MECHANICS OF SEDIMENT TRANSPORT

PROCEEDINGS OF EUROMECH 156 / MECHANICS OF SEDIMENT TRANSPORT
ISTANBUL / 12-14 JULY 1982

Mechanics of Sediment Transport

Edited by

B.MUTLU SUMER
Technical University of Istanbul

A.MÜLLER
ETH-Hönggerberg, Zürich

A.A.BALKEMA / ROTTERDAM / 1983

The texts of the various papers in this volume were set individually by typists under the supervision of each of the authors concerned.

ISBN 90 6191 221 0

Distributed in USA & Canada by: MBS, 99 Main Street, Salem, NH 03079, USA

Printed in the Netherlands

Table of contents

Preface

The European Mechanics Committee has included a Euromech colloquium on "The Mechanics of Sediment Transport" into its program of 1982. It was held in Istanbul during July 12-14, 1982. Fifty-five researchers from twelve countries were invited to discuss, in the light of modern fluid mechanics, sediment transport problems of current interest.

A better understanding of the complex feedback system between flow, bed forms and transported sediment is essential to improve engineering predictions of sediment transport problems. It was the objective of the colloquium to bring together scientists and engineers to cover these aspects of sediment transport. Forty-one papers including three introductory survey lecturers were presented in six different sessions. On the first day, turbulent flow structures related to sediment transport and particle dynamics were discussed. During the second day, the stability aspects of initiation, formation and behavior of ripples and dunes were investigated together with sediment transport under wave action. On the last day, studies on sediment transport in suspension and sediment transport in steep channels were presented. Also included in this session were discussions on degradation and particle sorting processes.

The very good response to the first announcement, both in quantity and quality, prompted us to look for a publisher of a volume comprised of the papers presented at the colloquium. The editors would like to thank all authors who have contributed to this volume. Their enthusiasm and willingness to participate are gratefully acknowledged.

The editors would also like to extend their thanks to the staff of the Hydraulics group (Technical University of Istanbul) for their help in the organization of the colloquium. The support of the hosting organization, the faculty of Civil Engineering of the Technical University of Istanbul, of the Scientific and Technical Research Council of Turkey and the ST and FA Co., are gratefully acknowledged.

B. Mutlu Sumer
Andreas Muller

1 Flow structure as related to sediment transport

Euromech 156: Mechanics of Sediment Transport / Istanbul / 12-14 July 1982

The influence of boundary layer turbulence on the mechanics of sediment transport*

A.J.GRASS
University College, London, UK

ABSTRACT: A summary of new knowledge relating to the structure of boundary layer turbulence is given. Particular emphasis is placed on recent research by the author and other investigators which has made use of elaborate flow visualisation and conditional sampling techniques in quantitative analysis. This work has confirmed the presence of coherent eddy patterns embedded in the flow, which appear to take the general form of combinations of contorted single or amalgamated groupings of vortex loop type structures.These play a crucial role in the processes of turbulence generation and the transport of momentum and contaminants such as bed sediment. The primary influence on the mechanics of the progressive modes of sediment transport, is illustrated by reference to specific research studies carried out by the author concerned with the initial instability,bed load transport and suspended load transport of fine bed sand. Links between these investigations and the transport of sediment by superimposed waves and currents are also briefly discussed.

1.INTRODUCTION

An essential precursor to any rational study of the wide ranging and frequently dominant effects of flow turbulence on sediment transport phenomena, is a comprehensive knowledge and understanding of the characteristics and mechanics of boundary layer turbulence itself. In the mid-sixties when the author first started work in this field, progress with sediment transport studies was greatly inhibited by the fact that relevant turbulence knowledge was rather limited and almost exclusively based on measurements in air flow boundary layers. The critical need for instrumentation for use in water, led to the subsequent introduction of hot film and laser doppler anemometers along with sophisticated flow visualisation techniques. These developments, linked with very substantial improvements in data processing methods including the use of increasingly efficient conditional sampling techniques, has over the past ten years inevitably produced a rapid growth in our knowledge of boundary layer turbulence in water, air and other fluid flows.

Many of the resulting investigations have placed emphasis on measurements in the near bed interaction zone which is particularly relevant in studies of initial bed sediment instability and bed load transport. Of perhaps even greater significance, has been the discovery of the so-called 'bursting' phenomenon and associated coherent flow structures in the boundary layer turbulence generation process which provide the primary mechanism for sediment suspension as discussed below.Excellent general reviews of many of these recent developments in boundary layer turbulent research have been produced by Laufer(1975) and by Willmarth (1975).A similarly compre-hensive survey of conditional sampling methods used in the study and identification of ordered flow structures has been given by Antonia (1981).

The present paper is not intended as a detailed review. Its main purpose is to describe some of the general trends and significant findings in boundary layer turbulence research over the past decade and illustrate the application of this new information in the study of sediment transport processes by reference to specific investigations in which the author has been directly involved. For the reasons outlined above, the latter research studies commenced with an investigation of the

* This paper was presented by invitation at the colloquium as a survey lecture.

structure of boundary layer turbulence in an open channel water flow. This fundamental fluid mechanics work has been continued up to the present time, running in parallel with linked sediment transport investigations.

The paper therefore starts with a summary of relevant new knowledge relating to boundary layer turbulence structure. Particular emphasis is placed on very recent flow visualisation observations by the author and his colleagues and by a number of other investigators which begin to form a consistent picture of the turbulence structure. A particularly significant advance in the work at University College London, has been the development of a technique for simultaneous monitoring of the instantaneous fluctuating velocity vectors at points throughout large three-dimensional blocks of boundary layer flow space. This has allowed calculation of vorticity vectors and, for the first time in boundary layer turbulence research, the plotting of instantaneous three-dimensional vortex lines embedded in the shear flow. The striking simplicity of the untangled looping form of these vortex lines strongly confirms the coherent nature of the associated vortex structures and their dominant influence on the instantaneous state of the encompassing regions of the flow field. These vortex structures and their self-induced patterns of migration, are a key element in the turbulence bursting process and the linked transport of mass, momentum and contaminants, such as bed sediment, across the boundary layer.

The direct influence of turbulence on sediment transport mechanics is illustrated by reference to parallel studies relating to three important stages in the transport of fine bed sand. Namely, initial bed particle motion, bed load transport and suspension transport. The first two of these investigations demonstrate how improved knowledge of the near bed flow field and in particular the associated characteristics of the fluctuating bed shear stresses induced by the turbulence action, can be applied in a quantitative analysis of certain sediment transport phenomena.A more qualitative study of the mechanics of sediment suspension and its direct links with the coherent vortex structures identified in the recent turbulence investigations, is described in the third investigation. The sediment suspension action of vortex structures in a turbulent boundary layer are in many respects qualitatively similar to the suspending action of vortices produced by the oscillatory fluid motion induced by surface waves over sediment ripples on the bed. By considering this bed level wave agitation as pseudo turbulence, a method is suggested for correlating measured suspended sediment transport rates produced by combined waves and tidal currents. An example of the application of this method to field data is presented.

2. BOUNDARY LAYER TURBULENCE STRUCTURE

As pointed out by Laufer (1975), it is somewhat ironic that one of the most interesting and significant discoveries in turbulent boundary layer research over the past two decades, namely, the so-called 'bursting' phenomenon identified at Stanford University by Kline and his co-workers, resulted from the imaginative use of simple flow visualisation methods rather than sophisticated modern electronic instrumentation.

Kline et al (1967) gave a detailed description of the streaky nature of the flow structure in the viscous sublayer on a smooth wall with its alternating narrow, elongated zones of high and low velocity occurring randomly in time and space. This sublayer flow pattern appears to result from counter-rotating streamwise vortex pairs (Bakewell and Lumley 1967 , Blackwelder and Eckelmann 1979). Kline et al linked the sublayer structure to the bursting process which they discribed as a randomly occurring event comprising gradual local lift up of the low speed streaks, sudden oscillation, bursting and ejection. This event is followed by an inrush or sweep event described by Corino and Brodkey (1969) and by Grass (1971) based on observations from similar visualisation studies of the boundary layer flow structure.

The ejection and inrush events form part of a randomly occurring yet linked cyclical process of turbulence generation in the boundary layer which has been shown to be responsible for most of the turbulence energy production and the major contribution to the Reynolds stress momentum transport in the wall layers. Grass also showed that the bursting event with its linked ejection of low momentum fluid from the wall region out into the main body of the flow, and the subsequent inrush event with its return high momentum fluid, were common features of both smooth and rough wall flows. During bursting and ejection events, the visualised erupting flow structure appears on slow motion films produced by Grass (1967) to literally 'rear up' from the bed. This behaviour is particularly pronounced in the case of the violent bursting and resulting highly dist -urbed flow field over the rough boundary as illustrated in the film frame reproduced in figure 1a. The interval between the lines of hydrogen bubble tracer blocks leaving the vertical generating wire in figure 1 is 1/50 second, with a mean flow velocity of 145mm/second and a vertical field of view of 35mm out of a total open channel flow depth of 50mm (see Grass (1971) for further details).

behind the burst as shown in figure 1a,b. The contrast between the highly disturbed flow within the burst structure and the well ordered, quiescent, almost laminar flow following up behind the burst struct -ure can be clearly seen in figure 1a,b. Grass (1971) also observed that the eject-ion events occurring during bursting and the subsequent inrush or sweep events appeared to influence the entire depth of the boundary layer.

(a) (b)

Fig.1 Visualised turbulence bursting(a) and associated transverse vortex (b) in the boundary layer of a rough wall open channel flow (Grass,1967).

The instantaneous velocity profiles app -roximately marked by the simultaneously generated lines of bubble tracer blocks in figure 1a, indicate a general velocity defect relative to the mean during burst-ing. This is to be expected due to the lin -ked upwelling and ejection of low momentum fluid from the bed region. The velocity profile takes up a more uniform distribut -ion throughout the depth with velocity magnitudes generally greater than the mean, particularly in the bed region, during the accelerating inrush phase following up

Lighthill(1963) drew attention to the fact that the only apparent mechanism by which the strong concentration of mean and fluc-tuating vorticity can be maintained at the wall in turbulent boundary layers, is by the stretching and compression of vortex lines resulting from the inflow and outflow of fluid close to the boundary.The bursting phenomenon and the associated ejection and inrush events appear to provide the neces -sary mechanism for this wall layer vorti -city concentration.

In attempting to explain the visualised

flow structure and the bursting process in the wall region of a smooth boundary flow, Kline et al (1967)and Kim et al(1971),invo-ked the concept of an initially locally compressed transversevortex filament,raised slightly outwards from the wall in the lift up region of a low speed streak. The filam-ent is thus subjected to subsequent stream-wise stretching in the highly sheared flow and evolves into a vortex loop. This process is illustrated in figure 2, taken from Hinze (1975), who in discussing the bursting mechanism also drew attention to certain similarities with the initial breakdown process in laminar to turbulent boundary layer transition.

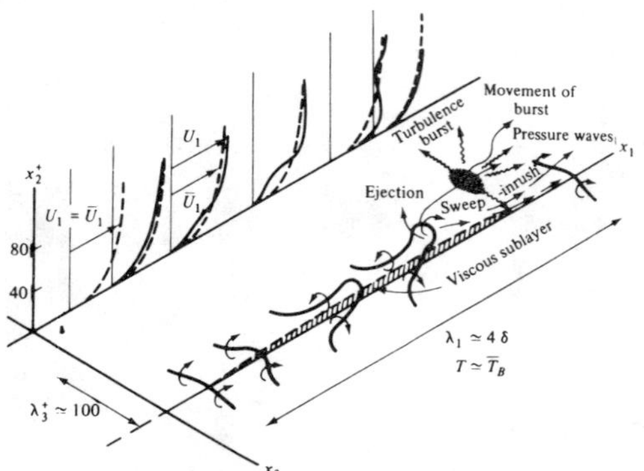

Fig 2 Conceptual model of the bursting process in turbulent boundary layers (Hinze,1975).

The mutual interaction of the streamwise limbs of the vortex loop structure illust-rated in figure 2 cause the loop to move away from the wall into regions of ever increasing streamwise velocity.This further enhances the streamwise stretching and hence the streamwise vorticity and the resulting self-induced outward velocity of the loop structure particularly in its tip region. Willmarth(1975), gives an interesting account of the mutual induction and hyper-bolic path taken by counter-rotating vortex pairs in the vicinity of a wall, based on observations of the behaviour of trailing vortices behind aircraft.

An outward ejection of low momentum fluid takes place between the streamwise limbs of the loops (figure 2) which makes a major contribution to the Reynolds shear stress. This low momentum fluid also creates a local inflexion, and a presumably dynamically instable free shear layer, in the local streamwise velocity profile as shown in figure 2. Kline et al (1967),suggested that the resultant instability of the flow surrounding the tip of the vortex loop, produced the observed sudden oscillation followed by bursting and ejection.Some un-certainty remains however concerning the exact details of the complex interaction between the vortex loop structures and the sheared flow through which they erupt.

Vortex loops,which are also variously referred to as'horseshoe'or'hairpin'vortic-es,were first postulated as a possible co-herent eddy structure in turbulent boundary layers by Theodorsen (1952). The concept was also incorporated into a vortex loop model of wall turbulence by Black(1966). Powerful new evidence in support of the actual existence of vortex loops as the predominant form of eddy structure in tur-bulent boundary layers,has come from the excellent flow visualisation studies recen-tly reported by Perry et al(1981) and by Head and Bandyopadhyay(1981).These studies were concerned respectively with the flow structure in turbulent spots and in fully turbulent boundary layers. Further defini-tive evidence has been produced in the rec-ently completed investigation by Grass & Stuart (1983) who have succeeded in plotting three-dimensional vortex loops from direct simultaneous measurements of the spacial distribution of the three-dimensional vort-icity vector in the boundary layer of an open channel flow. The latter tests are discussed in more detail below with examples of the results.

The streamwise streaky flow structure associated with turbulent spots have been convincingly explained by Perry et al(1981) as the 'footprints'generated by the stream-wise trailing limbs of the interacting and self-propagating vortex loop structures. Head & Bandyopadhyay (1981) in discussing the vortex loop structures, clearly obser-ved in their smoke visualisation studies of the boundary layer, drew attention to the coincidence in spanwise dimension of their observed loops and the spacing of the streamwise streaks in the viscous sub-layer of developed boundary layers and in turbulent spots. Perry et al (1981) also make the same observations.

The coincidence of these observed span-wise scales is clearly of great significance. Head & Bandyopadhyay(1981) conclude by inference that the burst sequence illustr-ated in figure 2 can be reasonably, though not with complete certainty, equated to the outward drift of vortex loops observed in their experiments. Perry et al (1981),on the basis of their observations,postulate a close connection between the flow struc-ture of turbulent spots and of fully turbulent boundary layers.

On the evidence, there appears to be some common mode of instability in the highly sheared viscous wall region of turbulent spots and fully turbulent boundary layers, which triggers the self-propagating echelon groupings of vortex loop structures observed by Perry et al (1981). Brown and Thomas (1977) suggest the possibility of a Taylor-Gortler type of instability based on their observations that the convected frame streamlines close to the wall are convex to the boundary in a particular zone under detected large-scale structures in the boundary layer. Coles (1978)has suggested a similar type of mechanism for the streamwise counter-rotating vortex structures in the wall region.

The spanwise spacing of these wall streaks is scaled, as apparently is the span dimension of the vortex loop structures, by the wall viscous length scale, having a mean value of approximately 100 ν/u_τ The streaky nature of the viscous sublayer on the smooth bed of a water channel flow, visualised by transported fine sand particles, is illustrated in figure 3. This photograph which covers a field of view width of approximately 800 ν/u_τ was taken as part of the investigation reported by Grass (1971). It is apparent from inspection of figure 3, that large fluctuations in bed shear stress occur, a characteristic which has important implications in relation to the transport of bed sediment as discussed below.

The study of and search for ordered coherent structures in boundary layer turbulence has proved more difficult than, for example, in the case of two-dimensional mixing layers, for two main reasons. Firstly, the flow structures are inherently three-dimensional. Secondly, there are two primary length scales involved, namely,the wall region length scale which takes the form of the viscous length scale on a smooth wall or the physical roughness scale on a fully rough wall, and a length scale determined by the total thickness of the boundary layer. Corresponding to these two length scales, observations suggest that there are also two types of flow structure. There is a wall region structure associated with the bursting phenomenon and which the recent research discussed above suggests takes the form of stretched vortex loops. Secondly, there is a large scale structure which appears to take the form of a large slowly rotating vortex type motion.

Many laboratory studies of turbulence structure have been conducted, usually for very good reasons, at relatively low Reynolds numbers. Under these conditions the scales of the two types of flow structure have often been of the same order of magnitude. This factor has generated its own difficulties and confusions in interpretation of the observations, particularly in attempts to separate the respective mechanisms involved in the formation and maintenance of the two interacting structural formations. These problems are now reasonably widely recognised and hopefully future investigations will place a greater emphasis on the study of flow structure at the high Reynolds numbers relevant to many geophysical flows.

Fig.3 Streaky structure of the viscous sublayer visualised by transported fine sand particles (Grass.1971).Flow direction ↑

Direct evidence of the existence of quasi-ordered large scale structures in the intermittent outer regions of the boundary layer was provided by the visualisation observations of Fiedler and Head (1966). Much earlier, clear photographic evidence of the presence of large transverse vortex structures in the boundary layer of a channel flow, taken with a camera moving with the flow, were obtained by Nikuradse (1929). (See Schlichting (1979:556).Willmarth(1975)reports interesting observations of similar large transverse swirling eddies at very high Reynolds numbers in the thick boundary layer formed along the side of a large ship.

The study of these observed large scale structures has been pursued energetically over the past decade or so using both flow visualisation and the various forms of conditional sampling described by Antonia (1981). A conceptual picture of the flow patterns inferred from studies by Kaplan and Laufer (1969), Kovasnay et al(1970), Blackwelder and Kovasnay (1972) and by Falco (1974), was presented by Laufer (1975) which is reproduced in figure 4.

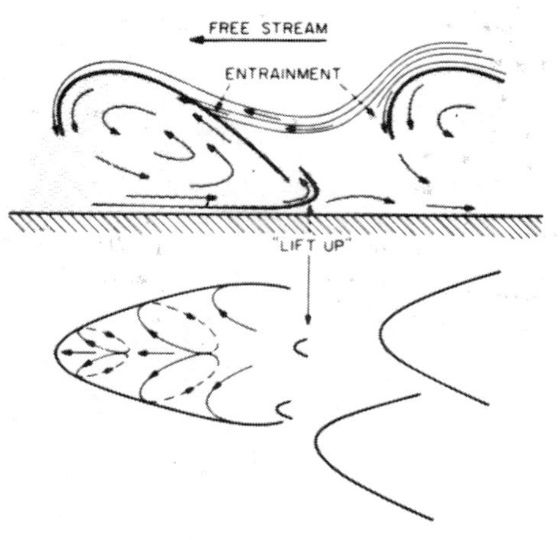

Fig.4 Conceptual model of the large coherent structures in a turbulent boundary layer (Laufer,1975).

Laufer aptly described the general form of the quasi-ordered structure as a large scale,three-dimensional vorticity 'lump' rolling over the wall vortex sheet pulling up with it small scale vorticity.This description is consistent with the later findings from studies by Blackwelder and Kaplan(1976), Brown and Thomas (1977)and by Chen and Blackwelder (1978). The conceptual picture shown in figure 4 is very similar to the structural flow pattern suggested by Brown and Thomas (1977) who described the three-dimensional structure as a large horseshoe vortex. As pointed out by Laufer, the plan view of the structure shown in figure 4 and the general three-dimensional form of the large scale structure, remains somewhat speculative. New investigations such as that reported by Chambers et al(1982) using two-dimensional arrays of measuring probes, should help to resolve some of these uncertainties.

The suggested link between the passage of large scale structures over a point on the wall, and the triggering of wall layer bursting which lifts high concentration (large vector magnitude) vorticity into the outer flow from the wall generation zone, was inspired by an important finding of Rao et al(1971). Namely, that the period between bursting activity scales with the outer flow parameters of free stream or mean velocity and the boundary layer thickness.

Uncertainties remain as to the exact nature of the link between the wall and outer layer structures and in particular the form of the triggering mechanism for the bursting process. Willmarth(1975) and others have suggested that it is possibly adverse pressure gradients associated with the passage of the large scale vortical structures which trigger the bursting by the production of vorticity of opposite sign to the mean vorticity, and the generation of unstable free shear layers above the severely retarded fluid close to the wall. The alternative suggestion by Brown and Thomas (1977)(discussed above), that the bursting is the result of Taylor-Gortler type instability occurring under a region of the large scale structure where the wall streamlines are convex to the wall (which is likely to coincide with the region of wall flow retarded by the adverse pressure gradients), appears to link in with the pressure gradient hypothesis.

It is interesting to note that the transverse vortex which can be seen in the bed region in figure 1b, results from the rapid roll-up of a local free shear layer generated close to the bed following the passage of the highly disturbed bursting zone shown in figure 1a. The visualised portion of the vortex presumably represents a cross-section through the tip area of a small loop structure forming in the bed zone. The film frames reproduced in figure 1 are 0.13 seconds apart which represents a convected flow distance of approximately 19mm.

Larger scale transverse vortices are apparent in the velocity vector representation of the flow structure patterns shown in figure 5. This diagram was produced from velocity data measured in the recent study by Grass and Stuart(1983) of the three-dimensional characteristics of coherent structures in the fully develo -ped turbulent boundary layer of an open channel flow over a rough bed. The flow depth in these tests was 50mm in a channel 0.5m wide and the depth averaged flow velocity at the centre of the channel was 144mm/second.

Velocity measurements were made using an adaption of the hydrogen bubble tracer method previously described by Grass(1971). The single bubble generating wire used in the earlier tests was replaced with a grid of orthogonal wires in a vertical plane transverse to the flow. This produced a pulsed, regular grid pattern of bubble tracers comprising 132 node points. By employing a specially designed, stereoscopic photographic recording system, it proved possible using this technique to record sequences of 132 regularly spaced, simultaneous three-dimensional velocity vectors across the grid plane which spanned the flow depth and covered a 50mm x 50mm square area in the centre of the channel.

Using Taylor's 'frozen eddy' hypothesis, it was possible to construct an approximate picture of the instantaneous three-dimensional velocity field throughout blocks of the flow space swept through the measuring grid. This velocity data was then used to compute corresponding distributions of three-dimensional vorticity which were in turn used to plot the position and establish the three-dimensional geometry of the vortex lines throughout the swept flow blocks.

The velocity vector plot example shown in figure 5, represents a vertical slice through the flow along the channel centre line. It gives a convected view impression of the instantaneous streamline patterns as seen by an observer travelling with the flow at the depth averaged velocity. Large transverse vortex structures, inclined internal shear layers, and ejection and inrush zones are clearly visible in the flow structure. A series of plots of the type illustrated in figure 5 were produced and examined to identify regions where interesting events or ordered flow structures were present. This procedure represented a simple form of visual conditional sampling. A complete three-dimensional analysis was then carried out for the selected flow blocks in order to construct the vortex line distributions associated with particular identified structural features such as well defined transverse vortices. A typical example of one of these vortex line plots is shown in figure 6. This clearly takes the form

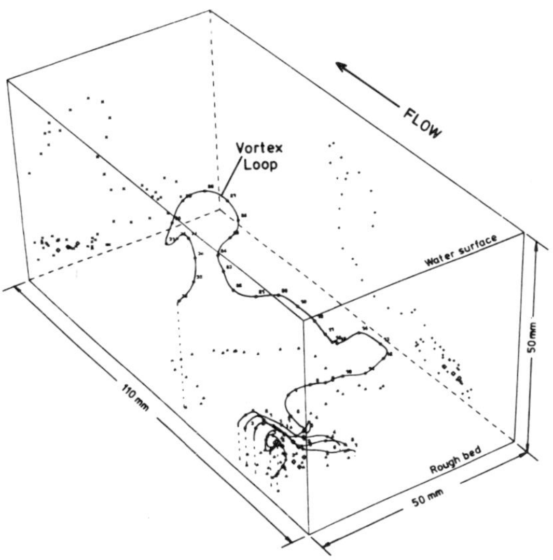

Fig.6 Inclined, three-dimensional vortex loop structure plotted from measured spacial distributions of the three-dimensional vorticity vector in an open channel boundary layer (Grass and Stuart)

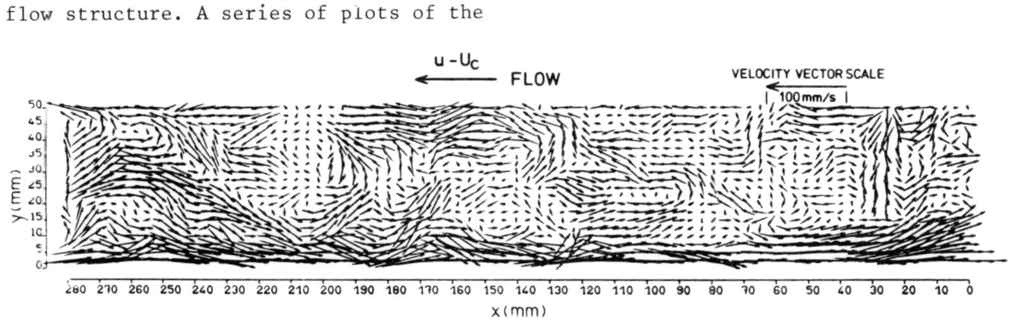

Fig 5. Transverse vortices and inclined shear layers in an open channel boundary layer, visualised by velocity vectors in a convected reference frame (Grass and Stuart).

of a large scale, inclined vortex loop structure. Projections of this loop structure onto three orthogonal faces of the block, also shown in figure 5, give a good impression of the loop geometry and its relative orientation.

These results represent the first direct confirmation, based on three-dimensional quantitative measurements, of the presence of vortex loop structures embedded in turbulent boundary layer flows. Indeed, to the authors knowledge, this is the first time in fluid mechanics research generally that quasi-instantaneous three-dimensional vortex lines have been plotted from physical experimental measurements. The observations are thus of considerable significance and the method clearly represents a powerful new tool in the study of complex three-dimensional flow phenomena.

There remains considerable uncertainty regarding the mechanism by which the large scale flow structures at high Reynolds numbers are formed. Head and Bandyopadhyay (1981) suggest, based on their visualisation observations, that stretched vortex loops or hairpin vortices are a major constituent of turbulent boundary layers at all Reynolds numbers. For the higher Reynolds number flows studied ($Re_\theta \simeq 10{,}000$), they suggest that the large scale features are made up of agglomerations of many small scale, stretched hairpin vortices which originated at the wall, and have presumably been drawn up during turbulence bursting.

Head and Bandyopadhyay(1981) also make the interesting observation that whilst vortices of opposite sign constituting the legs of the small span (wall scale) hairpins grouped in the large scale structure, may diffuse into each other and cancel, no such mechanism exists for destruction of the transverse vorticity in the hairpin tips. They therefore suggest that at really high Reynolds numbers, possibly only the tips of the hairpins survive and remain active in the outer flow regions. The enclosing circulation round such a grouping of surviving vortex tips would certainly be consistent with the observed slowly rotating motion of the large scale structures.

It appears possible that there may be some process of coalescence inside these clumps of like signed, transverse vortex elements similar to the vortex pairing mechanism involved in the growth of turbulent mixing layers, described by Winant and Browand(1974). Coalescence of discrete vortex systems is clearly a fertile area for future research.

At low Reynolds numbers, transverse vortices are a commonly observed feature in the outer flow structure as discussed for example by Falco(1974) and Praturi and Brodkey (1979). Head and Bandyopadhyay (1981) suggest that these vortices are simply slices through the tips of single vortex loops which form both the wall and outer flow structural elements at low Reynolds numbers.

In a recent article on the dynamics of vorticity, Saffman concluded that the increasing size and speed of modern computers has given theoreticians a new and potent tool in the study of vortex interactions. He also points out that the experimental discovery of coherent structures in turbulence is providing a practical impetus for their investigations. Such theoretical studies, particularly relating to three-dimensional vortex interactions, may help to fill in some of the gaps in existing knowledge discussed above. As in the past though, it appears likely that in the short term at least, major advances are most likely to come from further careful experimentation particularly at high Reynolds numbers.

Lighthill(1979) has described vorticity as containing all the 'memory' in a fluid flow. Because of the property that vortex lines are continuous and travel with the flow, they represent powerful flow tracers and historical indicators of where surrounding regions of a flow originate. For example the tip region of the vortex loop structure shown in figure 5 is linked back to the bed generation region by the trailing vortex lines.

The stretching of vortex lines, as takes place for example along the trailing limbs of the loop structures discussed above, represents a primary mechanism whereby energy is extracted from the mean flow, and the boundary extends its restraining influence out into the body of the flow. The study of vorticity and vortices in the context of turbulence and other fluid flow phenomena is thus of crucial importance. This central role has perhaps been best summed-up by Kuchemann(1965) who described vortices as "the sinews and muscles of fluid motions".

3.TURBULENCE AND SEDIMENT TRANSPORT

The brief survey of recently acquired knowledge and increasing understanding of the turbulence structure of boundary layers given in section 2, also serves to strongly underline the great complexity of the processes involved. The interaction between such flows and entrained bed sediment, adds further formidable dimensions of complexity in the resulting two phase flow regime. It is not really surprising therefore, that in spite of the enormous amounts of research effort expended on the study of sediment transport processes over the last century, the general state of our fundamental knowledge of much of the physics remains in a relatively primitive state. This reflects particularly in the lack of credible theorectical prediction models.

However, progress continues to be made at least in our qualitative understanding of certain aspects of sediment transport mechanics. In recent times, this can undoubtedly be linked to the parallel improvements in knowledge of the turbulent flows forming the primary transporting agent. It is not the intention here to attempt to review the many recent carefully conducted sediment transport studies which have contributed to this new understanding. Valuable relevant survey articles have been presented by Vanoni(1975) and by Jackson(1976) for example. These authors draw attention to the central role played by turbulence in sediment transport mechanics which is also a recurring theme of emphasis by Raudkivi(1976).

The primary purpose of this section is to illustrate,albeit in a rather restricted fashion, this wide ranging influence of flow turbulence, by briefly describing a number of specific sediment transport investigations in which the author has been directly involved and relating them to the preceding discussion of turbulence structure. These studies deal in sequence with the progressive stages of transport of fine bed sand from initial movement, through bed load transport, to final suspension transport by a turbulent boundary layer flow. The bed material used in all these tests was a fine Mersey sand, approximate mean diameter 0.14mm, which has the simplifying advantage of producing an hydraulically smooth boundary under flat bed transport conditions in water flows.

3.1 Initial movement of fine bed sand

It is clear from visual inspection of figure 3, showing the movement of fine sand in the viscous sublayer over a smooth surface, that the bed is subjected to a widely fluctuating range of instantaneous shear stresses, both temporally and spacially. This is quantitatively confirmed in figure 7, reproduced from Grass(1970) which shows instantaneous velocity profiles through low and high speed streak zones in the region immediately adjacent to a bed of fine sand and corresponding to ejection and inrush phases in the turbulence generation cycle. These velocity profiles were measured using hydrogen bubble tracers generated on a vertical wire passing through the bed. The visualised flow field and the corresponding bed sand motion were simultaneously recorded by high speed motion photography. Sample film frames illustrating the state of the flow at the instants when the velocity profiles were measured are also shown in figure 7.

The probability density function for the fluctuating bed shear stress on a smooth wall, shown on figure 8, assumes a positive skew form recorded by several investigators including, for example,Grass (1971),

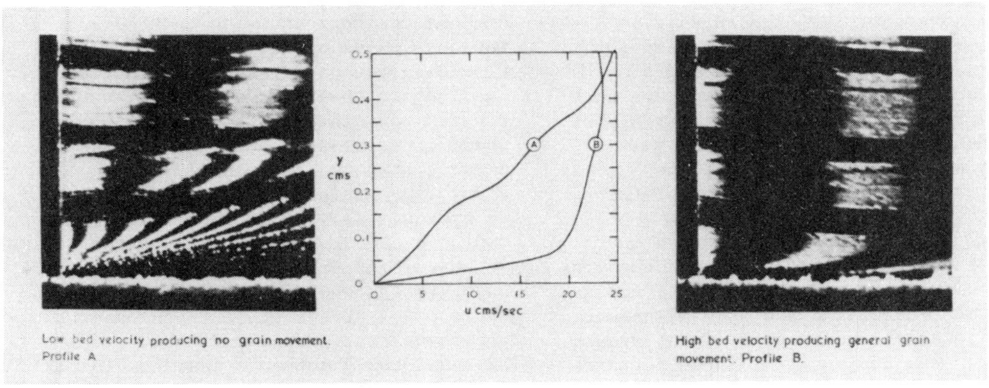

Fig.7 Visualised and measured velocity profiles through low and high speed streaks in the viscous sublayer over a flat bed of fine sand producing widely fluctuating local rates of bed load sand transport (Grass,1970).

Eckelmannand Reichardt(1972) and Blinco and Simons(1974).

These observations of fluctuating shear stress and induced bed sediment transport demonstrate that the problem of defining critical flow conditions for initial movem-ent of bed material is essentially statis-tical. No well defined mean flow condit-ion exists for which the bed sediment suddenly becomes unstable and begins to move. The process is a gradual one which commences when the largest fluctuating shear stresses applied to the bed by a particular flow exceed the critical values required to move the most susceptible exposed sand grains on the bed surface.

The novel concept that each individual sediment particle on the bed surface can be ascribed a certain critical bed shear stress value, τ_c, which, if it is exceeded then the particle moves, was introduced by Grass(1970) in his study of initial insta-bility of fine bed sand. For a particular fluid viscosity, the initial instability characteristics of the bed sediment can then be described by a probability density function distribution, as shown in figure 8, which for an non-armoured, flat bed condition, becomes an intrinsic property of the sediment similar to its settling velocity distribution. Initial movement charactersitics defined in this way can thus be applied in different flow condi-tions including unsteady flows such as those produced close the sea bed by surface waves.

Grass(1970) estimated the first and second moments of these τ_c distributions by recording the instantaneous bed shear stresses, measured from the local velocity gradient in the sublayer above particular individual sand grains, at the instant of initial movement. The high speed film records referred to above were used for these measurements. Critical conditions for steady unidirectional flows were then defined by locating the flow shear stress distribution so as to produce a small overlap with the lower tail of the characteristic, τ_c, distribution of the bed sand as shown in figure 8.

The motion film records produced in the above study in conjunction with the visc-ous sublayer visualisation films, emphas-ised the fluctuating and streaky nature of the localised zones of sediment trans-port over the bed surface. These patterns will inevitably produce a random sequence of local erosion and accretion on an initially perfectly flat bed, resulting in the statistical (inhibited random walk) growth of elongated surface undulations. These mounds eventually cause the flow to

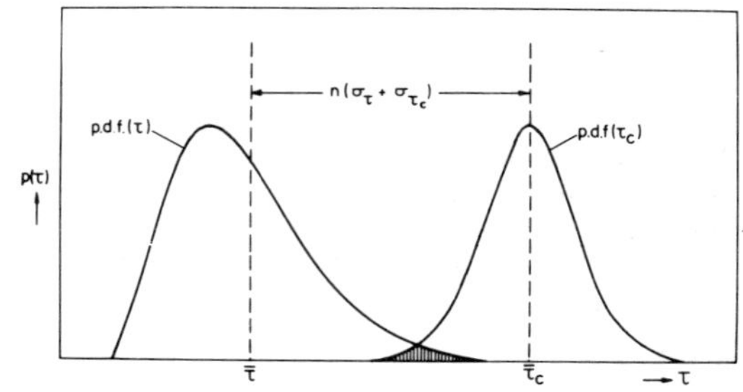

Fig.8. Critical flow conditions for initial movement of bed sand defined by the over-lap between the distributions of bed shear stress (τ) and the critical shear stress (τ_c) for the movement of individual sand grains in the surface layer (Grass 1970).

separate triggering the spontaneous formation of bed ripples as discussed by Grass (1970) and studied in detail by Williams and Kemp(1971). Boundary layer turbulence was thus linked through these basic studies to the formation of ripples on a flat bed of fine sand in the absence of any other flow disturbance.

3.2 Bed load transport of fine sand

Close examination of the films produced during the initial bed movement investig-ations and in particular those showing simultaneous flow visualisation and bed particle transport, confirmed the intuitive notion that there is a close correlation between the local instantaneous velocities in the viscous sublayer very close to the bed and the corresponding induced instan-taneous rates of bed sand transport. Providing the bed distance is sufficiently small, these sublayer velocities are also closely correlated with the fluctuating shear stresses acting on the bed as demonstrated by Eckelmann(1974).

For fine grained bed sediment producing an hydraulically smooth boundary condition, fluid forces are transferred to the sedime-nt particles primarily by viscous shear. Under these conditions and based on the above observations of flat bed sand trans-port behaviour, Grass and Ayoub (1982)have advanced the hypothesis that the local instantaneous rate of bed load transport produced by a particular instantaneous value of fluctuating bed shear stress, τ, can be represented by the transport rate, $q_{s\ell}(\tau)$ induced by a steady laminar flow with the same bed shear stress and fluid viscosity.

The net sediment transport rate under turbulent flow conditions, q_{st}, or more generally under arbitrary unsteady flow conditions, can then be calculated using the following equation:

$$q_{st} = \int_{\tau_{min}}^{\tau_{max}} p(\tau) q_{s\ell}(\tau) d\tau \qquad (1)$$

where $p(\tau)$ is the probability density function of the fluctuating bed shear stress, τ.

The hypothesis expressed in equation (1) was tested by comparing measured bed load transport rates produced by a flat plate turbulent boundary layer with the corresponding values calculated using equation (1) substituting $q_{s\ell}$ (τ) values provided by measurements in a steady laminar flow between two closely spaced parallel plates.

The results of these tests which largely confirm the general validity of equation (1) and the underlying concepts, are presented in figure 9. The potential value of this newly suggested method for calculating net bed load transport rates, lies once again in the fact that it can be applied to a wide range of different types of flow.

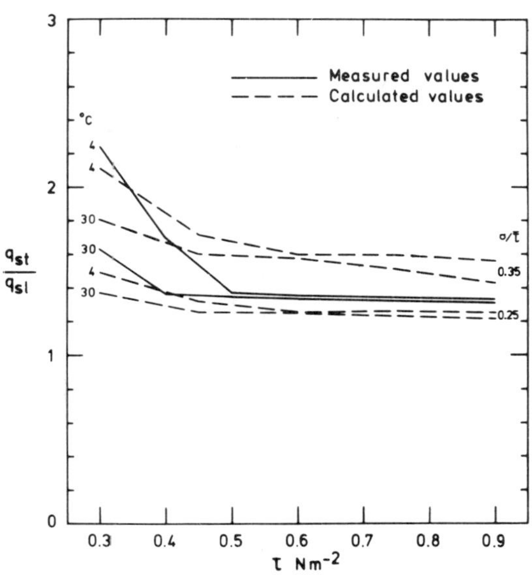

Fig.9 Ratio of bed load sediment transport rates produced by turbulent and laminar flows with the same average bed shear stress (Grass and Ayoub, 1982).

These might include for example the oscillatory flow induced at the bed by surface waves, or the developing boundary layer downstream of the re-attachment point in separated flow over bed sediment ripples.

3.3 Mechanics of sediment suspension

As part of the bed load sediment transport study described in section 3.2 above, Grass and Ayoub (1982) measured the transport rates induced by the turbulent boundary layer developed on a flat plate towed through still water. It was found in these tests that for plate velocities in excess of approximately 0.5 metres/second the fine Mersey sand used as the bed material began to lift into suspension off the flat bed. Under suitable slit lighting these suspended sand particles provided excellent visualisation of the bursting process and eddy structure in the boundary layer. This observation prompted setting up the study of supsension mechanics reported by Grass (1974).

The suspension process was filmed with a high speed motion camera. The small relative velocities in the outer regions of the boundary layer with respect to the stationary camera, produced very graphic pictures of the slowly translating flow structures. A typical frame from the resulting films is shown in figure 10. The form of the coherent flow structures visible in the outer regions of the boundary layer closely resemble the structures visualised by smoke in Head and Bandyopadhyay's(1981) investigation at similar Reynolds numbers (Re_θ=2300 in the sand suspension test) The transverse rotating form of the grouped clusters of sand particles evident in the outer layers and the characteristics of their eruption trajectories as they rise up from the near bed region, is entirely consistent with the hypothesis that the sand particles were entrained by and marked the progress of single vortex loop structures lifting off the bed as described in section 2.

By identifying clearly marked flow structures at the outer edges of the boundary layer, their trajectories could be tracked back down to the bed by analysing the film records frame by frame in a reverse direction. This enabled averaged streamwise and vertical ejection velocities to be estimated for the suspended clusters of sand particles as they travelled outwards across the boundary layer. The results are shown in Figure 11. Streamwise ejection velocities were somewhat smaller than the overall mean velocities which is consistent with ejections being associated

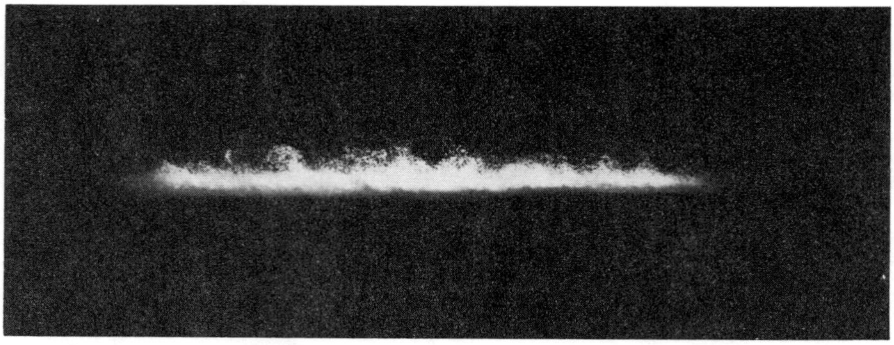

Fig.10 Sand suspended by ejection events in a flat plate turbulent boundary layer visualising the coherent flow structures (Grass,1974).

with low momentum bed fluid. The vertical ejection profile shows an interesting well defined peak at a position approximately 0.4 of the boundary layer thickness above the bed. It appears likely that this peak is related to the characterisitics of the self-induced velocities of vortex loop structures as they lift into and are stretched by the outer flow. This suggestion is however somewhat speculative and confirmation must await further study of the interactive vortex dynamics involved. The ejection velocities shown in figure 11 are consistent with similar measurements obtained from the trajectories of larger particles suspended in a open channel flow reported by Sumer and Oguz (1978)and Sumer and Deigaard (1981).

The sand suspension tests reported by Grass(1974) thus conclusively linked the mechanism of sediment suspension with the bursting and ejection processes in the turbulent boundary layer. The newly established knowledge and ideas relating to vortex loop structures, discussed in section 2 above, are also entirely consistent with the observed characteristics of the suspension process. Flow events capable of entraining low momentum fluid in the wall layer and ejecting and transporting it out across the thickness of the boundary layer will clearly similarly transport a contaminant such as fine sediment mixed up with the wall layer fluid. The observed vortex loop structures and the primary mechanism of sediment suspension can thus be reasonably equated.

The superposition of waves on unidirectional currents, has been shown to significantly enhance the bed region turbulence intensities in spite of the fact that the induced oscillatory flow at the bed is initially irrotational (see for example Kemp and Simons(1982)).This increase in the general level of turbulence agitation at the bed level results from the sensitivi-ty of the boundary layer bursting process to the imposed wave pressure gradients over smooth walls, and to the formation and lifting of separational vortices over

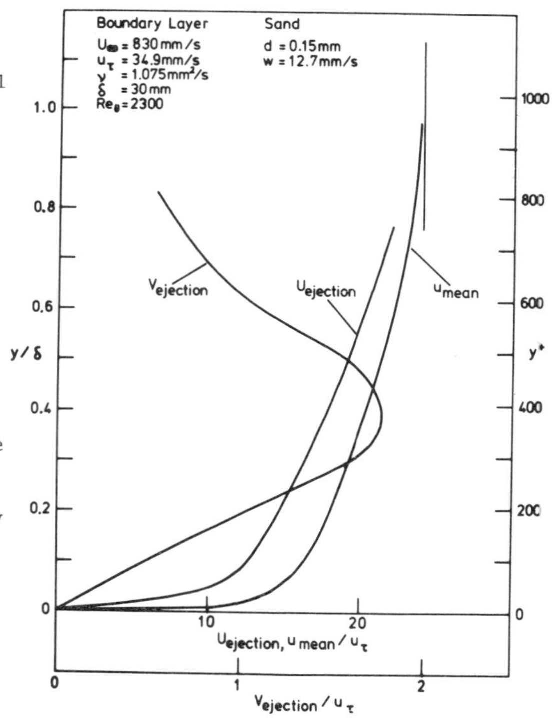

Fig.11 Streamwise and vertical velocity profiles recorded by sand particles suspended by ejection events in a flat plate turbulent boundary layer(Grass,1974).

rough beds. Increases in bed level turbulence mixing is inevitably linked with a corresponding increase in the concentration of sediment in the 'reservoir' bed layer, and an accompanying increase in the total suspended sediment transport rate as this bed concentration is distributed vertically throughout the depth by the main stream turbulence.

The formidable complexity of the combined two phase flow, coupled with the randomness of natural field conditions, rules out fully theoretical modelling of these processes at the present time. Practical prediction methods must inevitably be based therefore on directly measured field data. A simple method of correlating measured sediment transport rates under combined waves and currents and of comparing these rates with those obtained under current only conditions has been recently proposed by Grass (1981). This method is based on the simple concept that the wave induced bed motion can be viewed as 'pseudo' turbulence as far as the entrainment of sediment is concerned. Arguments are presented in support of this conclusion which involve the physical analogies between the bed sediment entrainment processes in waves and currents compared with the current only case and also the relative efficiency with which bed friction, particularly over rough rippled beds, converts irrotational wave motion into rotational, sediment entraining vortical motion.

The predicted ratio between the suspended sediment transport rates produced by combined waves and currents and the current only, are plotted in figure 12. The results are compared with good quality field data measured in the Maplin Sands area of the Thames Estuary by the Hydraulics Research Station, Wallingford as reported by Owen and Thorn (1978). Bearing in mind the complexity of the two phase flow under discussion, the theoretical predictions show a remarkable 'envelope' agreement with the scattered field data.

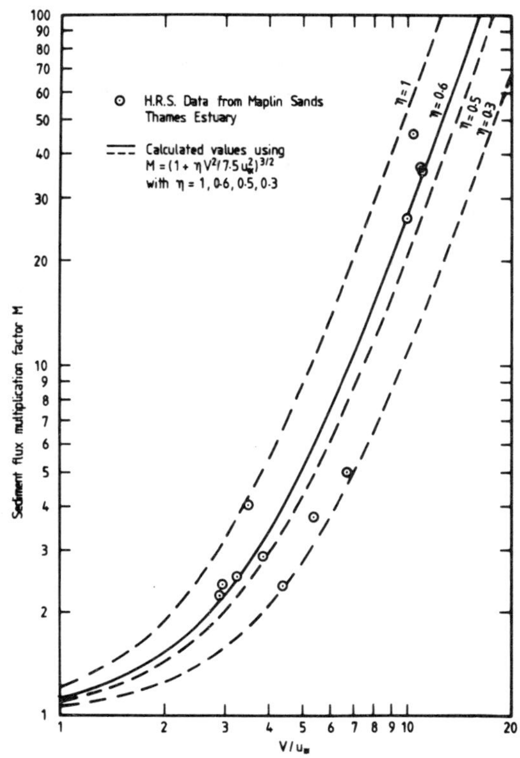

Fig.12. Large increases in suspended sediment transport rate produced by waves superimposed on tidal or other currents. (Grass,1981).

4. CONCLUSIONS

Over the past two decades, considerable progress has been made in improving our knowledge and understanding of turbulent boundary layers. A consistent picture of the coherent and deterministic turbulence structure, which is certainly far removed from the random chaos originally perceived, is beginning to emerge.

These developments in boundary layer research have been linked to parallel improvements in our understanding of sediment transport processes as illustrated by the examples given in section 3 above. Although this improving knowledge can, in a limited number of simple cases, be used in a direct quantitative manner in deriving sediment transport prediction methods, as shown above, the major value of fundamental research lies in improving qualitative understanding of the physical processes involved. In the continuing attempt to improve semi -empirical prediction methods for design purposes and in the application of these methods, such knowledge can never be a disadvantage.

REFERENCES

Antonia,R.A. 1981. Conditional sampling in turbulence measurement.Ann.Rev.Fluid Mech. 13:131-156.

Bakewell,H.P. & Lumley,J.L. 1967. Viscous sublayer and adjacent wall region in

turbulent pipe flow. Phys.Fluids 10:1880-1889.
Black,T.J. 1966. Some practical applications of a new theory of wall turbulence.Proc. 1966. Heat Transfer and Fluid Mech.Inst. Stanford University Press.
Blackwelder,R.J.& Kovasznay,L.S.G.1972. Time scales and correlations in a turbulent boundary layer. Phys.Fluids 15: 1545.
Blackwelder,R.F. & Kaplan,R.E. 1976. On the wall structure of the turbulent bound -ary layer. J.Fluid Mech.76:89-112.
Blackwelder,R.F. & Eckelmann,H.1979.Stream -wise vortices associated with the bursting phenomenon.J.Fluid Mech.94: 577-594.
Blinco,P.H. & Simons,D.B. 1974. Characteristics of turbulent boundary shear stress J.Eng.Mech.Div. ASCE.EM2:203-220.
Brown,G.L. & Thomas,S.W. 1977. Large structure in a turbulent boundary layer. Phys.Fluids 20: S243-S253.
Chambers,A.J. Antonia,R.A. Brown,L.W.B. & Raupach.M.R. 1982. Study of the large structure in a turbulent boundary layer over a rough surface. Program and Summaries Book of Euromech 156,Tech.Univ.of Istanbul.
Chen,C.P. & Blackwelder,R.F. 1978. Large scale motion in a turbulent boundary layer: a study using temperature contamination. J.Fluid Mech.89:1-31.
Coles,D.1978. A model for flow in the viscous sublayer. Workshop on coherent structure of turbulent boundary layers, p.462 Lehigh University.
Corino,E.R. & Brodkey,R.S. 1969. A visual investigation of the wall region in turbulent flow. J.Fluid Mech.37:1-30.
Eckelmann,H & Reichardt,H.1972. An experimental investigation in a turbulent channel flow with a thick viscous sublayer. Proc.Symp. on turbulence in liquids, Univ.of Missouri-Rolla.
Ecklemann,H.1974. The structure of the viscous sublayer and the adjacent wall region in a turbulent channel flow. J.Fluid Mech. 65: 439-459.
Falco,R.E. 1974. Some comments on turbulent boundary layer structure inferred from the movements of a passive contaminant A.I.A.A. Paper 74-99
Fielder,H. & Head,M.R. 1966. Intermittency measurements in a turbulent boundary layer. J.Fluid Mech.25:719-735.
Grass,A.J. 1967. Boundary layer turbulence in open channel flow. Ph.D.Thesis.Univ. of London.
Grass,A.J. 1970. Initial instability of fine bed sand. J.Hyd.Div.ASCE.HY3. 619-632.
Grass,A.J. 1971. Structural features of turbulent flow over smooth and rough boundaries.J.Fluid Mech. 50.233-255.
Grass,A.J.1974. Transport of fine sand on a flat bed: Trubulence and suspension mechanics.Proc.Euromech 48, Tech.Univ. of Denmark, Copenhagen.
Grass,A.J. 1981. Sediment transport by waves and currents.SERC London Centre for Marine Technology Report No.FL29. October 1981.
Grass,A.J. & Ayoub,R.N.M. 1982. Bed load transport of fine sand by laminar and turbulent flow. Proc.18th ICCE.Cape Town.
Grass,A.J. & Stuart,R.J. 1983. Three-dimensional vortex lines and coherent flow structures in boundary layer turbulence. Publication in preparation.
Head,M.R. & Bandyopadhyay,P. 1981. New aspects of turbulent boundary-layer structure. J.Fluid Mech.107: 297-338.
Hinze,J.O. 1975. Turbulence 2nd Ed. McGraw-Hill.
Jackson,R.G. 1976. Sedimentological and fluid dynamic implications of the turbulent bursting phenomenon in geophysical flows. J.Fluid Mech. 77: 531-560.
Kaplan,R.E. & Laufer,J. 1969. The intermittently turbulent region of the boundary layer. Proc.Int.Congr. Mech.12:236.
Kemp,P.H. & Simons,S.R. 1982. The interaction between waves and a turbulent current: waves propagating with the current. J.Fluid Mech. 116:227-250.
Kim,H.T. Kline,S.J. & Reynolds,W.C. 1971. The production of turbulence near a smooth wall in a turbulent boundary layer. J.Fluid Mech.50: 133-160.
Kline,S.J. Reynolds,W.C. Schraub,F.A. & Runstadler, P.W. 1967. The structure of turbulent boundary layers. J.Fluid Mech. 30:741-773.
Kovasnay,L.S.G. 1970. The turbulent boundary layer, Ann.Rev. Fluid.Mech.2: 95-112.
Kuchemann, D. 1965. Report on the IUTAM Symposium on concentrated vortex motions in fluids. J.Fluid Mech. 21:1-20.
Laufer,J. 1975. New trends in experimental turbulence research. Ann.Rev.Fluid Mech. 7:307-326.
Lighthill,M.J. 1963. Laminar boundary layers.Ed.L.Rosenhead. Clarendon Press. Oxford.
Lighthill,M.J. 1979. Waves and hydrodynamic loading. Proc.BOSS. 79 London.
Nikuradse,J.1929. Kinematographische Aufnahme einer turbulenten Stromung. ZAMM 9: 495-496.
Owen,M.W. & Thorn,M.F.C. 1978. Effect of waves on sand transport by currents. Proc. 16th ICCE Hamburg.Paper No.76
Perry,A.E., Lim,T.T. & Teh,E.W. 1981. A visual study of turbulent spots. J.Fluid Mech. 104: 387-405.

Praturi,A.K.& Brodkey .R.S.1978. A stereoscopic visual study of coherent structures in turbulent shear flow. J.Fluid Mech.89: 251-272.
Rao,K.N. Narasimha,R. & Badri Narayanan, M.A. 1971. The bursting phenomenon in a turbulent boundary layer. J.Fluid Mech. 48:339-352.
Raudkivi,A.J.1976 Loose boundary hydraulics. 2nd Ed. Pergamon.
Saffman,P.G. 1981. Dynamics of vorticity. J.Fluid Mech.106:49-58.
Schlichting,H. 1979. Boundary layer theory 7th Ed. McGraw-Hill.
Sumer,B.M. & Oguz, B. 1978. Particle motions near the bottom in turbulent flow in an open channel. J.Fluid Mech.86:109-127.
Sumer,B.M. & Deigaard,R. 1981. Particle motions near the bottom in turbulent flow in an open channel. Part 2. J.Fluid Mech. 109:311-337.
Theodorsen,T. 1952 Mechanism of turbulence Proc. 2nd Midwestern Conf. on Fluid Mech. Ohio State Univ.
Vanoni,V.A. 1975. River dynamics.Adv. in Appl. Mech.15:1-87.
Williams,P.B. & Kemp. P.H. 1971. Initiation of ripples on flat sediment beds. J.Hyd. Div. ASCE. HY4. 97: 505-520.
Willmarth,W.W.1975. Structure of turbulence in boundary layers.Adv.in Appl.Mech.15: 159-254.
Winant, C.D. & Browand F.K. 1974. Vortex pairing: the mechanism of turbulent mixing layer growth at moderate Reynolds number. J.Fluid Mech. 63:237-255.

Turbulent structure in an open-channel flow

H.O.ANWAR & R.ATKINS
Hydraulics Research Station Ltd., Wallingford, UK

SYNOPSIS: Instantaneous velocity components have been measured in the vertical and the mean flow directions in a two-dimensional open-channel flow. The results indicate the existence of large-scale structures superimposed on background small structures of isotropic behaviour. From the measured data the statistical properties of these structures, such as their contribution to the Reynolds shear stress, duration periods, frequencies, and the time periods between their occurrence have been determined throughout the water depth. From the measured data the auto-correlation functions and the energy spectra of the horizontal and vertical velocity fluctuations and their product have been evaluated.

INTRODUCTION

It is now known (1,4,6,8,10,11) that the turbulence in a plain boundary layer shear flow is dominated by the intermittent occurrence of a repetitive sequence of hydrodynamic events, characterised by 'bursting events'. These intermittent events consist of large-scale relatively organised structures of a coherent nature superimposed on background small-scale structures which exhibit isotropic behaviour. The contribution of the large-scale structure to the Reynolds shear stress is large. The results of the measurements presented here are intended to describe the characteristic behaviour of these events and their statistical values, which have been determined throughout the water depth in a two-dimensional shear flow.

The experiments have been conducted in a flume flow over a hydraulically smooth bed. The instantaneous velocity components in the mean flow and the vertical directions have been measured for two different values of the mean velocity.

From the measured data the mean velocity profiles, the auto-correlation functions and the statistical parameters of the bursting events, such as the mean time periods between events, their duration and frequency together with the energy spectra have been determined. It is interesting to note that there are quiescent time periods during which only the small-scale structure are active.

EXPERIMENT AND EQUIPMENT

The experimental study has been conducted in a flume with smooth boundaries. The flume was 27 m long, 0.6 m wide and 0.4 m deep. The water depth was 0.18 m, and a uniform two-dimensional flow has been confirmed by velocity and the Reynolds shear stress profiles measured at the working section, 19 m downstream from the entrance.

The instantaneous velocity components in the longitudinal and vertical directions have been measured using a 22 mm diameter discus type electromagnetic current meter obtained from Colnbrook Instrument Development Ltd (2). The mean velocity profiles in the flow directions have also been measured independently using a miniature propeller current meter and a good agreement was obtained between these two methods (2,3).

The slope of water surface has been measured with a series of twin-wire wave probes (3,5). The bed shear stress has been measured directly using a standard Preston tube, (2,3). The surface slopes and fitted logarithmic profiles to the measured velocity distributions have also been employed to determine the bed shear stress.

The experiments have been conducted at

two different reference velocities, $\overline{U}_r$ = 0.5 and 0.75 m/s, the reference velocity was measured near the free surface.

RESULTS

The arithmetical mean velocity $\overline{U}$ and the root mean square of the velocity fluctuations, $(\overline{u^2})^{\frac{1}{2}}$, in the mean flow direction have been evaluated from the digital data recorded on a magnetic tape as follows:

$$\overline{U} = \frac{1}{N}\sum_{i=1}^{N} U_i, \quad (\overline{u^2})^{\frac{1}{2}} = \{\frac{1}{N}\sum_{i=1}^{N}(U_i-\overline{U})^2\}^{1/2} \quad (1)$$

where U_i is the instantaneous longitudinal velocity. Similar expressions have been used for velocity in the vertical direction.

A logarithmic velocity profile expressed in the following form has been fitted to the measured velocity distribution by least squares regression at 95% confidence limits, i.e.:

$$\frac{\overline{U}}{U_*} = \frac{1}{K}\ln\frac{yU_*}{\nu} + B \quad (2)$$

where U_* is the friction velocity $(=\sqrt{\tau_o/\rho}$, τ_o being the bed shear stress), y is the height above the bed, K is von Karman's constant, taken as 0.41 and ν the kinematic viscosity of water. In Eqn.(2) B is an empirical constant representing the change in $\overline{U}$ across the viscous layer. A value for B of 6.8 has been obtained from the present data.

It will be assumed that in a boundary layer shear flow there exists fluid volumes, as a coherent structure, (1,6), with an average dimension equal to the eddy size. These eddy-like structures, similar to the macroscales in a turbulent shear flow, can be defined by their length- and time-scales as suggested by G.I.Taylor (15). For instance the Eulerian definition of the integral time-scale L_t at a point is defined by:

$$L_t = \int_o^{\infty} R(t)dt \quad (3)$$

where

$$R(t) = \int_o^{T} \frac{p(t_o)p(t_o+t)}{\overline{p^2}(t)} dt \quad (4)$$

in which t is the delay time and p is the turbulent fluctuation. It should be noted that the auto-correlation function, R(t), provides parameters of the macroscale structure, but not the structure of the turbulent flow.

A typical graphical representation of the auto-correlation functions for 5000 sampling points of u and the velocity fluctuation in the vertical direction v and their product uv are shown in Fig.1 by using discrete forms of Eqns.(3) and (4). The results indicate that the high correlation, for close spacing, approaches zero at the limit of the large eddies.

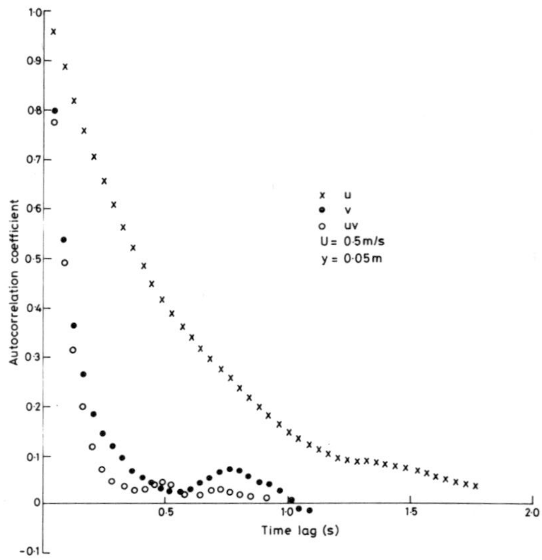

Fig.1. Auto-correlation functions of velocity fluctuation u in the mean flow direction and v in the vertical direction and their product uv.

The corresponding length scales or eddy sizes have been determined by calculating the area under each curve, to obtain L_t, and then multiplying by the advection velocity $\overline{U}$. The results in non-dimensional form are shown in Fig.2, where h = 0.18m is the water depth. It is to be noted that L_v (see Fig.2b) represents a horizontal dimension of the coherent eddy-like structure moving vertically and L_{uv} (Fig.2a) is a dimension of the large eddies responsible for the Reynolds shear stress and momentum transport. In Fig.2 are also shown the results of others for comparison. Fig.2 indicates that L_v and L_{uv} increase with increasing y in the region where $\frac{y}{h} < 0.5$ and remain almost unchanged when $\frac{y}{h} > 0.5$. Furthermore the results in Fig. 2 show that $0.2 \lesssim \frac{L_v}{L_u} \lesssim 0.3$. Although this scale-ratio increases from 0.2 near the bed to 0.3 near the free

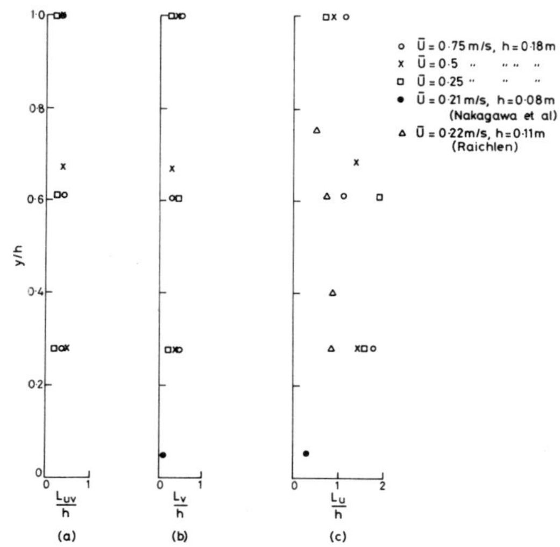

Fig.2. Eddy-dimensions (a) for the momentum transport, (b) and (c) for the vertical and horizontal movements of fluid elements.

surface, the flow is far from being isotropic, this will be discussed later. It is interesting to note that the horizontal dimension of the large-scale structures responsible for the vertical movement of fluid volumes and the momentum transport described by L_v and L_{uv} are less than ½ h, and not the water depth as would be expected.

The results also indicate that the time scales L_t, obtained from the v and uv auto-correlation coefficients, are related directly to the average period of bursting events of the flow. This will be shown later. It was pointed out previously that the results shown in Fig.2(a-c) indicate the dimensions of the large-scale structures, but not their contribution to the Reynolds shear stress. In order to determine these a computer program has been developed to identify specific features in the uv product time series as follows in relation to the directions of u and v:

v>0	u<0	Ejection event
v<0	u>0	Inrush event
v>0	u>0	Weak outward interaction event
v<0	u<0	Weak inward interaction event

The ejection events indicate the movement of fluid volume away from the boundary towards the outer region and the inrush events represent the motion towards the boundary from the outer region.

Fig.3 shows an example time series of 8 secs of the uv product together with u and v components. Fig. 3 further presents the way in which the events identified above contribute to the Reynolds stress in a highly intermittent manner; the ejection and inrush events make large, positive contributions to the Reynolds shear stress whereas those of the interactions are weak and negative, these will be discussed later.

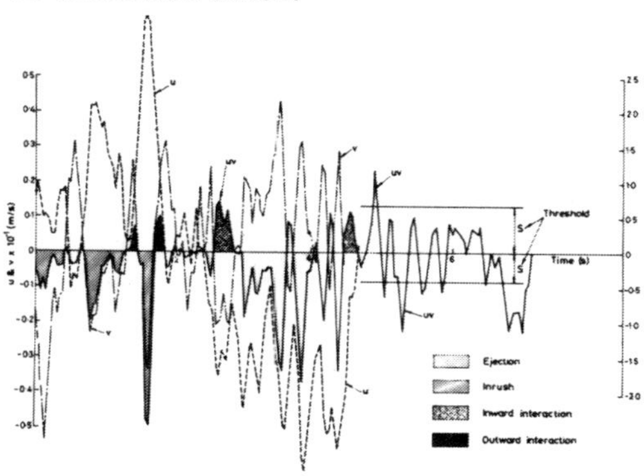

Fig.3. Typical time series of u, v and of uv product demonstrating (I) ejection of fluid elements away from the boundary (II) inrush events, the movement of fluid elements towards the boundary and (III) the weak interaction events of isotropic behaviour.

The bursting events classified above may provide a mechanism for the intermittent nature of sand grains being picked up by turbulent flow, a process which could be related to the ejection of fluid from near the bed into the high momentum region of the flow. Furthermore the time series of the uv product (see Fig.3) closely resembles the motion of a particle suspended in turbulent shear flow as studied by Sumer et al. (13,14).

It is evident from the above that the bursting events contribute greatly to the mechanism of turbulence and to the related hydrodynamic parameters. The computer program has been employed to evaluate various statistical parameters from the turbulent shear flow. The computations have been carried out for various values of thresholds, S, which are amplitudes in the uv-time series measured from the uv-zero-line in both the positive and negative directions (see Fig.3). By using the threshold S it is possible to compute the contribution of an event, say ejection at a level S. The evaluations have been terminated when the average contributed value of the event became statistically invariant. The threshold S was made dimensionless using σ the root-mean-square value of the uv-fluctuation.

THE REYNOLDS SHEAR STRESS

The contribution of the four above mentioned events, depicted in Fig. 3, to the Reynolds shear stress has been computed from the uv-time series. The computed average values for ejection, inrush and the interaction events have been normalised by $\overline{uv}$, and Fig.4(a) shows a typical set of results of this computation. This figure shows that the contribution of the ejection events to the Reynolds shear stress is significantly higher than that of the inrush events, i.e. about 70% due to ejection and about 50% from the inrush events at $\frac{S}{\sigma} = 0$.

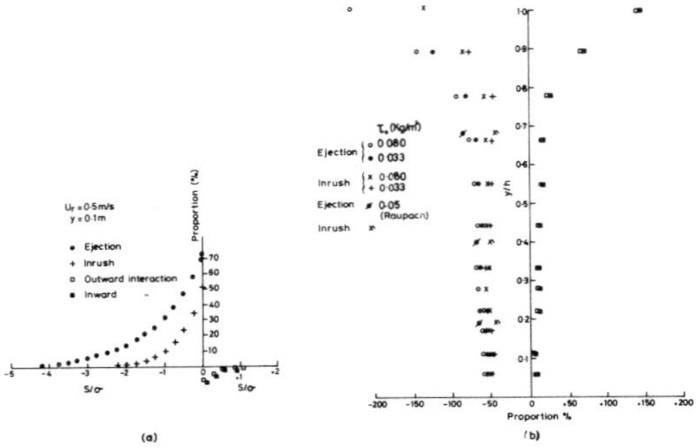

Fig.4. Contribution of events (a) over the non-dimensional threshold range S/σ and (b) throughout the water depth at S/σ = 0.

Fig.4(a) also demonstrates that the contribution of the interaction events to the Reynolds shear stress are small and negative. Fig.4(b) shows the maximum contributions to the Reynolds shear stress from the ejection and inrush events through the depth at $\frac{S}{\sigma} = 0$ and normalised by $-\overline{uv}$. It is evident that the contributions from the ejection events increases towards the free-surface, and to a lesser extent this is true for the inrush events. It has been found that the contributions of these events are reversed in the viscous layer where $\overset{+}{y} = \frac{yU_*}{\nu} \lesssim 15$, (4,12). From the above results it can be concluded that the momentum transport process is strongly asymmetric over the $\frac{S}{\sigma}$-range (see Fig.4 (a,b) and also throughout the depth h.). Fig.4(b) further shows that the contributions from the ejection and inrush events increase with increasing bed shear stress $\tau_o = \rho U_*^2$. In Fig.4(b) is also shown the results of Raupach (11) obtained in air as test fluid.

FREQUENCY OF EVENTS

The frequencies of the ejection, inrush and interaction events have been determined at various y, and the results for the non-dimensional threshold $\frac{S}{\sigma}$ are shown in Fig.5. Fig.5(a) shows that the frequency of ejection is higher than that of the inrush when $\frac{S}{\sigma} < -0.5$ and the reverse is true when $\frac{S}{\sigma}$ approaches zero, while the frequencies of the two interaction events remain much the same over the threshold range. With the existing electromagnetic current meter it has been found that the velocities are being affected by the rigid boundary when $y \lesssim 40$ mm, (2,3), hence on the basis of the argument given previously it is reasonable to assume that the inrush frequencies will be higher than that of ejection in $\overset{+}{y} \lesssim 15$ region. Fig.5(b) shows that the frequencies, at $\frac{S}{\sigma} = 0$, of the ejection and inrush events increase towards the free surface. It is interesting to note that the frequencies of these events become equal at $\frac{S}{\sigma} \approx -0.5$ (see Fig.5(a))

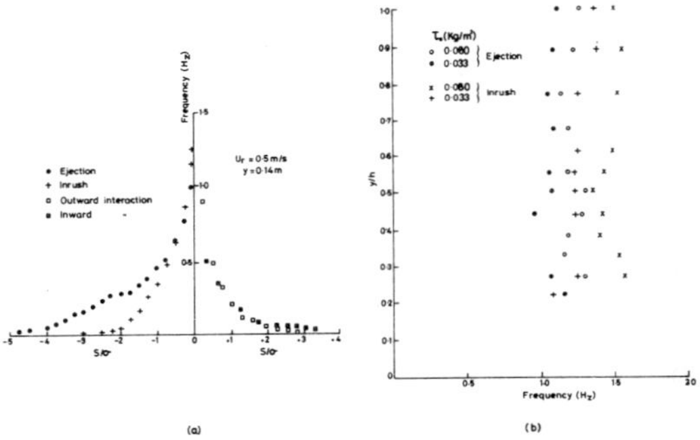

Fig.5. Frequency of events (a) over non-dimensional threshold range S/σ and (b) throughout the water depth at S/σ = 0.

and the frequency of the inrush events is higher than that of the ejection events. Furthermore Fig.5(b) shows that the difference frequency between ejection and inrush increased towards the free surface.

DURATION OF EVENTS

It has been shown previously that the momentum-transporting process comprises fluctuating events generally of a well-defined sequence, namely ejection followed by inrush and then by a period of interaction events. From the uv product time-series the fraction of time during which the uv signals become active for specified events has been evaluated, and the results

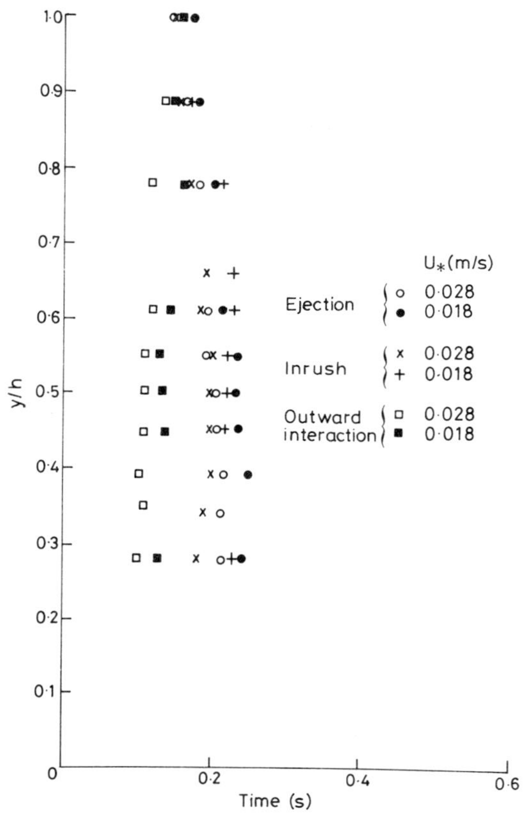

Fig.6. Duration periods of events at non-dimensional threshold $S/\sigma = 0$.

for ejection and inrush at various y are shown in Fig.6. The results show that the mean duration periods of the ejection event is longest in the bed region (y = 40 mm) and it decreased towards the free surface. Moreover the results show that the duration of the inrush events remains almost unchanged in the central region and decreases towards both the free surface and the bed. Fig.6 further shows that the mean duration periods of the interaction events are largest near the free surface and reduces towards the bed.

The mean duration periods between ejection and inrush events have been computed using the following expression, (1):

$$T = \frac{\text{Record time length} - (T_{ej} + T_{in})}{1 + \text{number of ejection and inrush events}}$$

The results are shown in Table 1 together with the integral time-scale $T_{L_{uv}}$ and T_{L_v}, calculated previously. Comparison indicates that $T_{L_{uv}} \approx T_{L_v} \approx \frac{1}{2}T$.

Table 1. Mean duration of periods T between ejection and inrush and time-scale $T_{L_{uv}}$ and T_{L_v} obtained from auto-correlation functions (see Fig.1).

	$\bar{U}_r$ = 0.5 (m/s)		$\bar{U}_r$ = 0.75 (m/s)		
y/h	1	0.28	1	0.61	0.28
T(s)	0.24	0.195	0.23	0.17	0.15
$T_{L_{uv}}$ (s)	0.125	0.120	0.09	0.1	0.075
T_{L_v} (s)	0.1	0.125	0.1	0.085	0.09

It can be concluded that ejection and inrush events represent the large-scale structures responsible for the transport of momentum among large-scale eddies.

It has been suggested (9,12) that the time period T is related directly to the main flow parameters, i.e. the boundary layer thickness δ assumed here to be the water depth h and $\bar{U}_r$ so that

$$\alpha = \frac{T\bar{U}_r}{h}$$

in which α is an empirical constant. The results of the present investigation shown in Fig.7 indicate that α varies throughout the water depth; it increases towards the free surface and remains almost unchanged in the mid-water depth. The values of α shown in Fig.7 are much lower than the range 3 to 7 found from records of the velocity fluctuation u close to bed (12).

ENERGY SPECTRA

The energy spectrum of turbulence describes the distribution of energy among the eddies of different sizes. The spectrum has been evaluated using a Fast Fourier transform analysis to obtain one-dimensional wave number spectrum, $E_{pj}(K)$, where p is turbulent velocity fluctuation defined previously, j is the jth record, and the wave number $K = 2\pi n\bar{U}^{-1}$ at a frequency n. The wave number K was made dimensionless using the water depth h($K^* = Kh$) and the spectra by $\int_o^\infty E_p(K)dK$ the total variance.

Typical energy spectra of u and v and their product uv in their normalised forms are shown in Figs.8-10. The figures

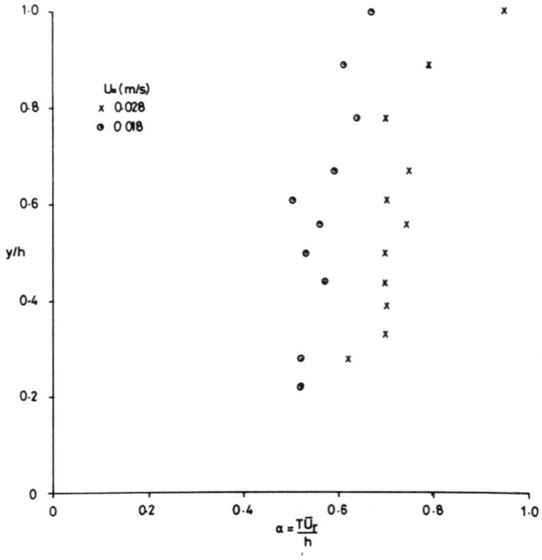

Fig.7. Variation of constant α throughout the water depth.

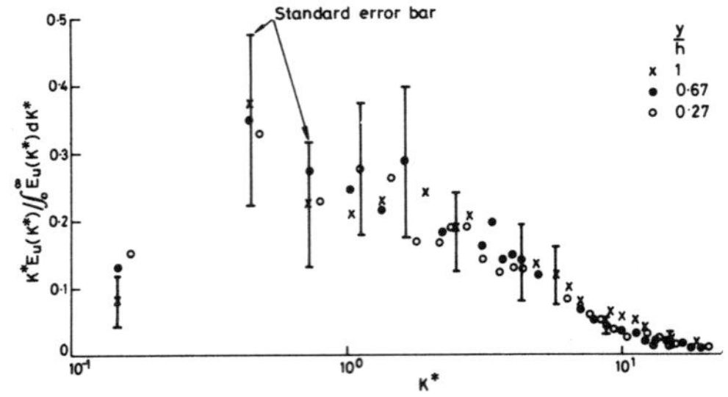

Fig.8. Normalised energy spectra of horizontal velocity fluctuation.

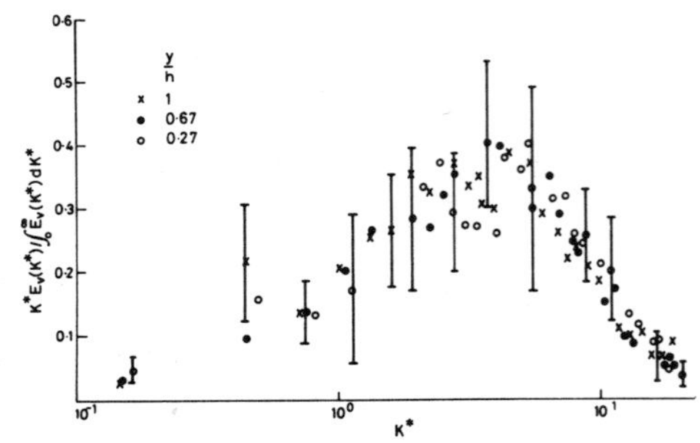

Fig.9. Normalised energy spectra of vertical velocity fluctuation.

show that the normalised spectra are almost coincident over the entire range of K^*, hence the above scaling is acceptable. Furthermore it appears that the shapes of the spectra for v and uv are identical and their peaks occur at almost the same values of K^*. This means that the eddies of the same sizes are responsible for the vertical movement of the fluid elements and the momentum transport.

The u and v spectra have shown a region, in the frequency range $1 \lesssim n \lesssim 5$, in which the spectra appear to follow a $-\frac{5}{3}$ law, but it was found, by evaluating $E_v(K)/E_u(K)$ in this region, that this ratio differs greatly from the expected value $\frac{4}{3}$, usually associated with isotropic flow. Hence it can be concluded that the $-\frac{5}{3}$ law does not necessarily mean that turbulent flow is isotropic when the Reynolds number appears high. The Reynolds numbers of the present investigation, based on $\bar{U}_r$ and h, were about 1×10^5, viz, not high enough to produce an isotropic turbulent flow.

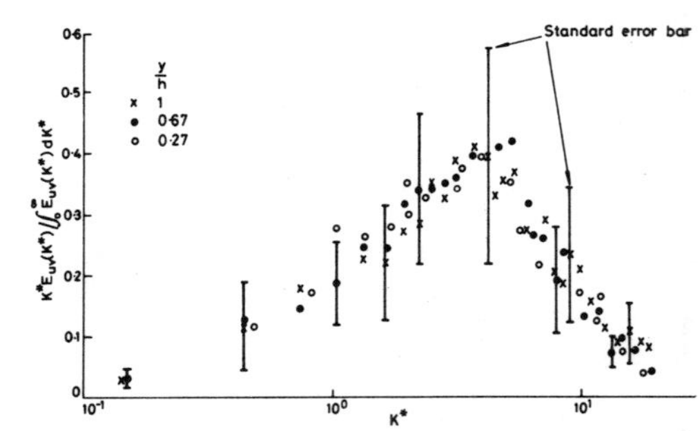

Fig.10. Normalised co-spectra of horizontal and vertical fluctuations.

ACKNOWLEDGEMENTS

The work described here was carried out as part of the research programme of the Hydraulics Research Station, and is published with the permission of the Director of Hydraulics Research Station.

REFERENCES

1. Anwar, H.O. 1981. A study of the turbulent structure in a tidal flow. Estuarine, Coastal and Shelf Science. 13, 373-387.
2. Atkins, R. 1980. Turbulence measurements using a small electromagnetic current meter in open channel flows. Hydraulics Research Station, Wallingford. Report No. IT 196.
3. Anwar, H.O. & Atkins, R. 1980. Turbulence measurements in simulated tidal flow. Proc. ASCE, 106, HY8, 1273-1289.
4. Brodkey, R.S., Wallace, J.M. & Eckelmann, H. 1974. Some properties of truncated turbulence signals in bounded shear flow. Journal of Fluid mechanics. 63, 209-224.
5. Fryer, D.K. & Thomas, M.W.S. 1975. A linear twin-wire probe for measuring water waves. Journal of Physics E: Scientific Instruments, 8, 405-408.
6. Gordon, C.M. 1974. Intermittent momentum transport in a geophysical boundary layer. Nature, 248, 392-394.
7. Grass, A.J. 1971. Structural features of turbulent flow over smooth and rough boundaries. Journal of Fluid Mechanics. 50, 233-255.
8. Heathershaw, A.D. 1979. The turbulent structure of the bottom boundary layer in a tidal current. Geophysical Journal of the Royal Astronomical Society. 58, 395-430.
9. Jackson, R.G. 1976. Sedimentological and fluid-dynamic applications of the turbulent bursting phenomenon in geophysical flows. Journal of Fluid Mechanics. 77, 531-560.
10. Nakagawa, H. & Nezu, J. 1981. Structure of space-time correlations of bursting phenomena in an open-channel flow. Journal of Fluid Mechanics. 104, 1-43.
11. Raupach, M.R. 1981. Conditional statistics of Reynolds stress in rough-wall and smooth-wall turbulent boundary layers. Journal of Fluid Mechanics. 108, 363-382.
12. Rao, K.N., Narasimha, R. & Badri Narayanan, M.A. 1971. The bursting phenomenon in a turbulent boundary layer. Journal of Fluid Mechanics. 48, 339-352.
13. Sumer, B.M. & Oguz, B. 1978. Particle motions near the bottom in turbulent flow in an open channel. Journal of Fluid Mechanics. 86, 109-127.
14. Sumer, B.M. & Deigaard, R. 1981. Particle motions near the bottom in turbulent flow in an open channel. Part 2. Journal of Fluid Mechanics. 109, 311-337.
15. Taylor, G.I. 1953. Statistical theory of turbulence. Proc. Royal Society. A151, 421-454.

Euromech 156: Mechanics of Sediment Transport / Istanbul / 12-14 July 1982

Sediment entrainment by a turbulent spot: A progress report

F.K.BROWAND, D.OSTER & D.PLOCHER
University of Southern California, Los Angeles, USA
D.McLAUGHLIN
Dynamics Technology, Torrance, Calif., USA

ABSTRACT: A turbulent spot is allowed to pass over a bed containing artificial sediment. Particles entrained into the flow are tracked using a digital camera.

Introduction

A turbulent spot might be thought of as an isolated patch of turbulence in an otherwise laminar boundary layer. Spots have been studied since the mid fifties, when first noted by Emmons, but many of the interesting properties of spots have come to light only recently. For example, spots appear to be composed of substructures, vortex features, arranged in a rough pattern (Wygnanski 1981). Spots "grow" by adding (or generating) new vortex features. Although much study is needed to make a complete correspondence, the vortex features in spots appear similar to the features observed in fully turbulent boundary layers. If the spot is indeed the prototypal turbulent boundary layer, much can be learned by studying the entrainment properties of spots. Here, there are several distinct advantages. There is a beginning and an end to each spot, and the vortex sub-structures tend to have an organized alignment which is similar from spot to spot. This pattern can be exploited by sampling with respect to the passage of each spot front. In this way, an ensemble can be constructed which represents the contribution to entrainment of the individual vortex structures. On the basis of this information, we ultimately understand the process of entrainment in a fully turbulent boundary layer. Our spot experiment is in the early stages. The following paragraphs describe the major features of the experimental apparatus and the data logging technique.

Experimental apparatus

Turbulent spots are produced in a laminar boundary layer by puffing at intervals through a small hole near the leading edge of a flat plate mounted in a free-surface water channel. The arrangement is shown schematically in Figure 1. Once generated, the individual spot grows in a characteristic fashion, and eventually passes over a circular portion of the plate containing an artificial sediment of plastic beads. During the spot passage, the plastic beads are lifted upward into a vertical light sheet, where position and velocity are determined by means of a digital camera controlled by a microcomputer. After spot passage, the flow returns to a laminar state, the bed of erodible material is made flush with the surface of the flat plate by incrementing a piston and screw mechanism in the supply tube, and the sequence is repeated -- perhaps 500-600 times.

The flow velocity and boundary layer thickness are largely dictated by the nature of the experiment. The undisturbed flow must remain laminar, and this places an upper limit upon the obtainable Reynolds number, or equivalently, upon the product of velocity and boundary layer thickness. The smallest free stream velocity which will give reasonable values for wall shear stress is of the order of 20-25 cm/sec. This in turn fixes the boundary layer thickness (and the maximum vertical scale in the turbulent spot). Spots occur spontaneously in our channel at a transition Reynolds number, $Re_x = Ux/\nu = 7\times10^5$, corresponding to a distance of 270 cm from the plate leading edge at a free stream velocity of 25 cm/sec. (Sidewall contamination may be responsible for this transition. In the best of circumstances, natural transition could be extended to about $1\text{-}1.5\times10^6$.) The plastic bead erodible bed is presently located at 135 cm from the leading edge, and the spot is examined in the downstream region be-

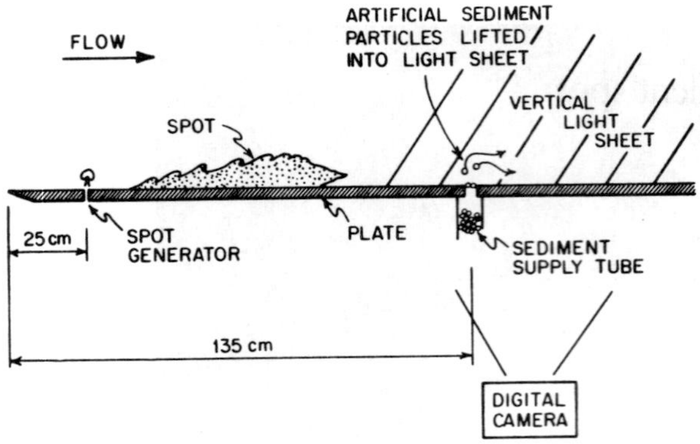

Fig. 1. Sketch of experiment.

tween 120 cm and 200 cm -- the corresponding Reynolds numbers are $3\text{-}5 \times 10^5$. The equivalent displacement thickness Reynolds number for the laminar boundary layer, $U\delta^*/\nu = 1.72(Ux/\nu)^{\frac{1}{2}}$, lies in the range 1000-1200. The spot generator is located 25 cm from the plate leading edge. The displacement thickness Reynolds number here is 430 -- just at the critical Reynolds number for instability.

The boundary layer plate has a width of 91 cm, and is mounted 25 cm below the free surface and 8 cm above the channel bottom. The pressure difference above and below the plate can be controlled by opening and closing the downstream gates shown in Figure 2. The channel is operated with a slightly negative pressure below the plate, and this has two advantages. First, it allows the attaching streamline at the leading edge of the plate to be favorably positioned to avoid separation over the upper, working surface. Second, the pressure difference allows the corner flow between the upper plate surface and the sidewall to be removed by means of a suction slot -- in turn greatly increasing the downstream distance available to observe the growing spot. In practice, the suction slot is set to produce a flow toward the corner having a limiting streamline slope of about 1 in 10 near the corner. This slight horizontal divergence in the boundary layer near the sidewalls has no noticeable effect on the flow in the central 30 cm where the spots are produced.

A turbulent spot is visualized by means of a unique laser fluorescence technique. The photograph in Figure 3 was taken by releasing a thin sheet of fluorescein over

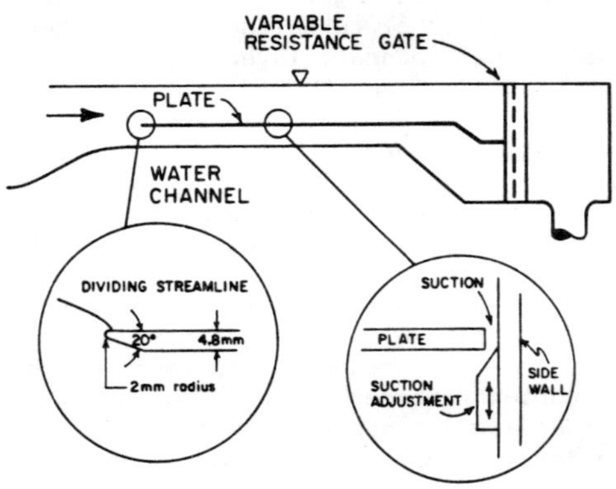

Fig. 2. Detail of boundary layer plate.

the surface of the plate just below a horizontal sheet of laser light. When the turbulent spot passes, fluorescein is lifted into the laser sheet and the fluorescence is photographed from above. In plan view, the spot measures approximately 25 cm by 40 cm. The vertical extent of the spot is about 3 cm (three times the thickness of the laminar boundary layer), and the plane of the laser light in Figure 3 is roughly 15% of the spot height above the wall. The spot side view is taken using ordinary red food coloring and backlighting.

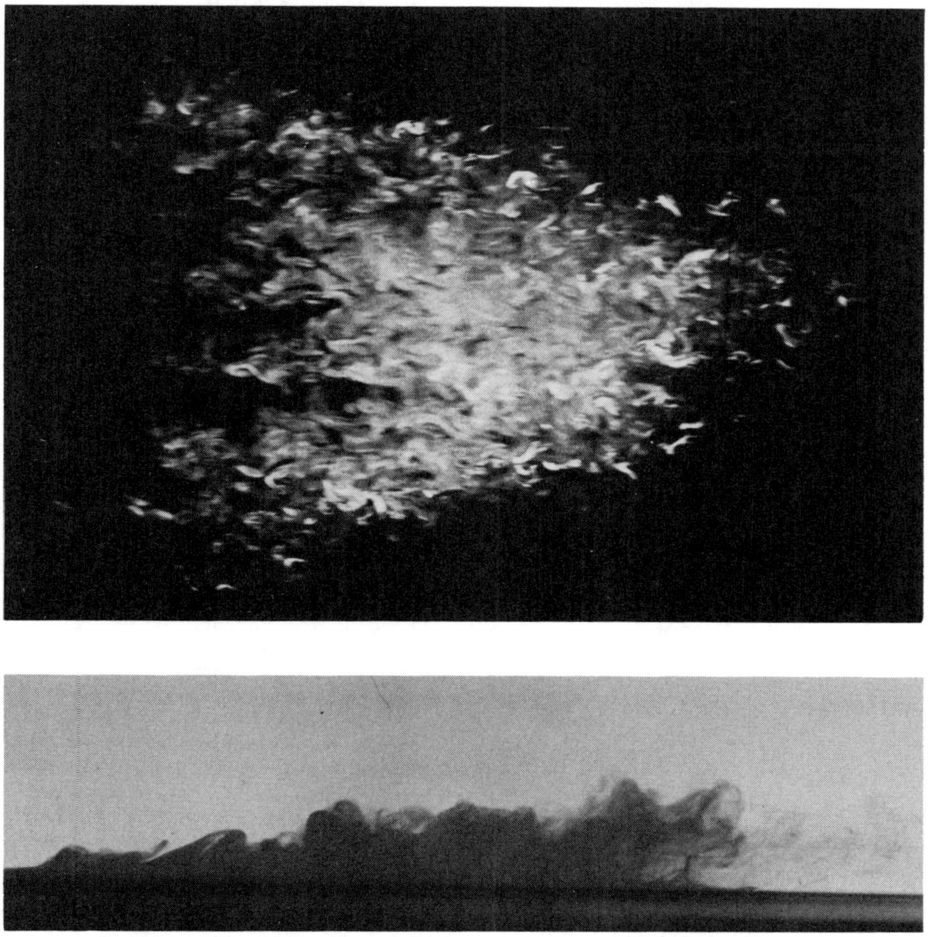

Fig. 3. Visualization of typical spot.

The wall shear velocity within the turbulent spot is in the range 1-2 cm/sec for free stream velocities in the range 20-30 cm/sec. This represents a small shear stress, and the fall velocity of the artificial sediment must be suitably scaled. The particles are raw polystyrene beads commonly used in the manufacture of styrofoam. Large quantities of beads can be obtained in the size range 300µm-1500µm and having various weights. Furthermore, the effective weight can be changed by slight heating. We are presently experimenting with beads in the range 500µm-1000µm, having fall velocities in the range 0.5-1.5 cm/sec. (The viscous wall layer is approximately 1 mm in thickness, and although the particles are a substantial fraction of this dimension, the smooth wall regime probably still obtains.)

Beads must be sorted while maintaining close tolerance on size and fall velocity. As a further complication, the beads absorb water and change weight slightly when immersed. The simplest procedure is to sort by size using ordinary sieves. Typical examples of fall-time probability distributions for sieve sorted samples are shown in Figure 4 as a function of immersion time. The samples have a standard deviation in fall velocity (fall velocity is inversely proportional to fall time) of about $\pm 5\%$, which seems to improve slightly with increased immersion time. It appears from these results that a total change of about 30% in fall velocity will occur after approximately 60 days. We are also designing a bead-dispenser-and-collector apparatus to be placed in the water channel, which will sort beads directly according to fall velocity. Our initial estimates using direct sorting is that the standard deviation can be improved to about ±2%.

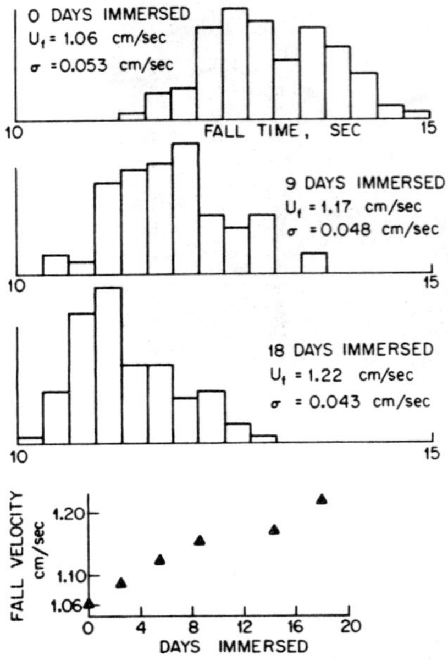

Fig. 4. Distributions of particle fall times for sorted samples.

Data acquisition system

The data acquisition system is controlled by a DEC LSI 11/23 microprocessor. The various elements are shown schematically in Figure 5. The 11/23 has 256 K bytes of core memory and a data storage capability of 21 M bytes (two RL02 discs). The Tektronics terminal allows sophisticated graphics, including the use of 3-dimensional plotting routines. Sensor outputs, such as hot-film voltages for turbulent velocity measurement, are input into the computer through a 12 bit A to D converter manufactured by Data Translation. A single channel can be digitized at a rate of 125 K Hz. For multiple channels, the aggregate rate drops. For example, 10 data channels can be digitized at a rate of 5 K Hz per channel.

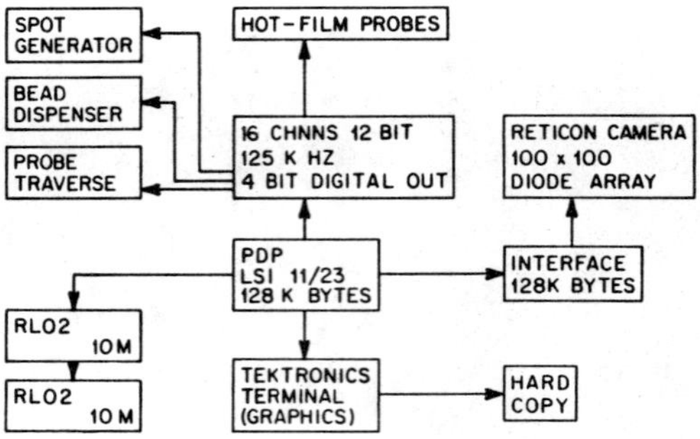

Fig. 5. Block diagram of data acquisition system.

The novel portion of the acquisition system is the digital camera manufactured by Reticon Corporation. It consists of a photodiode matrix containing 10,000 elements. The array measures 0.6x0.6 cm, and is assembled on the surface of an integrated circuit. Reflected light from the entrained sediment particles is imaged on the diode array using conventional camera optics. Twelve digital images (12x10000, 8 bit bytes) are stored in memory locations on the interface board. Within limits, the camera has independently controlled shuttering and framing rates, with a maximum framing rate of 180 pictures per second. When the spot has passed, the particle position information in each frame is transferred to disc storage, and the system is readied for another spot passage. Since only 10-20 particles are anticipated to be in the field of view at any one time, there is a considerable compression of the data actually written to disc. The camera interface, which implements commands from the computer and controls output data flow, is the only non-standard item in the acquisition system. A series of hardware and software problems with the interface have slowed our progress considerably. We hope to have these difficulties resolved shortly.

Deductions about turbulent entrainment of bed material will require many spots and many particles -- that is, large amounts of data. Our goal is to achieve an interactive system with simultaneous data acquisition and processing. The computer will control most aspects of the experiment including probe positioning, generation of turbulent spots, and the resupply of sediment particles. The last two figures are examples of our real-time data processing capability applied to hot-film velocity measurement. The hot-film output is first calibrated against a small propellor meter located in the free stream. Channel velocity is incremented by changing the motor RPM until a sufficient number of calibration points are obtained. The computer then displays the result of a least squares, 4th order polynomial fit; the coefficients are stored if the fit is acceptable. The whole procedure takes perhaps 15 minutes. Figure 6 shows two laminar boundary layer profiles -- one recorded at the location of the spot generator (25 cm from the leading edge), and one recorded 15 cm upstream from the position of the artificial sediment bed (120 cm from the leading edge). These profiles show good agreement with the Blasius boundary layer solution.

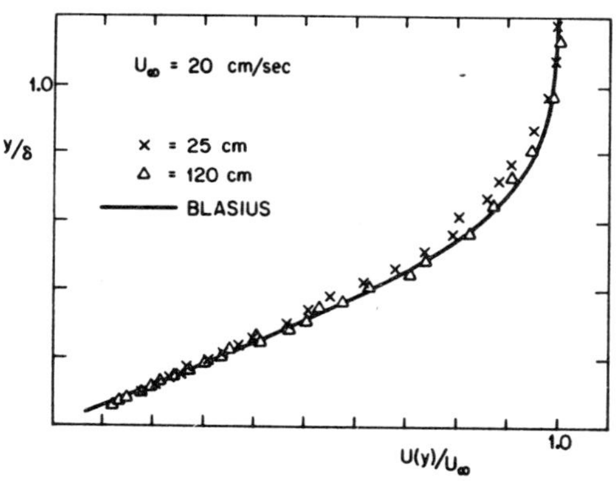

Fig. 6. Measured boundary layer profiles.

Turbulent spots are produced by a 50 m sec computer pulse initiated every 20 seconds. The computer waits 12 seconds for the spot to arrive at the downstream location; the hot-film output is then digitized for an eight second interval (8192 data points). The number of spots to be produced is entered in the computer, and the velocity trace for each spot passage is automatically displayed. Otherwise, the experiment will proceed unattended. The first three traces in Figure 7 are individual spots -- each different from one another (especially the second one), but having characteristic features in common. The fourth trace represents an ensemble of 100 spots. The ensemble average is comparable to data obtained by other investigators (c.f. Cantwell, Coles and Dimotakis 1978). Much more of the internal spot structure could be educed by using an elaborate averaging technique to take into account the slightly different arrival times and the different sizes of individual spots (Wygnanski 1981).

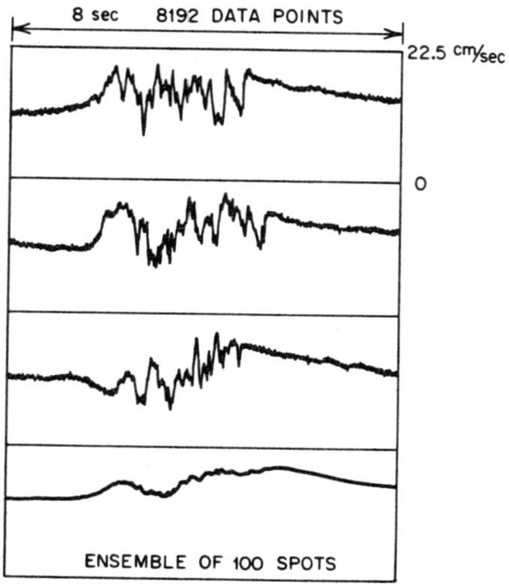

Fig. 7. Velocity signatures of turbulent spots.

Concluding remarks

We hope to promote an understanding of the relationship between the entrainment of bed material and the fluid mechanic structure of turbulent spots (and, by inference, turbulent boundary layers). There is a duality here which can be exploited. We bring to bear upon the entrainment problem the present knowledge of the fluid mechanics of spots. But particle entrainment can also be used as a technique for exploring the structure of turbulent spots. Many realizations are needed to demonstrate correlations or causal relationships in turbulent flow. Thus, we are prepared to process large quantities of particle position and velocity data.

The mapping of entrainment in spots will be accomplished by controlling two parameters:

(a) the size and location of the erodible bed with respect to the spot generator;

(b) the diameter and fall velocity of the artificial sediment.

Controlling particle fall velocity with respect to the wall shear velocity, allows precise adjustment of entrainment. For lighter particles, which fall more slowly, many portions of the spot will contribute to entrainment. For slightly heavier particles, only the most energetic portions of the spot will be capable of lifting particles. These energetic regions can be related to spot structure.

To be of general use, the experimental observations must provide a predictive capability. Initially, this capability may take the form of scaling arguments, but ultimately, we hope the results can be incorporated into a numerical computation scheme. The suspended load in a turbulent flow has conventionally been calculated from a time-mean concentration distribution obtained by integrating the convective/diffusion equation. This equation contains the turbulent diffusivity of sediment which is related by Reynolds Analogy arguments to the diffusivity of fluid momentum. We will try to extend this model concept to describe the spatially non-uniform field within a turbulent spot by making use of the ensemble measurements to provide the spatial distribution of mean flow and the appropriate second order turbulence quantities. We will, of course, have a substantial data base of particle motion and concentration measurements to compare with our numerical results.

Acknowledgements

This work is supported by the National Science Foundation under Grants CME 80-03427 and CME 79-10730.

References

Cantwell, B., Coles, D. & Dimotakis, P. 1978. Structure and entrainment of a turbulent spot, J. Fluid Mech (87): 641-672.

Wygnanski, Israel, 1981. Transition and turbulent spots, Turbulence Symposium Proceedings, Ed., G.K. Patterson and J.L. Zakin, Department of Chemical Engineering, University of Missouri-Rolla.

Boundary shear stress distributions in open channel and closed conduit flows

D.W.KNIGHT & H.S.PATEL
University of Birmingham, UK
J.D.DEMETRIOU
National Technical University of Athens, Greece
M.E.HAMED
Al-Azhar University, Egypt

ABSTRACT: The results of some laboratory experiments are reported concerning the distribution of boundary shear stresses in smooth open channels and closed conduits with a rectangular cross section. Perturbations in these distributions are explained in terms of secondary flows and the formation of pairs of contra-rotating cells adjacent to the long side of the rectangle. The proportion of the total shear force which acts on the channel or duct walls is shown to vary systematically with aspect ratio, but not in accordance to the ratio of wall to overall perimeter lengths. Equations are presented for the variation of the wall shear force, mean wall and bed shear stresses, and centreline bed shear stress with aspect ratio. The differences and similarities between open channel and closed conduit flows at comparable aspect ratios are noted.

1 INTRODUCTION

The distribution of boundary shear stresses in open channel and closed conduit ducts of arbitrary cross section remains an analytically intractable problem due to the nature of the governing equations. This is particularly unfortunate as regards sediment mechanics, on account of the common use of the local boundary shear stress or shear velocity as a primary variable. Moreover, certain aspects of sediment behaviour are extremely sensitive to small changes in the boundary shear stress. For example, sediment transport rates normally depend upon the shear velocity to the fifth or sixth power. For these reasons, any information concerning the detailed behaviour of boundary shear stress distributions is of considerable fundamental and practical interest.

The time averaged Reynolds equation in the x-direction is given by equation (1), where {UVW} are the mean velocities in the {xyz} directions and {uvw} are the turbulent fluctuating components. The longitudinal pressure gradient, $\partial p/\partial x$, (or resolved weight force in open channel flow) is thus seen to be expended against either turbulent shear forces or in driving the secondary motion (transverse velocity components V and W). Differentiation of the corresponding Reynolds equations in the y and z directions gives the mean vorticity equation for ξ in the x direction, equation (2), in which $\xi = \partial W/\partial y - \partial V/\partial z$, $\eta = \partial U/\partial z - \partial W/\partial x$ and $\zeta = \partial V/\partial x - \partial U/\partial y$.

$$U\frac{\partial U}{\partial x} + V\frac{\partial U}{\partial y} + W\frac{\partial U}{\partial z} = -\frac{1}{\rho}\frac{\partial p}{\partial x} - \left(\frac{\partial \overline{u^2}}{\partial x} + \frac{\partial \overline{uv}}{\partial y} + \frac{\partial \overline{uw}}{\partial z}\right) + \nu\left(\frac{\partial^2 U}{\partial x^2} + \frac{\partial^2 U}{\partial y^2} + \frac{\partial^2 U}{\partial z^2}\right) \quad ......(1)$$

$$\underbrace{U\frac{\partial \xi}{\partial x} + V\frac{\partial \xi}{\partial y} + W\frac{\partial \xi}{\partial z}}_{I} = \underbrace{\nu\left(\frac{\partial^2 \xi}{\partial x^2} + \frac{\partial^2 \xi}{\partial y^2} + \frac{\partial^2 \xi}{\partial z^2}\right)}_{II} + \underbrace{\xi\frac{\partial U}{\partial x}}_{III} + \underbrace{\eta\frac{\partial U}{\partial y} + \zeta\frac{\partial U}{\partial z}}_{IV} + \underbrace{\frac{\partial}{\partial x}\left(\frac{\partial \overline{uv}}{\partial z} - \frac{\partial \overline{uw}}{\partial y}\right)}_{V}$$

$$+ \underbrace{\frac{\partial^2}{\partial y \partial z}(\overline{v^2} - \overline{w^2})}_{VI} + \underbrace{\left(\frac{\partial^2}{\partial z^2} - \frac{\partial^2}{\partial y^2}\right)\overline{vw}}_{VII} \quad ......(2)$$

The formation of secondary flow cells (distribution of ξ) in corner regions has been studied extensively in recent years on account of its importance in calibrating three-dimensional turbulence models (e.g. Brundrett & Baines 1964, Gessner 1973, Melling & Whitelaw 1976, Gosman & Rapley 1980). The physical significance of the various terms in equations (1) & (2) has been demonstrated by a number of different investigators (e.g. Einstein & Li 1958, Tracy 1965, Perkins 1970). In equation (2), the terms on the left hand side, I, represent the convection of streamwise vorticity by the primary and secondary flows, and on the right hand side the terms in II account for the viscous diffusion of the streamwise vorticity, term III represents streamwise vortex stretching term IV the production of streamwise vorticity, and terms V to VII the production of vorticity by turbulence. Because the region of diffusion generally occurs nearer a boundary than the region of production, convection of vorticity is assumed to take place between the two regions, indicating the importance of terms I, II and VI. The corresponding equation for ζ may be shown to be a more suitable equation for investigating the mechanisms which generate secondary flow on account of its terms being an order of magnitude greater than those for ξ. Whatever vorticity component is examined however, it is clear that the transverse gradients of the Reynolds stress components which act wholly in planes normal to the primary flow direction, i.e. $\overline{v^2}$, $\overline{w^2}$ and $\overline{vw}$ (terms VI and VII in equation (2)), are primarily responsible for the production of axial flow vorticity.

Little attention has been paid to the influence of streamwise vorticity, although its effect upon boundary shear stresses has been postulated for some time (Perkins 1970). The formation of multi-cellular secondary flows adjacent to the long side in rectangular sections is usually ignored in most hydraulic flows, but as will be shown later, these secondary flows influence the boundary shear stress distributions to a considerable extent. See Figure 1. The experiments described in the following sections were aimed at evaluating this influence in both open channel and closed conduit flows. On account of the influence of the free surface and the different secondary flow structures, the two flows are not directly comparable at corresponding aspect ratios. Although the relevance of the results to sediment mechanics is emphasized, it should be recognized that they have a similar bearing upon dispersion and resistance studies.

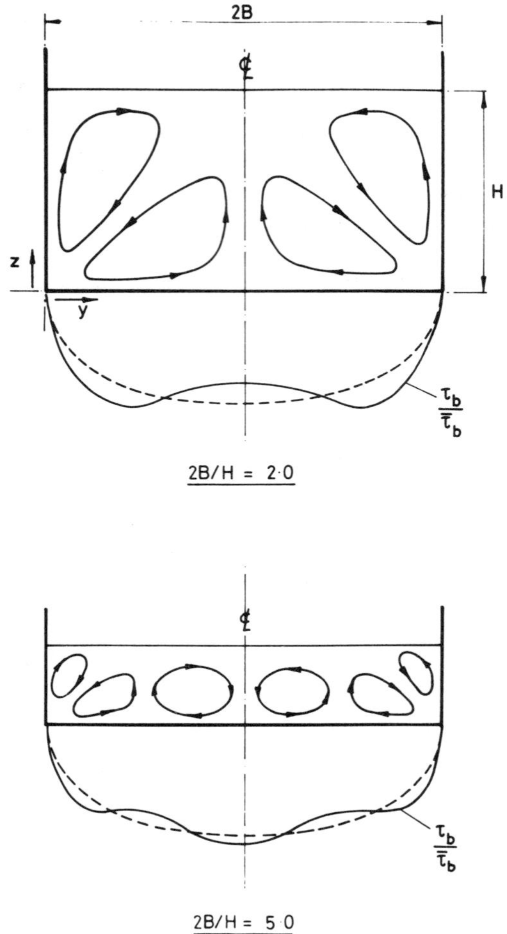

Fig.1. Influence of secondary flow cells on boundary shear stresses.

2 EXPERIMENTAL PROCEDURE

Experiments were conducted in several hydraulic flumes and a wind tunnel facility over a range of aspect ratios. In each case, boundary shear stress distributions were measured with various sized Preston tubes, using standard calibration equations (Patel 1965).

2.1 Open channel flow studies

A 610mm wide perspex flume, 15m in length, and set at a bed slope of 9.66×10^{-4} was constructed in such a way that its sidewalls could be adjusted to give four widths 610, 381, 152 and 70mm. A second flume 460mm wide and set at a bed slope of 9.58 x 10^{-4} was used to supplement the results from the four individual flumes described earlier. Stage discharge curves were obtained for each flume under uniform flow conditions by a series of preliminary experiments. The measured boundary shear stress distributions were averaged to give a section mean value, $\bar{\tau}_o$, which was then compared with the energy gradient mean value, $\bar{\tau}_e$ (= ρgRS_f). The mean error between $\bar{\tau}_o$ and $\bar{\tau}_e$ was + 1.5% with a standard deviation of 4.4%. The individual shear stress readings and distributions were subsequently adjusted to give the correct mean energy slope value. Further details are given elsewhere (Knight Demetriou & Hamed 1983a, Knight, Demetriou & Hamed 1983b).

2.2 Closed conduit flow studies

A 17m long test rig was constructed in such a way that a 9.25m long section of it could be adjusted to give a wind tunnel with a working cross section of variable dimensions, ranging between 165 x 165mm and 400 x 40mm. Two variable speed contra-rotating fans sucked air through the rectangular duct at various flow rates which were measured by means of an orifice meter. Pressure tappings along the tunnel allowed the longitudinal pressure gradient to be measured accurately. The measured boundary shear stress distributions were again averaged to give the section mean value, $\bar{\tau}_o$ which was then compared with the energy gradient mean value, $\bar{\tau}_e$ ($=R\partial p/\partial x$). The mean error between $\bar{\tau}_o$ and $\bar{\tau}_e$ was -2.3% with a standard deviation of 6.4%. Individual shear stress readings and distributions were adjusted in a similar manner to that for open channel flow. Further details are given elsewhere (Knight & Patel 1983).

3 EXPERIMENTAL RESULTS

In order to compare the open channel and closed conduit data, the dimensions of the working section of the wind tunnel were taken to be 2B x 2H, with the long side, 2B, denoted as the 'bed' and the short side 2H, denoted as the 'wall'. The aspect ratio, B/H, is therefore half the value normally associated with open channel flow 2B/H.

3.1 Boundary shear stress distributions

Values of the local wall or bed shear stress, τ_w and τ_b, were non-dimensionalized by the mean wall and bed shear stresses, $\bar{\tau}_w$ and $\bar{\tau}_b$, and plotted against z/H or y/B respectively. A typical set of bed shear stress distributions is shown in Figure 2 for aspect ratios, 2B/H, between 2.0 and 7.73. One of the most interesting features of these distributions, is their correspondence with the distributions shown in Figure 1, thus indicating that the secondary flow behaviour may be detected. It is evident from Figure 2 that various perturbations in the distributions are present in both open channel and closed conduit results, which on the basis of Figure 1, correspond to secondary flows either towards or away from the bed. As the aspect ratio increases, each additional

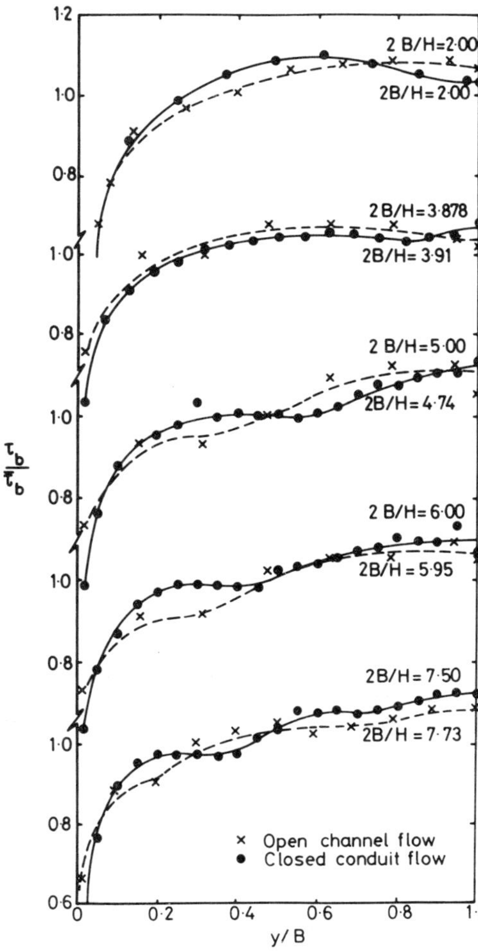

Fig.2. Bed shear stress distributions

pair of contra-rotating secondary flow cells introduced about the centreline, causes a reversal of the mean secondary flow normal to the bed at the centreline, and additional perturbations in the distributions elsewhere. The closed conduit results exhibit somewhat larger perturbations than the open channel flow results. A more detailed analysis of these perturbations (Knight & Patel 1983) reveals the presence of up to 28 secondary flow cells in the full section of the wind tunnel. Similar perturbations have been observed by other investigators (Leutheusser 1963) for an aspect ratio of 6.0. The influence of streamwise vorticity, ξ, on boundary shear stress distributions is thus shown to be considerable.

3.2 Mean boundary shear stresses

The percentage of the total shear force acting on different elements of the boundary was evaluated via the following equations:

Closed conduit flow

$$SF_w = \bar{\tau}_w \, 2H \quad \ldots\ldots\ldots\ldots \quad (3)$$

$$SF_b = \bar{\tau}_b \, 2B \quad \ldots\ldots\ldots\ldots \quad (4)$$

$$SF_T = 2SF_w + 2SF_b \quad \ldots\ldots\ldots\ldots \quad (5)$$

$$\%SF_w = 200SF_w/SF_T \quad \ldots\ldots\ldots \quad (6)$$

Open channel flow

$$SF_w = \bar{\tau}_w \, 2H \quad \ldots\ldots\ldots\ldots \quad (7)$$

$$SF_b = \bar{\tau}_b \, 2B \quad \ldots\ldots\ldots\ldots \quad (8)$$

$$SF_T = SF_w + SF_b \quad \ldots\ldots\ldots\ldots \quad (9)$$

$$\%SF_w = 100SF_w/SF_T \quad \ldots\ldots\ldots \quad (10)$$

where SF_w is the shear force acting on one wall of the closed conduit or on both walls of the open channel, SF_b is the shear force acting on one side of the closed conduit or on the bed of the open channel, and SF_T refers to the total shear force in either case.

The percentage of the total shear force acting on the walls, $\%SF_w$, was determined for each experiment and plotted against aspect ratio. The open channel flow results together with all other known open channel flow data is shown in Figure 3. The closed conduit flow data was similarly dealt with. Noting the exponential form of both curves, a relationship between $\%SF_w$ and 2B/H was found by linear regression to be

$$\%SF_w = e^{\alpha} \quad \ldots\ldots\ldots\ldots \quad (11)$$

where

$$\alpha = -3.253 \log_{10}\left(\frac{2B}{H} + 3\right) + 6.189 \quad (12)$$

for closed conduit flow, and

$$\alpha = -3.230 \log_{10}\left(\frac{2B}{H} + 3\right) + 6.146 \quad (13)$$

for open channel flow.

The percentage of the total shear force which would be expected to act on the walls, if it were assumed that the shear force was directly related to the ratio of wall lengths to overall boundary length, is given by

$$\%SF_w = 100/(1 + B/H) \quad \ldots\ldots\ldots\ldots \quad (14)$$

The open channel and closed conduit results are shown together with equation (14) in Figure 4.

The mean wall and bed shear stresses are related to the percentage of the shear force acting on the walls by the equations

$$\frac{\bar{\tau}_w}{\rho g H S_f} = 0.01 \, \%SF_w \left(\frac{B}{H}\right) \quad \ldots\ldots\ldots\ldots \quad (15)$$

$$\frac{\bar{\tau}_w}{\rho g R S_f} = 0.01 \, \%SF_w \left(1 + \frac{B}{H}\right) \quad \ldots\ldots \quad (16)$$

$$\frac{\bar{\tau}_b}{\rho g H S_f} = 1 - 0.01 \, \%SF_w \quad \ldots\ldots\ldots\ldots \quad (17)$$

$$\frac{\bar{\tau}_b}{\rho g R S_f} = (1 - 0.01 \, \%SF_w)\left(1 + \frac{H}{B}\right) \quad \ldots \quad (18)$$

Thus having obtained a relationship between $\%SF_w$ and 2B/H, equations (15) to (18) may be used to determine the mean wall or bed shear stress, $\bar{\tau}_w$ and $\bar{\tau}_b$, and the mean bed shear velocity, $\bar{u}_{*b}$ $(= \sqrt{(\bar{\tau}_b/\rho)})$. These relationships are shown in Figures 5 and 6, and indicate again slight differences between the open channel and closed conduit results. In the interest of clarity, curves generated by combining equation (14) with equations (15) to (18) have been omitted.

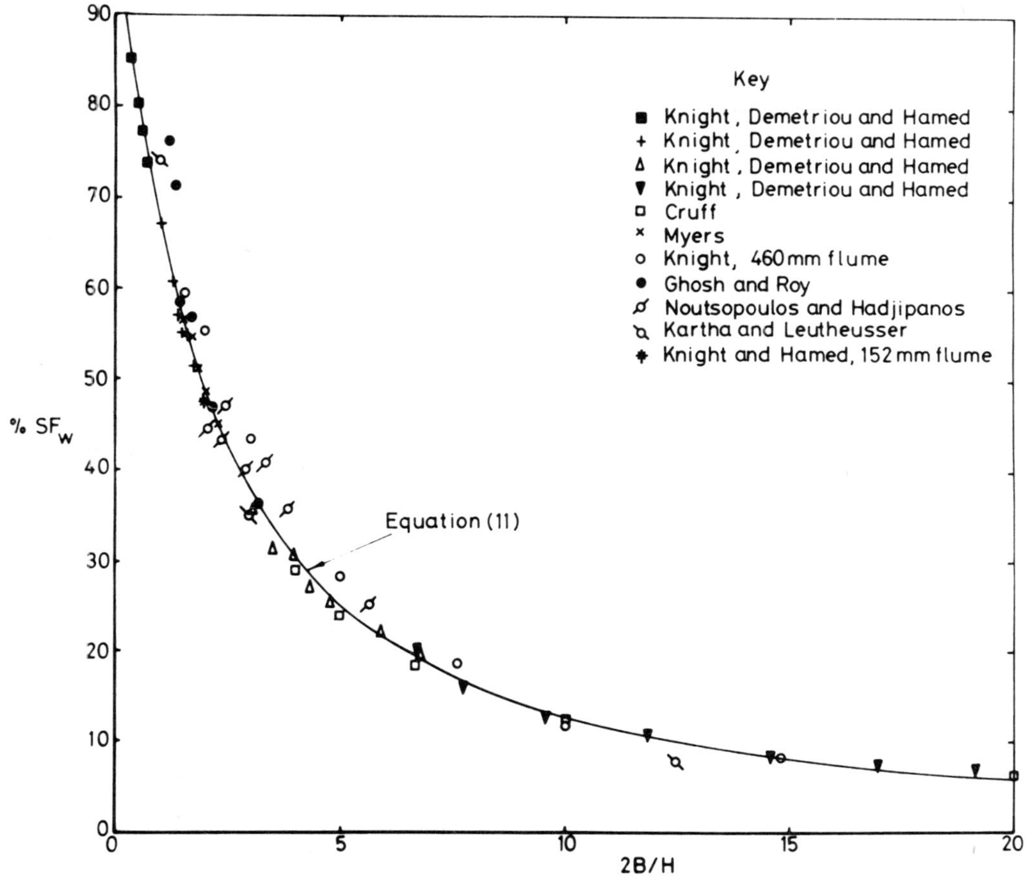

Fig. 3 Variation in the percentage of the total shear force carried by the walls in open channel flow.

3.3 Maximum boundary shear stresses

On account of secondary flow behaviour, the maximum bed shear stress does not necessarily occur at the centreline, but at some intermediate position between the centreline and the corner, See Figure 1. For aspect ratios greater than about 4.0 the maximum stress occurs at the centreline position. Because the value of the maximum bed shear stress is of such importance in sediment mechanics, the distributions shown in Figure 2, together with several other distributions from additional experiments, were analysed to give the variation of the bed centreline shear stress, $\tau_{b℄}$ with aspect ratio. The experimental data was fitted by the equations

$$\frac{\tau_{b℄}}{\tau_b} = \frac{1.025\left[1 - e^{-\left(\frac{2B}{3H}\right)^{0.95}}\right]}{1 - 0.01e^{\alpha}} \quad \ldots (19)$$

for closed conduit flow, and

$$\frac{\tau_{b℄}}{\tau_b} = \frac{1.025\left[1 - e^{-\left(\frac{2B}{3H}\right)^{0.85}}\right]}{1 - 0.01e^{\alpha}} \quad \ldots (20)$$

for open channel flow. Equations (12) and (13) should be used for the appropriate values of α in each case. The empirically derived equations, (19) and (20), are shown in Figure 7.

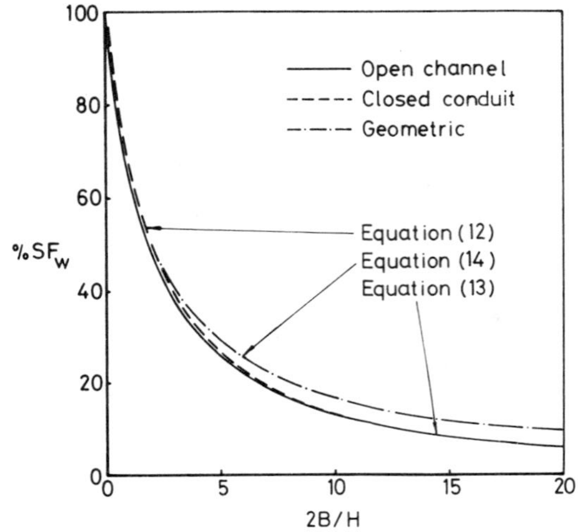

Fig.4. Comparison between open channel and closed conduit $\%SF_w$ results.

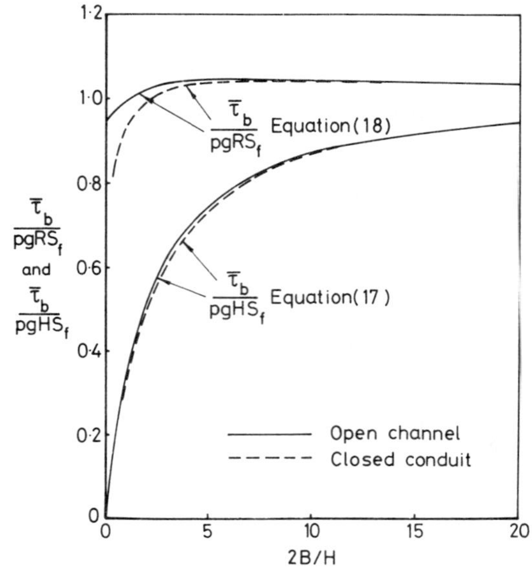

Fig.6 Variation of mean bed shear stresses with aspect ratio.

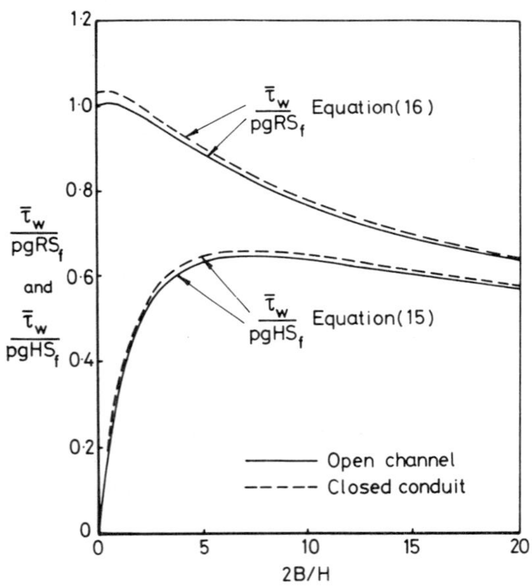

Fig. 5. Variation of mean wall shear stresses with aspect ratio.

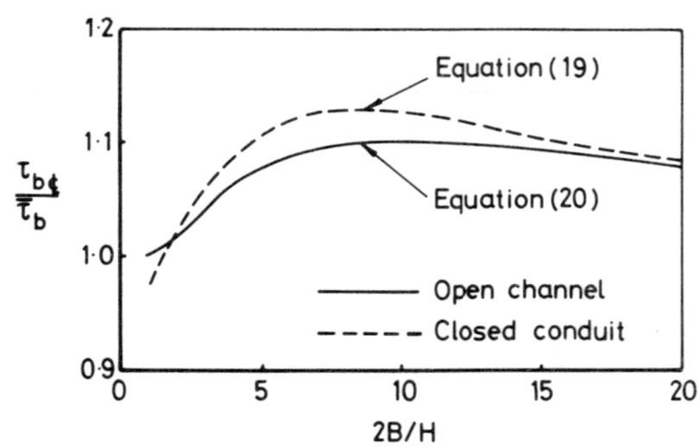

Fig.7. Variation of bed centreline shear stresses with aspect ratio.

3.4 Range of validity.

The empirically derived functions, equations (11) to (13) and (15) to (20), are necessarily limited by the conditions under which they were obtained. The closed conduit results are limited to $0.2 < 2B/H < 20.0$, $4.82 \times 10^4 < 4UR/\nu < 1.07 \times 10^5$, and the open channel flow results to $0.3 < 2B/H < 50.0$, $2.1 \times 10^4 < 4UR/\nu < 3.7 \times 10^5$, and $0.2 < U/\surd(gH) < 0.57$.

4 DISCUSSION OF RESULTS

The boundary shear stress distributions in both closed and open channel conduits with a rectangular cross section have been shown to be influenced by secondary flow cells. The number of perturbations in the shear stress distributions may be correlated with positions where the net secondary flow is either normally towards or away from the boundary. At an aspect ratio of 2B/H = 2.0, these perturbations are at their greatest, causing the maximum shear stress to occur at a position some distance from the centreline. As the aspect ratio increases, the number of perturbations increases and the amplitudes decrease. At aspect ratios greater than about 12.0, the perturbations become so small that they are unable to be detected by the Preston tube technique.

Although there are similarities between open channel flow and flow in a similar shaped closed conduit, small differences have been shown to exist between the two types of flow. That such small differences should exist is not surprising when one considers the influence of the free surface in open channel flow. The depression of the filament of maximum velocity below the free surface at low aspect ratios is a well known phenomenon. Clearly the secondary flow cells adjacent to the bed in open channel flow are reduced relative to closed conduit flow, and influence the mean wall and bed shear stresses accordingly. The patterns of secondary flow cells shown in Figure 1 demonstrate that in addition to influencing the boundary shear stresses, they may well influence the lateral and vertical diffusion coefficients. This may have certain implications for suspended sediments.

The two equations describing the variation of $\%SF_w$ with 2B/H in open channel and closed conduit flows are of some considerable practical significance in that they allow the shear forces on certain boundary elements to be identified. This is particularly useful in flume studies where the mean bed shear stress variation with aspect ratio is often required. The results are however restricted to the case in which both the bed and the walls are hydrodynamically smooth. The case in which the bed is differentially roughened with respect to the walls has been examined by the senior author and some tentative guidelines given (Knight, 1981). Flows in non-rectangular sections have also been investigated (Knight & Demetriou, 1982).

The centreline bed shear stress has been shown to vary systematically with aspect ratio. According to Figure 7, the ratio of bed centreline shear stress to mean bed shear stress reaches a maximum of 1.13 in closed conduit flow and 1.10 in open channel flow at 2B/H = 8.5. Since the maximum bed shear stress coincides with the centreline bed shear stress for 2B/H > 4, then from a design point of view it is suggested that the maximum bed shear stress ratios should be conservatively estimated as 1.13 and 1.10 for 2B/H < 8.5, and that equations(19)and(20) should be used for 2B/H > 8.5.

5 CONCLUSIONS

Multi cellular secondary flows have been shown to have some considerable influence upon boundary shear stress distributions in both closed conduit and open channel flows with rectangular sections. Perturbations in the shear stresses have been linked with secondary flows either normally towards or away from the boundary.

The proportion of the total shear force which acts on the boundary walls has been shown to vary systematically with aspect ratio, but in a different way to that expected on the grounds of geometric scaling. Certain differences have been shown to exist between open channel and closed conduit flows.

Equations have been proposed for the mean wall and bed shear stress variations with aspect ratio. The bed centreline shear stress has been shown to be different from the maximum bed shear stress for aspect ratios between 1 and 4. The aspect ratios covered in the study range from 0.2 to 20.

ACKNOWLEDGEMENTS

The experimental work was undertaken in the Civil Engineering Department at the University of Birmingham as part of a

Science and Engineering Research Council research program entitled "Momentum transfer in asymmetrical sections with varying boundary roughness". The writers are grateful to Mr. R.I.Cooper and Mr. M. Vanderstam for their expert technical assistance in the construction of the various flumes and wind tunnel.

REFERENCES

Brundrett, E. & Baines, W.D. 1964, The production and diffusion of vorticity in duct flow, J.Fluid Mech.19:375-394.

Einstein, H.A. & Li, H. 1958, Secondary currents in straight channels. Trans Amer.Geophysical Union, 39: 1085-1088.

Gessner, F.B. 1973, The origin of secondary flow in turbulent flow along a corner. J. Fluid Mech 58(1): 1-25.

Gosman, A.D. & Rapley, C.W. 1980, Fully developed flow in passages of arbitrary cross section. In C.Taylor & K.Morgan (ed.) Recent Advances in Numerical Methods in Fluids, p.335-399. Swansea: Pineridge.

Knight, D.W. 1981, Boundary shear in smooth and rough channels. J.Hyd.Div.,Proc.ASCE. 107(HY7): 839-851.

Knight, D.W. & Demetriou, J.D. 1982, Flood plain and main channel flow interaction, (submitted for publication).

Knight, D.W., Demetriou, J.D. & Hamed, M.E. 1983a, Boundary shear in smooth rectangular channels, (submitted for publication)

Knight, D.W., Demetriou, J.D. & Hamed, M.E. 1983b, Boundary shear stresses in rectangular flumes, (submitted for publication).

Knight, D.W., & Patel, H.S., 1983, Boundary shear stress distributions in closed conduit flows with a rectangular cross section,(submitted for publication)

Leutheusser, H.J.1963, Turbulent flow in rectangular ducts. J.Hyd.Div., Proc.ASCE. 89(HY3):1-19.

Melling, A. & Whitelaw, J.H. 1976, Turbulent flow in a rectangular duct. J.Fluid Mech. 78(2): 289-315.

Patel, V.C. 1965, Calibration of the Preston tube and limitations on its use in pressure gradients. J.Fluid Mech. 23(1): 185-208. ,

Perkins, H.J. 1970, The formation of streamwise vorticity in turbulent flow. J. Fluid Mech. 44(4): 721-740.

Tracy, H.J. 1965, Turbulent flow in a three dimensional channel. J.Hyd.Div.,Proc. ASCE, 91(HY6): 9-35.

Visualization of the mixing layer behind dunes

A.MÜLLER & A.GYR
Swiss Federal Institute of Technology, Zürich

ABSTRACT: In an open channel flow over a dune bed, a mixing layer is formed between the recirculating flow behind each dune and the main flow. A progress report on a visualization study of this flow situation is given. Dye was added to the flow which was illuminated by a light sheet. The motion was recorded on video tape. The shear layer which originates at the crest of the dune develops Kelvin Helmholtz instabilities. By vortex pairing, the vorticity is concentrated in large vortices. As soon as such a vortex tube is bent, it starts to interact with itself by Biot Savart forces. A vortex loop bent in the downstream direction moves upward as a boil or an internal boil. A vortex loop bent in the upstream direction moves downwards and starts to scour in the trough of the dune. By interaction with its mirror vortex, such a vortex can be forced upstream and its vorticity is accumulated in a circular vortex roller.

1 INTRODUCTION

An alluvial river is capable of adapting its friction factor to water and sediment discharge by changing the shape of the bed forms (Alam, Kennedy, 1969). In this feedback system, the interaction of flow and bed forms is the most evident. Scour and deposition of the local flow determine the evulution of the bed forms which themselves determine the flow as a boundary condition.

The flow in the lee of a migrating sand dune is an example of this interaction. The flow behaviour in the time mean is described in the literature (see e.g. Engelund, Fredsøe, 1982) as a flow separation at the crest and a reattachment in the trough such that an elongated bottom roller is formed. Above the roller is a zone of violent free turbulence. In the converging zone, a boundary layer starts to grow (fig. 1)

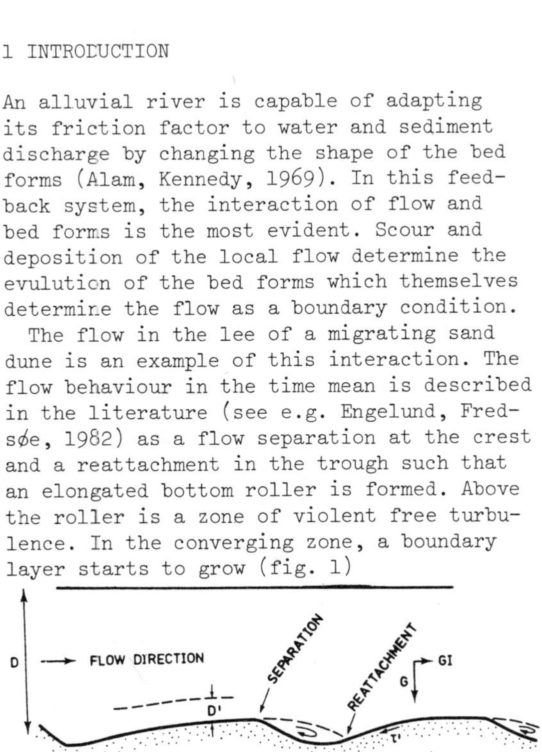

Fig. 1 Sketch of the mean flow over a dune bed (Engelund, Fredsøe, 1982) (L: dune length, D: mean flow depth)

Boils which can be observed at the free surface and a highly intermittent scour indicate that the flow contains organized structures which are three dimensional even behind two dimensional dunes. To learn more of the origin and the role of such structures, a flow visualization study was started. A light sheet was used to illuminate different planes in the flow behind a two dimensional dune. The motion of the marked fluid was recorded on a video tape. This paper is a progress report, where first interpretations of the visualized time resolved flow are given.

2 EXPERIMENTAL CONDITIONS

To simplify the experimental conditions, two dimensional dunes were formed in a .6 m wide, 6 m long working section of a laboratory flume. These dunes of length L = 1.5 m and a height H = .1 m were built, using a uniform sand with a grain size diameter of .7 mm. The dunes were then exposed to a flow of .064 m^3/sec.m to let their shape evolve naturally. This procedure led to a mean flow depth D = .21 m, such that the ratio L/D was 7.5. At the crest of the dune, the flow depth was .16 m and the mean velocity 40 cm/sec. The sediment discharge

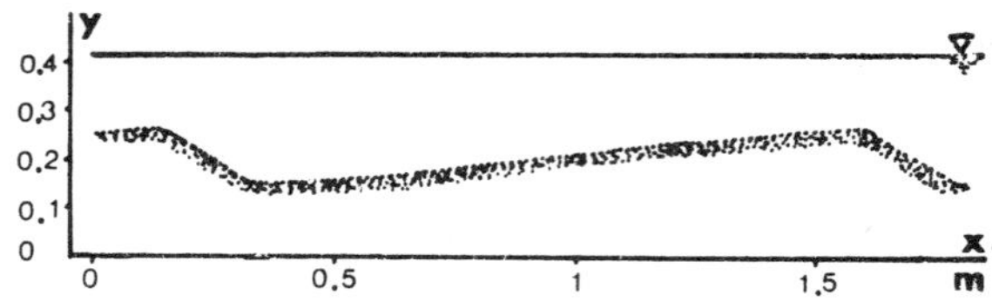

Fig. 2 Longitudinal profile of the two dimensional dune, as it was obtained in the present study

was estimated to be 3 cm^3/sec from a measured dune migration velocity of $5 \cdot 10^{-3}$ cm/sec. Fig. 2 shows the bed profile together with the measured free surface profile, as it was established for the flow visualization experiments. As soon as the bed became three dimensional, the original shape of the dunes was restored and again exposed to the flow.

A light sheet of approximately 10 mm width was used to illuminate different cross sections of the flow. Side views of the flow at one lateral position are given by a vertical sheet in flow direction. Plane views at different levels were obtained by a horizontal sheet across the channel. Fluoresceine was used as dye. It was injected in rather high concentration at one point in the trough of the dune, where it was mixed into the recirculating flow and transported into the mixing layer. In this way, the boundary between the main flow and the recirculating flow was marked. In some cases, colour was injected at the crest through hypodermic needles to mark the zone of vortex pairing of the shear layer.

The motion of the dyed fluid was recorded on a video film. Single frames were copied onto transparencies typically every .2 sec. In figure 5, e.g. five such frames are overlaid which allowed an analysis of the instantaneous motion as it was visible in the two dimensional light sheet. Size and spacing of the structures were measured and flow velocities were computed from the displacement of dye boundaries. These velocities have to be interpreted with some care, as a displacement of a dye pocket normal to the plane of the light sheet can give rise to a displacement of the visible dye boundary in the illuminated plane.

Additional qualitative information was gained by direct information of the experiments. The motion within a dye pocket is not resolved on the video film, but it is visible by eye if the velocities are not too high.

3 INTERPRETATION OF THE VISUALIZED FLOW

In figure 3, a typical instantaneous vertical section of the investigated flow field is depicted, as it was recorded on the video film. We divided the flow in different regions which are characterized by their dominant vortex dynamic, as indicated in fig. 3.

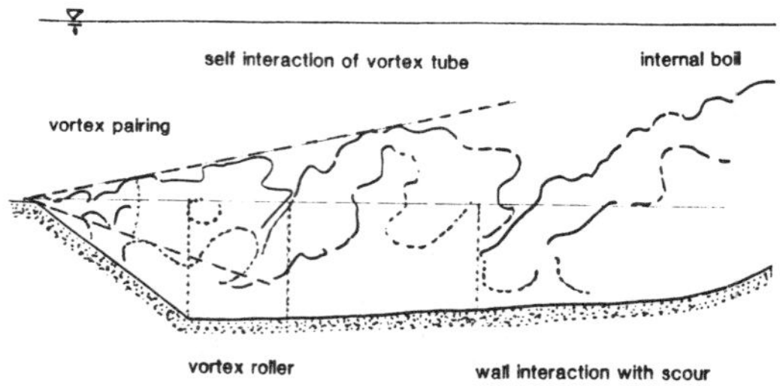

Fig. 3 Typical side view of the flow showing the boundary between the marked fluid and the main stream. The different flow regions are characterized by their dominant vortex dynamics

The free shear layer which originates at the crest of the dune develops Kelvin-Helmholtz instabilities. The vorticity is concentrated into discrete vortices which pass through pairing cycles and form large vortical structures. This zone of vortex pairing extends to about 1.5 dune heights H downstream of the crest. If such a vortex tube is bent (fig. 4) locally, it starts to interact with itself. If it is bent in the downstream direction, it will rise and pump low speed fluid towards the free surface.

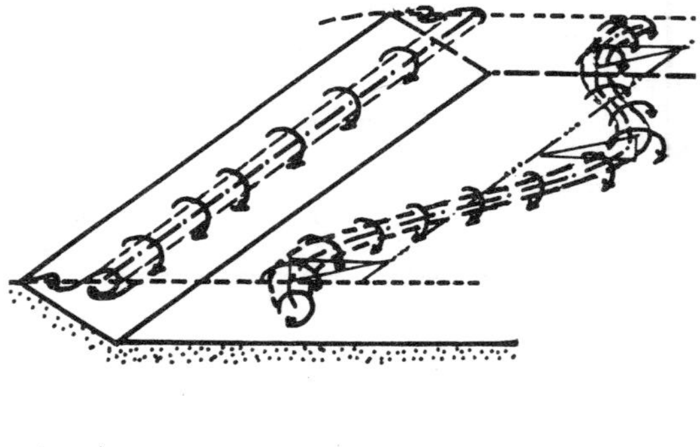

Fig. 4 Sketch of the circulation around a bent vortex tube. The central vortex loop is lifted due to the self interaction of the axial vorticity. The opposite is true for the loop bent upstream.

It produces a boil or an internal "boil-like" structure. If the vortex tube is bent in the upstream direction, it will move towards the bed. Its interaction with the wall can be described with the aid of a mirror vortex which produces an upstream force. The vorticity of vortices, which move in the upstream direction, is concentrated in a circular vortex roller which stays in interaction with the initial shear layer and with the wall.

3.1. Vortex pairing in the shear layer

The flow over the crest of the dune is a boundary layer flow. Its mean lateral vorticity is concentrated near the bed, where it is produced by the wall shear stress. As soon as the flow separates, a shear layer develops between the main flow and the recirculating flow.

The velocity difference ΔU across the shear layer is close to 40 cm/sec and carries the vorticity distribution of the upstream boundary layer. Since the shear layer is unstable, vertical deformations start to grow and the vorticity is transported into discrete vortices. By viscous diffusion and entrainment, these vortices grow until stey start a pairing cycle (Roshko, 1976). During pairing, they gain additional fluid by engulfment.

Jimenez (1981) measured the increase of the area due to entrainment and engulfment. He found that the area of the newly developed vcrtex is about four times larger than the area of each of the original vortices. Since the circulation is preserved, the mean vorticity of the vortex is divided by a factor of two per cycle.

The spreading angle of the mixing layer is given by the ratio of the size of the first vortex and the wave length of the first instability. In order to estimate this wave length, it was assumed that the free shear layer was disturbed by the vertical turbulent fluctuations. Within a linear stability theory, each wave number is disturbed according to the corresponding density of the power spectrum and was allowed to grow for the finite time given by Acton's (1976) calculations. This estimate gives a spreading angle of 16° which is in good agreement with the observed values.

Figure 5 shows five consecutive steps of a pairing cycle, where three vortices are combined into one vortex. The centers of the vortices are convected with a velocity of $\Delta U/_2$, i.e. about 20 cm/sec., which agrees with the theory on two dimensional mixing layers. The lateral extent of the vortices could not be determined with the present visualization technique.

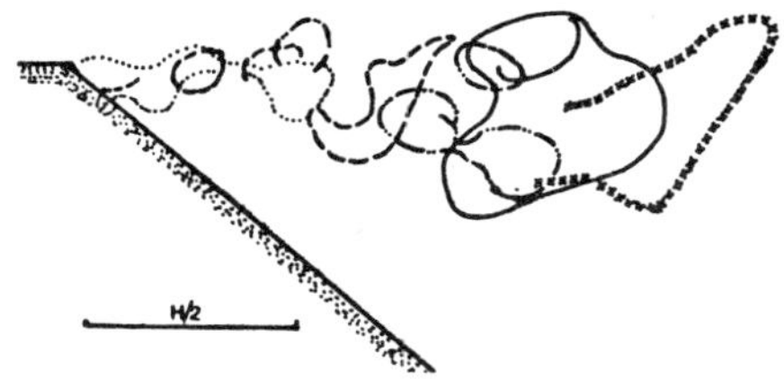

Fig. 5 Sequence of five frames of a side view showing the pairing of vortices in the initial shear layer.
(.... : t_o, ---- : t_o + .2 sec, -...- : t_o + .6 sec, ——— : t_o + .8 sec, xxxx : t_o + 1 sec)

Normally, the axis of the shear layer is close to horizontal. We observed exceptions of this behaviour during the initiation of a boil (see fig. 9), where an angle of about + 25° was measured. Intervals with a downward inclination of the shear layer were also observed. Both exceptions are discussed later in connection with boil events.

3.2 Self interaction of vortex tubes

In the pairing zone, the vorticity of the boundary layer is organized into discrete vortices. These vortices start to dominate the flow field by self interaction of the vortex tubes (fig. 4). This interaction of the vortices can be explained by Biot - Savart's law. The interaction of a vortex with a wall can be described in the same way by introducing its mirror vortex. It is a consequence of Biot Savart's law that a vortex tube of the shear layer which is bent in the downstream direction starts to rise and initiates an internal boil. In a plane view of such a boil, as it is shown in fig. 7a, the circulation due to the two branches of the loop was clearly visible during the experiment. For the same reason, a vortex loop moves towards the wall if it is bent in the upstream direction. This concept is compatible with the observation of the sand motion which occurs 3 to 4 H downstream of the crest. Typically, the sand motion is confined to a U-shaped region or in streamwise direction. The sand is transported in the direction normal to the loop of the U-shape. The phenomenon has a lifetime of less than a second. We interprete this sand movement as a wall interaction of a vortex loop.

Since the tip of the vortex loop has always the same positive sign of vorticity, the wall acts as a source of acceleration towards the crest. Therefore, this part of the vortex loop is decelerated close to the wall and sometimes it starts to move upstream.

3.3 Vortex roller

In the time mean, the recirculation of the separation zone is an elongated bottom roller. In the time resolved flow, we find that a fairly stable vortex is present most of the time, where the slope of the dune reaches the bottom of the trough. Fig. 6 shows a sequence of video frames with a vortex roller which is approximately circular with a diameter of H. Downstream of this roller, there is a zone of about 3 H length, where vortices of the shear layer interact with the wall, as described in the previous section. In this region, the flow direction is, therefore, alternating in time and also in the spanwise direction.

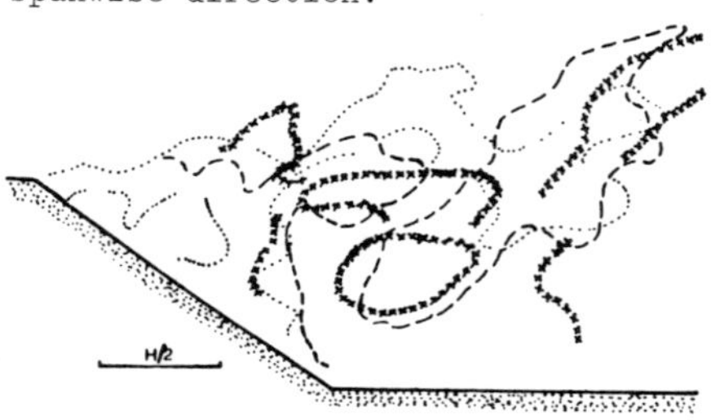

Fig. 6 A circular vortex roller persists adjacent to the corner of dune slope and trough. (.... : t_o, ---- : t_o + .4 sec, xxxx : t_o + .7 sec)

3.4 Internal boils

A rising vortex loop, bent in the downstream direction, pumps low speed fluid upwards. Therefore, an internal stagnation area is formed which forces the main stream to flow around this obstacle. The newly formed shear layer between the main stream and the low speed fluid becomes unstable and its vorticity is concentrated into secondary vortices which have the same sign of rotation as the primary vortices of the original vortex loop.

If such a structure hits the free surface, it is visible as a boil and in a later stage, it shows two primary and several secondary vortices at the free surface. In the visualized flow, similar structures, which we call internal boils, are visible. They decay before reaching the free surface.

Figure 7a shows a plane view of the flow .75 H below the water surface. The flow direction is upwards and the lower end of the figure is about 4 H downstream of the crest. Internal boils appear in two rows about .15 m apart. In the right row, three consecutive boils are depicted which entered the illuminated plane about 1 second apart. They are convected with a velocity of about 20 cm/sec. Their lateral extent is about .06 m or 1/10 of the channel width. The fact that they appear in rows indicates that the local deceleration of the main stream by the first boil triggers a subsequent internal boil at the same lateral position.

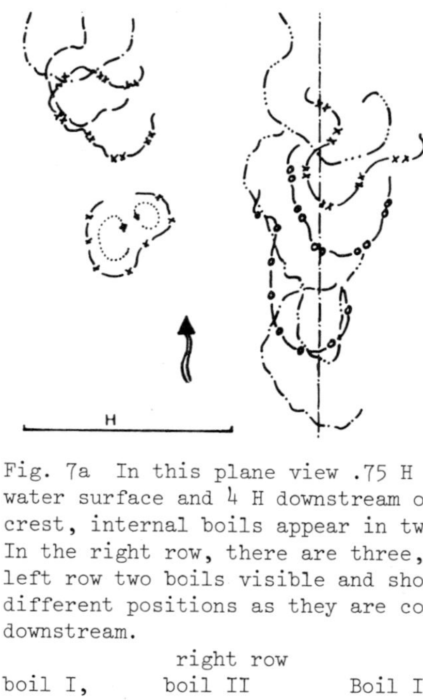

Fig. 7a In this plane view .75 H below the water surface and 4 H downstream of the crest, internal boils appear in two rows. In the right row, there are three, in the left row two boils visible and shown in different positions as they are convected downstream.

right row

boil I,	boil II	Boil III
-xx- : t_o	-o- : t_o + 1.2 sec	-.- : t_o + 2.1 sec
	-oo-: t_o + 1.6 sec	-..-: t_o + 2.4 sec
		-...-: t_o + 2.7 sec

left row

boil I	boil II
-x- : t_o	--- : t_o + 1.9 sec
-xx-: t_o + .2 sec	-.- : t_o + 2.1 sec

A sequence of vertical sections of an internal boil is depicted in fig. 7b. At the rear of the boil, the secondary vortices are visible. A dye pocket above the boil gives a velocity estimate for the main stream of 35 cm/sec.

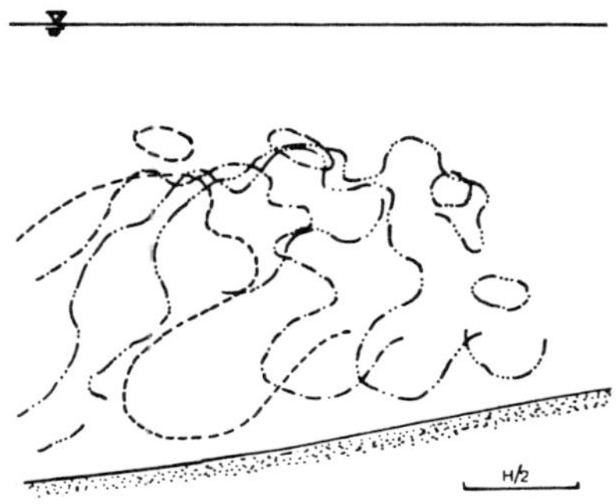

Fig. 7b Sequence of four frames showing a side view of an internal boil. The rear front shows the evolution of secondary vortices (---- : t_o, -.-.- : t_o + .2 sec, -..-.. : t_o + .4 sec, -...- : t_o + .6 sec)

Relatively rare is the evolution of a boil which reaches the free surface. Figure 8 is a sequence which starts with an upward diversion of the entire mixing layer. We have the impression that it is related to an instability of the vortex roller which starts to rise on the slope of the dune due to its mirror vortex. The sequence shows that marked low speed fluid starts to rise and reaches the free surface about four dune heights after the crest. Such a violent extrusion of low speed fluid must be compensated.

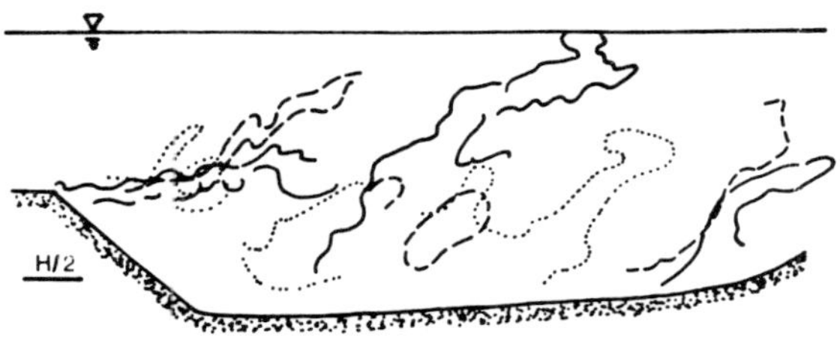

Fig. 8 Sequence of three frames showing the evolution of a boil (.... : t_o, ---- : t_o + .5 sec ―― : t_o +1.6 sec)

In our view, a downward diversion of the shear layer, as it was mentioned in section 3.1, is one way to supply fluid to a boil at a different lateral position. Another event which compensates for the fluid extruded by a boil is shown in fig. 9. A dye pocket which was released upstream at the free surface is transported by the main stream. On the crest, it has a velocity of 40 cm/sec and 5 H, downstream it gets decelerated to a velocity of 30 cm/sec. The lower part of the dye pocket is trapped by a structure of the mixing layer and is moving with a velocity of about 35 cm/sec. It reaches the bed about 5 H after the crest.

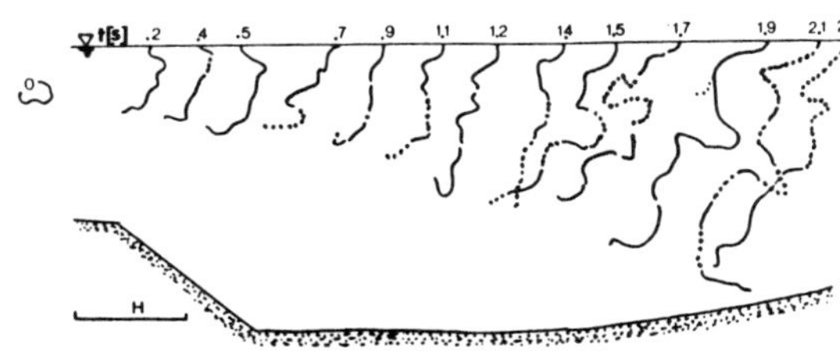

Fig. 9 Dye pocket released in the main stream and trapped by the mixing layer

4 CONCLUSIONS

1. The time resolved visualization of the flow in the lee of a two dimensional dune shows the development of vortices out of the initial shear layer.

2. The self interaction of these vortex tubes, which forms loops in the spanwise direction, leads to the formation of boils and internal boils and to vortices which scour and preserve the dune shape. It produces a three dimensional flow with a spanwise structure in the entire cross section.

3. A circular vortex roller is persistent most of the time in the separation region. It gains vorticity from vortices which are forced upstream along the wall due to their mirror vortex.

REFERENCES

Acton, E. 1976. The modeling of large eddies in a two dimensional shear layer. JFM 76, 561-592.

Alam, A.M.Z., Kennedy, J.F. 1969. Friction Factors for Flow in Sand-Bed Channels. ASCE J. of Hydraulics Div. 95, HY6, 1973-1992.

Engelund, F., Fredsøe, J. 1982. Sediment Ripples and Dunes. An. Rev. Fluid Mech. 14, 13-37.

Jimenez, J. 1981. Shear Layer Models and Computer Analysis of Data. Lecture Notes in Physics, No. 136, Springer Verlag, 1981.

Roshko, A. 1976. Structure of Turbulent Shear Flows: A New Look. AIAA J. 14, 1349-1357.

2 Single – particle dynamics

Resolution of equations governing the saltation motion in the air ($Re_{particles} < 1$)

G.SALAÜN-PENQUER & C.NASSAR
Institut de Mécanique des Fluides de Marseille, France

R.GUILLAUME
I.U.T. d'Aix-en-Provence, France

ABSTRACT : The transport of heavy particles by the wind is generally governed by three kinds of motion : suspension, saltation and creeping motion. The present theoretical work is a natural development of those of Bagnold , Kawamura and Owen and generalise their results. The equations of motion are solved in the case of horizontal or inclined sand surface ; in that last case the wind is supposed to be flowing (up or down) in a direction parallel to the line of the greatest slope.

NOTATIONS

a_o, a_1 constant of integration
d particle diameter
F_{ap} action of air on the particles
g gravity acceleration
G mass flux of particles across a vertical plane perpendicular to the fluid stream
h height of saltation layer
L jump length
m mass of one ideal particle
n number of particles per unit volume
n_1 number of rising particles per unit volume
n_2 number of falling particles per unit volume
u_τ friction velocity
V_o limit vertical speed of falling particles

$\bar{g} = (\frac{1}{18}\frac{\sigma}{\rho}\frac{d}{\nu})^2 gd$ dimensionless gravity acceleration

$\bar{G} = \frac{1}{3\pi\mu} G$ dimensionless mass flux

$\bar{n} = n\, d^3$

$\bar{u} = \frac{1}{18}\frac{\sigma}{\rho}\frac{d}{\nu} u$ dimensionless speed

α coefficient of proportionality connecting $v_1(o)$ to u_τ
β_c constant
ε eddy coefficient of viscosity
κ Von Karman's constant
μ dynamic viscosity
ν kinematic viscosity
ν_t eddy exchange coefficient
ρ volumic mass of air
σ volumic mass of particle
τ shearing stress
τ_s shearing stress closed by the sand surface.

INTRODUCTION

The transport of heavy particles deposited on a surface blown by the wind is a very complex problem ; various aspects of fluid mechanics are concerned, besides it is intimately connected with the flow turbulence ; that transfer is generally governed by three kinds of motion : suspension, saltation and creeping motion.

The present work is mainly theoretical, it concerns the saltation movement and naturally follows the spirit of Bagnold, Kawamura and Owen's works and generalise their results.

Bagnold's analysis allows to understand the main mechanisms which govern that kind of movement ; that author was the first to suppose that the saltation action on the wind is similar to that of a surface roughness ; he showed the existence of a critical friction velocity u_{τ_c} so that the saltation movement cannot appear if $u_\tau < u_{\tau_c}$. He proposes for valuing this speed, the always used expression :

$$u_{\tau_c} \simeq A\sqrt{\frac{\sigma g d}{\rho}} \quad (A = Cte) \quad (1)$$

Then he defines an ideal sand, the grains of which have all the same diameter d_e, called "equivalent diameter", and for that sand gives an expression of the mass flux of grains G across unit width of a plane perpendicular to fluid stream :

$$G = C\sqrt{\frac{d_e}{D}}\ \frac{\rho}{g}\ u_\tau^3 \quad (C=Cte) \quad (2)$$

Kawamura using Bagnold's hypotheses, showed that it was possible to establish, from the study of the saltation movement of a single sand particle, mathematical expressions of various characteristic quantities of that movement : jump length L, mass flow G, saltation layer height h... .He develops his calculation in the case where the wind speed in the saltation layer is constant and the aerodynamic strength on the particles obeys to Stokes' law ; moreover he supposes that the magnitude of the vertical components of two particles speeds, one up the other down, situated at the same level are equal.

Owen completes a few assumptions made by Bagnold and Kawamura and this allows him to establish a precise theory. As these predecessors did, he supposes that the saltation action on the general flow is similar to that of a surface roughness but he considers this equivalent roughness to be proportional to the height h. He also assumes the existence of a critical shearing stress τ_s close to the sand layer but he adds that this stress keeps a constant value whatever the wind speed (as long as this speed is less than the critical speed of suspension) :

$$\tau_s = \beta_e\ \sigma\ g\ d \quad (3)$$

Finally, he develops his theory considering a turbulent flow in the saltation layer ; the eddy viscosity keeps a constant value in this layer :

$$\nu_t = \kappa\ u_\tau\ h \quad (4)$$

Thanks to this last hypothesis he can give an explicit expression of the coupling equation connecting the grains motion to the air motion.

THEORETICAL CALCULATION

The theory that we propose here, is developped from the same hypotheses as those used by Owen, nevertheless the theory itself is different : on the one hand we have not taken the same form of approximation for the vertical component grain speed and we have used a different dependant variable for the integration : on the other hand we have chosen a different expression for the coefficient α relating the vertical component of the grain velocity $v_1(o)$ to the friction velocity u_τ :

$$v_1(o) = \alpha\ u_\tau \quad (5)$$

In order to help understanding our theoretical calculus, we give below the used hypotheses and which are not yet mentionned or stated sufficiently precisely.

- Flow movement

The wind speed U and its derivative of the first order $\frac{dU}{dy}$ are continuous at the top of the saltation layer.

The vertical velocity profile in the fluid outside the saltation layer obeys the law :

$$U = \frac{u_\tau}{\kappa}\ \text{Log}\ \left(\frac{y}{h}\right) + 10\ u_\tau \quad (6)$$

The horizontal pressure gradient may be neglected in comparison with the vertical gradient of shearing stress.

- Particles movement

We assume that the particles trajectories are all identical ; they are in vertical planes parallel to the mean flow direction.

Given that the initial direction of a particle in saltation cannot be anticipated, we shall assume that direction to be vertical.

The grains movement is studied, using a cartesian frame of reference fixed in the space ; we must also notice that the theoretical study of saltation amounts to a coupling equations system : the first ones concern the movement of grains particles, the other the fluid movement.

The equations of horizontal and vertical movement of one particle are :

$$m\ \frac{du_i}{dt} = 3\ \pi\mu\ d\ (U-u_i) \quad (7)$$

$$m\ \frac{dv_i}{dt} = -\ mg\ -\ 3\ \pi\mu\ d\ v_i \quad (8)$$

in the case of a rising particle i=1 and in that of a falling particle i=2.

The Navier-Stokes'equation in the saltation layer is reduced to :

$$\rho \frac{dU}{dt} = F + \varepsilon \frac{d^2U}{dy^2} = 0$$

It is easily demonstrated that $\frac{du_i}{dt}$ and $\frac{dv_i}{dt}$ are respectively connected to $\frac{du_i}{dy}$ and $\frac{dv_i}{dy}$ by the following relations :

$$\frac{du_i}{dt} = v_i \frac{du_i}{dy} \quad \text{and} \quad \frac{dv_i}{dt} = v_i \frac{dv_i}{dy} \qquad (9)$$

The equations (7) and (8) can be then written under dimensionless form :

$$\bar{v}_i \frac{d\bar{u}_i}{d\bar{y}} = \bar{U} - \bar{u}_i \qquad (10)$$

$$\bar{v}_i \frac{d\bar{v}_i}{d\bar{y}} = -\bar{g} - \bar{v}_i \qquad (11)$$

In the case of stationnary flow, the rising particles flux across any horizontal plane, situated at a level y in the saltation layer, is equal to the falling particles flux across that same plane ; besides the value of that flux may be proved to be independent of y :

$$\phi = n_1 v_1 = - n_2 v_2 = cte$$

That remark allows us to write the equations 10 and 11, in the case of a particles flux ϕ, under the form :

$$\bar{\phi} \frac{d\bar{u}_1}{d\bar{y}} = \bar{n}_1 (\bar{U} - \bar{u}_1) = \bar{F}_{1ap} \qquad (12)$$

$$-\bar{\phi} \frac{d\bar{u}_2}{d\bar{y}} = \bar{n}_2 (\bar{U} - \bar{u}_2) = \bar{F}_{2ap} \qquad (13)$$

$$\bar{v} \frac{d\bar{v}}{d\bar{y}} = -\bar{g} - \bar{v} \qquad (14)$$

If we observe that :

$$F = -F_{1ap} - F_{2ap}$$

We get immediately the explicit expression of the coupling equation, valid for both rising and falling particles. After integration and using the boundary conditions, we obtain :

$$\frac{\varepsilon}{3\pi\mu} \frac{d\bar{U}}{d\bar{y}} = \bar{\phi} (\bar{u}_1 - \bar{u}_2) + \frac{\varepsilon}{3\pi\mu} \left(\frac{d\bar{U}}{d\bar{y}}\right)_h \qquad (15)$$

We can observe that the integration of equation 11 is easy, it leads to :

$$\bar{h} - \bar{y} = \bar{v} - \bar{g} \, \mathrm{Log} \left(1 + \frac{\bar{v}}{\bar{g}}\right) \qquad (16)$$

It is impossible to find from this equation a strict explicit expression of $\bar{v}$, however and noting that $\bar{v} << \bar{g} \Rightarrow \frac{\bar{v}}{\bar{g}} << 1$, we can give an approximate expression of this quantity using an limited expansion of logarithm bounded at the third term. We find after all calculation :

$$\bar{v}_1 = 3\bar{g} \frac{\bar{x}}{1-\bar{x}} \quad (17) \quad \text{and} \quad \bar{v}_2 = -3\bar{g} \frac{\bar{x}}{1+\bar{x}} \quad (18)$$

$$\text{with} \quad \bar{x} = \frac{\sqrt{2\,\bar{g}\bar{h}}}{3\bar{g}} \left(1 - \frac{\bar{y}}{\bar{h}}\right)^{1/2}$$

That enables to reduces the initial system to :

$$-\frac{d\bar{u}_1}{d\bar{x}} = 3\,(1-\bar{x})(\bar{U}-\bar{u}_1) \qquad (20)$$

$$\frac{d\bar{u}_2}{d\bar{x}} = 3\,(1+\bar{x})(\bar{U}-\bar{u}_2) \qquad (21)$$

$$a \frac{d\bar{U}}{d\bar{x}} + b\bar{x} = \bar{x}\,(\bar{u}_2-\bar{u}_1) \qquad (22)$$

For integrating, put : $\bar{U}-\bar{u}_1 = X$, $\bar{U}-\bar{u}_2 = Y$, $z = \bar{u}_2-\bar{u}_1$, we get the following system (23):

$$\bar{u}_1 = \bar{U} - \frac{x+1}{2} z - \frac{1}{6} \frac{dz}{d\bar{x}}$$

$$\bar{u}_2 = \bar{U} - \frac{x-1}{2} z - \frac{1}{6} \frac{dz}{d\bar{x}}$$

$$z = \bar{u}_2 - \bar{u}_1 = \frac{a}{\bar{x}} \frac{d\bar{U}}{d\bar{x}} + b$$

In order to establish the differential equation of z, we shall derive once u_1 and u_2 with respect to $\bar{x}$ and substract the resulting equations one from the other ; we obtain :

$$\frac{1}{6} z'' + \bar{x}z' + \left(\frac{3\bar{x}^2}{2} - \frac{\bar{x}}{a} - 1\right)z = -\frac{b}{a}\,\bar{x} \qquad (24)$$

The solution of that equation is researched under the form :

$z = C(\bar{x})\, e^{3/2\,\bar{x}^2} + b$, that implies that the function C(x) must satisfy to the differential equation :

$$C'' - \frac{6}{a}\left(\bar{x} + \frac{3a}{2}\right)C = 6b\left(1 - \frac{3}{2}\bar{x}^2\right) e^{3/2\,\bar{x}^2} \qquad (25)$$

For integrating that last equation we

shall first substitute the exponential by its expansion limited to the third term (that is allowable since $\bar{x} << 1$) and make the change of variable :

$z = \left(\frac{6}{a_4}\right)^{1/3} \left(\bar{x} + \frac{3a}{2}\right)$; disregarding the term a^4 and those of higher order, we obtain :

$$C''-zC=6b\left(\frac{a}{6}\right)^{2/3}$$

$$\left[1-\frac{9}{8}\left(\frac{a}{6}\right)^{4/3}z^4+\frac{81}{2}\left(\frac{a}{2}\right)^2 z^3-\frac{3^7}{4}\left(\frac{a}{8}\right)^{8/3}z^2\right]$$

The general solution of that equation has been expressed under a serial form ; we obtain for z :

$$z=b+\Big\{a_o+a_1 z+\sum_1^{\infty}(a_oA_n+a_1B_{n+1}z)\,z^3$$

$$+3b\left(\frac{a}{6}\right)^{2/3}z^2\left(1+\sum_1^{\infty}C_{n+2}z^{3n}\right)+\frac{81}{2}\,b\left(\frac{a}{6}\right)^2$$

$$\left(\frac{z^3}{6}-\sum_1^{\infty}A_n z^{3n}\right)-\frac{9^4}{2}\,b\left(\frac{a}{6}\right)^{10/3}\sum_1^{\infty}B_{n+1}\,z^{3n+1}$$

$$+\,243\,b\left(\frac{a}{6}\right)^{8/3}\sum_1^{\infty}C_{n+2}\,z^{3n+2}\Big\}e^{-3/2\,\bar{x}^2} \qquad (26)$$

with : $A_n = \dfrac{1.4.7...(3n-2)}{(3n)!}$

$B_{n+1} = \dfrac{2.5.8...(3n-1)}{(3n+1)!}$

$C_{n+2} = \dfrac{3.6.9...3n}{(3n+2)!}$

Knowing z, we may determinate U by integration and therefore calculate u_1 and u_2.

$$U=U_h+9\left(\frac{a}{6}\right)^{4/3}\left[\frac{1}{a}-\frac{27}{8}a-\frac{9^5}{48}\left(\frac{a}{6}\right)^3\right]\phi_1-\left(\frac{a}{6}\right)^{2/3}$$

$$\left[\frac{1}{a}-\frac{81}{8}a+5\,\frac{9^5}{48}\left(\frac{a}{6}\right)^3\right]\phi_2-\frac{27}{24}\,a\left[1-\frac{5}{2}\,81\left(\frac{a}{6}\right)^2\right]\phi_3$$

$$+\frac{1}{4}\left(\frac{a}{6}\right)^{1/3}\left[1-\frac{15}{2}\,81\left(\frac{a}{6}\right)^2\right]\phi_4$$

$$+\,5\,\frac{27}{16}\left(\frac{a}{6}\right)^{5/3}\phi_5-\frac{3}{16}\left(\frac{a}{6}\right)\phi_6 \qquad (27)$$

with

$$\phi_i=(a_o-K_3)\left[F_i(z_o)-F_i(z)\right]+(a_1-K_4)$$

$$\left[G_{i+1}(z_o)-G_{i+1}(z)\right]+(K_2+K_5)$$

$$\left[H_{i+2}(z_o)-H_{i+2}(z)\right]$$

$$+K_3\left[\frac{z_o^i-z^i}{i}+\frac{z_o^{i+3}-z^{i+3}}{6(i+3)}\right]+\frac{K_4}{i+1}\left(z_o^{i+1}-z^{i+1}\right)$$

$$-\frac{K_5}{i+2}\left(z_o^{i+2}-z^{i+2}\right)$$

$$F_i = z^i\left[\frac{1}{i}+\sum_1^{\infty}\frac{A_n}{3n+i}\,z^{3n}\right]$$

$$G_i = z^i\left[\frac{1}{i}+\sum_1^{\infty}\frac{B_{n+1}}{3n+i}\,z^{3n}\right]$$

$$H_i = z^i\left[\frac{1}{i}+\sum_1^{\infty}\frac{C_{n+2}}{3n+i}\,z^{3n}\right]$$

$K_2 = 3b\left(\frac{a}{6}\right)^{2/3}$ $\qquad K_4 = \frac{9^4}{2}\,b\left(\frac{a}{6}\right)^{10/3}$

$K_3 = \frac{81}{2}\,b\left(\frac{a}{6}\right)^2$ $\qquad K_5 = 243b\left(\frac{a}{6}\right)^{8/3}$

PRACTICAL USE OF THE THEORETICAL RESULTS

The system of equations to solve is :

$$\bar{u}_1 = \bar{U} - \frac{\bar{x}+1}{2}\,z - \frac{1}{6}\frac{dz}{d\bar{x}} \qquad (21)$$

$$\bar{u}_2 = \bar{U} - \frac{\bar{x}-1}{2}\,z - \frac{1}{6}\frac{dz}{d\bar{x}} \qquad (22)$$

$$z = b + f_o\,(a_o,a_1,z)\;e^{-3/2\,\bar{x}^2} \qquad (26)$$

$$\bar{U} = \bar{U}_h + f_1\,(a_o,a_1,z) \qquad (27)$$

We must add to these equations the expressions of the constants a and b displayed in z :

$$a = \varepsilon/27\,\pi\,\mu\,\phi g$$

$$b = \varepsilon\left(\frac{d\bar{u}}{d\bar{y}}\right)_h/3\,\pi\,\mu\,\phi$$

Let us state precisely that the expression of the coefficient α connecting $v_1(o)$ to u_τ , is here :

$$\alpha = 2200\,d + 0.14\,\frac{v_o}{u_\tau} \quad \text{(d in meters)}$$

The study of the preceding system has been made taking various value of d and u_τ . The hypotheses and the boundaries conditions supply the supplementary equations allowing the determination of all quantities of which the movement depends. We must notice that it is interesting to calculate the jump length L and the mass transfer G, those both quantities are not independent but satisfy the obvious relation :

$$G = m\,\phi\,L$$

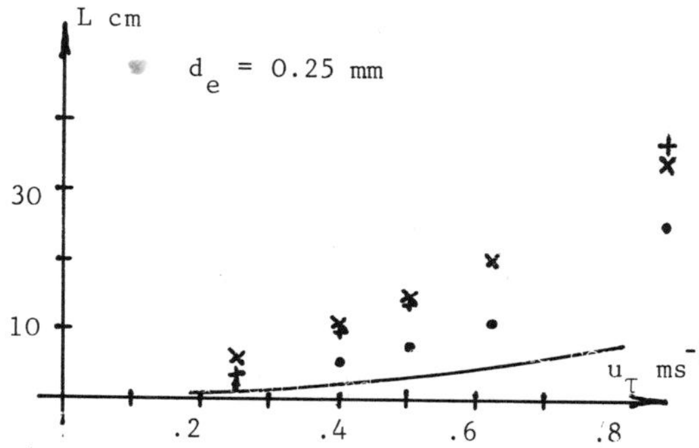

Fig.1.Jump length: - Kawamura's theory ; + Bagnold ; x Present theory ; . Approximate expression

The calculus of G amounts to reckoning L ; we have :

$$L = \int_0^h \left(\frac{u_1}{v_1} - \frac{u_2}{v_2}\right) dy \text{ which leads to :}$$

$$L = \frac{2\varepsilon}{m\,\phi\,v_o}\left[U_s - U_h + \left(\frac{dU}{dy}\right)_h h\right] - \frac{u_{2s}v_{2s}}{g}$$

Substituing in that expression v_{2s} by its expression in function of v_{1s} and provided that $U_s - U_h$ may be approximated by $-h\left(\frac{dU}{dy}\right)_h$ then it is easy to show, for the studied saltation conditions, we have (see fig. 1 to 3)

$$L \approx \frac{u_{2s}v_{1s}}{g} \rightarrow G = m\,\phi\,\frac{u_{2s}v_{1s}}{g}$$

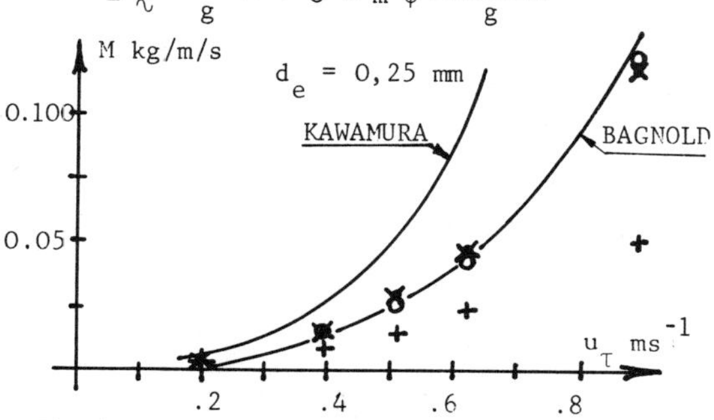

Fig.2. Mass flux of particles : + Owen's theory ; x Present theory ; . Approximate exp. ; o Bagnold's experiments

We can also observe that :

$$L = \frac{u_{2s}\,v_{1s}}{g} = \frac{v_{1s}\,v_{2s}}{g}\,\frac{u_{2s}}{v_{2s}} = \frac{v_{1s}\,v_{2s}}{g\,tg\,\gamma} \rightarrow$$

$$tg\,\gamma = \frac{g\,L}{v_{1s}\,v_{2s}}$$

Substituing in the last relation v_{1s} and v_{2s} by their respective expression in function of αu_τ and v_o, we find :

$$tg\,\gamma = -\frac{2h}{L}$$

To study the saltation movement over any plane sand surface we shall introduce the inclination δ of this plane with the horizontal direction. The sign of δ will be chosen (+) if the wind flows up the line of the greastest slope and (-) in the contrary.

The system formed by the equations 18-19-20 become then :

$$-\frac{d\bar{u}_1}{d\bar{x}} = 3\,(1-\bar{x})\,(\bar{U} - \bar{u}_1 - \bar{g}_1)$$

$$\frac{d\bar{u}_2}{d\bar{x}} = 3\,(1+\bar{x})\,(\bar{U} - \bar{u}_2 - \bar{g}_1)$$

$$a\,\frac{d\bar{U}}{d\bar{x}} + b\bar{x} = \bar{x}\,(\bar{u}_2 - \bar{u}_1) + 6\,\bar{g}_1\,\bar{x}^{-2}$$

$$\text{with } \bar{g}_1 = \left(\frac{1}{18}\,\frac{\sigma}{\rho}\,\frac{d}{\nu}\right)^2 g\,d\,\sin\delta$$

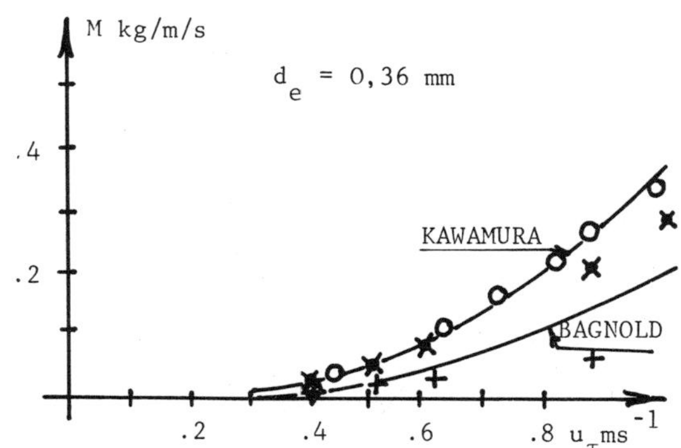

Fig.3. Mass flux of particles : + Owen's theory ; x Present theory ; . Approximate expression ; o Kawamura's experiments

The treatment of this last system is quite comparable to that of the preceeding one, it leads to the following z differential equation :

$$z'' + 6\bar{x}z' + 6\left(\frac{3}{2}\bar{x}^{2} - \frac{\bar{x}}{a} - 1\right)z =$$

$$- \frac{6}{a}\left(b - 6\bar{g}_1\bar{x}\right)$$

The solution of this equation has been researched under the form :

$$z = C(\bar{x})\, e^{-3/2\,\bar{x}^{2}} + b - 6\,\bar{g}_1\bar{x}$$

That leads, under the same approximations as in the previous case, to the following $C(\bar{x})$ differential equation :

$$C'' - \frac{6}{a}\left(\bar{x} + \frac{3a}{2}\right)C = 6b\left(1 - \frac{9}{8}\bar{x}^{-4}\right)$$

$$+ 54\,\bar{g}\,\bar{x}^{-3}\left(1 + \frac{3}{2}\bar{x}^{-2}\right)$$

The solution of this equation has been researched under a serial form which is slightly more complicated than that obtained in the case of horizontal sand surface.

The main results which we have obtained concern the mass flux and the angle γ of grains arrival, namely :

$$G = \alpha\frac{\rho u_{\tau}^{3}}{g\cos\delta}\left(1 - \beta_c\frac{\sigma g\, d\cos\delta}{\rho\, u_{\tau}^{2}}\right)$$

$$- \frac{\alpha\, u_{\tau}}{3g\cos\delta}\,\pi d^{3}\,\phi\,\sigma\sqrt{2gh\cos\delta}\;\; tg\delta$$

$$tg\gamma = -\,2h/L$$

We must also point out the fact that we do not know for the moment if β_c is function or not of δ.

REFERENCES

Bagnolds, R.A. 1941. The Physics of Blown Sand and Desert Dunes. Mathuen, London.

Bagnolds, R.A. 1956. The flow of cohesionless grains in fluids, **Phil.Trans.A**, Vol.249, p.235-297.

Bagnolds, R.A. 1954. Experiments on a gravity-free dispersion of large solid spheres in a Newtonian fluid under shear. Proc.Roy. Soc. A,225, 49.

Brocher, E. 1971. Etude théorique des phénomènes d'netraînement de produits pulvérulents par le vent (rapport interne).

Brocher, E. 1971. Calcul du poids de minerai et de charbon entraîné par le vent (rapport interne I.M.F.M.).

Comolet, R. 1976. Mécanique des fluides expérimentale. Tomes I et II,Masson et Cie.

Kawamura, R. 1951. Mouvement du sable sous l'effet du vent.Colloque International du C.N.R.S. à Alger (27-31 Mars).

Kells, L.M. 1947. Elementary differential equations,Mc Graw-Hill Book Cy,Inc.

Kennedy, J.F. 1963. The mechanics of dunes and antidunes in erodible-bed channels. Journal of Fluid Mechanics, February.

Owen, P.R. 1964. Saltation of uniform grains in air. Journal of Fluid Mechanics Vol.20, Part 2, pp.225-242.

Prandtl, L. Guide à travers la mécanique des fluides. Dunod.

Reynolds, A.J. 1976. A decade's investigation of the stability of Erodible stream beds. Nordic Hydrology, 7, 161-80.

Valiron, G. Théorie des fonctions. 1948 Masson et Cie.

Nassar, C. 1982. Résolution des équations régissant les phénomènes de saltation, Thèse de 3e Cycle I.M.F.M.

Euromech 156: Mechanics of Sediment Transport / Istanbul / 12-14 July 1982

On the relation between size and distance travelled for wind-driven sand grains – Results and discussion of a pilot experiment using coloured sand

OLE BARNDORFF-NIELSEN, JENS LEDET JENSEN & MICHAEL SØRENSEN
University of Aarhus, Denmark

ABSTRACT: An experiment is reported, which was conducted in order to obtain information on the stochastic process of distance travelled by a single sand grain in a heterogeneous sand bed under the action of the wind. The way a ridge of coloured grains was distributed down through a wind tunnel by the wind is studied. The main finding is that, for the size range considered, the distance travelled in unit time increases with grain size.

1 INTRODUCTION

When a sand grain is transported by the wind over a flat sand bed it passes through a sequence of periods of movement and rest. After being picked up from the sand bed by the action of turbulence or ejected by a descending grain, a saltating grain makes a number of jumps and then begins a new rest period, during which it may be sheltered from the wind by other grains. Contrary to this, a creeping grain rolls over the bed surface after being struck by a descending grain or by the action of the wind.

Precise simultaneous measurements of size and velocity of single sand grains, as these are transported by a wind blowing over a sand bed, are difficult to obtain due to the high velocities and the multitude of grains in transport. Instead of studying the momentary velocity, we therefore consider the distance travelled by a sand grain during a time interval.

The distance travelled by any single grain is determined by a large number of physical parameters. In modelling the travel behaviour it is both useful and natural to consider some of these as fixed during an experiment while the rest are supposed to vary randomly, so that the movement of each grain is considered a realisation of a stochastic process determined by the fixed parameters. The fixed parameters may, for instance, be mean wind speed, intensity of turbulence, size and shape distribution of the grains in the bed, and size of the particular grain whose movement is to be modelled; whereas grain shape and the turbulent velocity field may be taken as random. It is likely that the complete stochastic process is determined by a considerable number of parameters, and it is a large scale operation to investigate this in detail. The process is, however, essential for a comprehensive theory of the physics of blown sand and, in particular, it is a prerequisite for explaining the hyperbolic shape of the log-size distribution of sand samples from aeolian sand deposits (cf. Bagnold (1941), Bagnold and Barndorff-Nielsen (1980), Barndorff-Nielsen (1977), Barndorff-Nielsen et al. (1982), and Barndorff-Nielsen and Darroch (1981). If we restrict our interest to some main characteristics of the process we may, however, expect that only a small number of explaining parameters are needed, such as those mentioned above.

The experiment reported in the present paper is part of an investigation of the stochastic process of distance travelled and consisted in a study of the way a ridge of coloured heterogeneous sand is distributed down through a wind tunnel, by the wind, over a flat sand bed of the same kind of sand. The experiment had the character of a pilot experiment, performed to determine whether the methods employed are useful at all, and the experimental technique can therefore be improved considerably. As only one experiment has been conducted so far, the discussion given in the following is limited to the question of how distance travelled by a sand particle depends on the size of the particle.

The experiment and its analysis indicate

that - contrary to what might be expected - within the range of sizes observed the mean of the distance travelled in unit time increases with particle size.

Similar observations have, however, been made in the case of alluvial sediment transport, e.g. Gilbert (1914), Nevin (1946), and Byrne (1965). The stochastic processes governing the motion of bed load in alluvial sediment transport have been studied much more intensively than those governing aeolian sand transport, but since the physics of the two phenomena are different, the results from alluvial studies are not directly applicable to the case of windblown sand. The study of the bed load in alluvial streams by means of stochastic processes was initiated by H.A.Einstein (1937), who observed the motion of a collection of coloured stones in a flume. It was revived by Hubbell and Sayre (1964), who made flumes as well as field experiments using radioactive tracer particles. Since then a lot of experiments have been performed with radioactive tracers, e.g. Grigg (1970), Yang and Sayre (1971), and Hung and Shen (1979), but as most of these investigators aim at a determination of the form of certain probability laws defining the stochastic process, rather than at determining how the parameters involved in these laws depend on important physical parameters, none of them report results comparable to our results. However, Meland and Norrman (1969) studied the dependency of the average particle velocity in alluvial streams on the total transport, on the size distribution and on the shape of the grains, using coloured grains but in a way very different from ours. They reported that the average velocity increased with size to a mode point from which it decreased again.

In section 2 the material used and the experimental method are presented, and in section 3 the observations are analyzed statistically and a model explaining the data is proposed. This model and its connection to the stochastic process describing distance travelled by a single sand grain is studied mathematically in section 4. In particular, the two first moments of the probability law of our model are found, and are used to estimate the mean and variance of distance travelled in unit time, as functions of grain size. Finally the experimental method is discussed in section 5.

We wish to stress the tentative character of both the experiment and the analysis put forward in the sequel. We have argued rather freely and speculatively from the data, and much further experimental and theoretical work is needed to determine the extent to which the conclusions we suggest correspond to physical reality.

Further details on the experimental procedure and the analysis of the observations may be found in Barndorff-Nielsen, Jensen and Sørensen (1981).

2 EXPERIMENTAL MATERIAL AND METHODS

The experiment was carried out in a wind tunnel which is 14.5 m long, 0.35 m wide, and 0.5 m high. A sand from an aeolian sand deposit on Bønnerup Strand, on the north coast of Djursland in Denmark, was used.

Approximately 1 liter of the sand was coloured by 'Coomassie Brilliant Blue', dissolved in water, methanol and acetic acid in the proportions 5:5:1. The sand was kept in this dissolution for almost 24 hours and was then dried. It turned out that after a light manual separation only very few grains sticked together. Furthermore, one hour of sieving of the coloured grains showed that the colour sticks to the grains, even under the rather extreme physical conditions of sieving.

The bottom of the wind tunnel, from the intake to 10 m downwind, was covered by a 2 cm deep layer of sand. Prior to the experiment a wind 5 m/s was allowed to blow for half an hour. At the end of this period there was virtually no movement of the sand grains and only very small ripples had been formed. A ridge of coloured sand, 2 cm high and 9 cm wide, was placed 2.15 m from the intake of the wind tunnel, see figure 2.1. The shape and size of the ridge before and

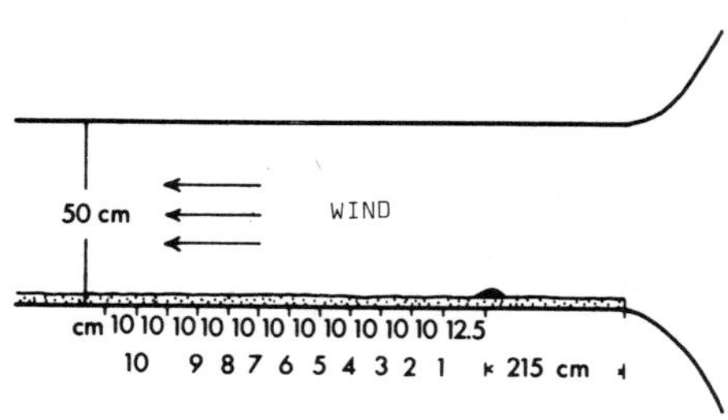

Figure 2.1. The experimental set up in the wind tunnel. The numbers 1-10 indicate the locations where the surface samples were taken.

after the experiment is shown in figure 2.2. The wind was then turned on, at a velocity of 7 m/s, and was allowed to blow for 2.5 minutes, at which stage the ridge had been partially covered by uncoloured grains. No measurements of the wind profile and turbulence were made. After the experiment samples were taken of the surface layer. At a number of locations along the tunnel, a 3-4 cm wide cross section of the bed surface was sprayed with an adhesive spray and thereby fixed. From each section the central part, of length approximately 15 cm, was taken up in small pieces constituting a sample, and on each side of this a 3-4 cm by 3-4 cm single piece was taken and placed on a flat glass tray. In the sequel we refer to these two kinds of samples as central part samples and glass tray samples, respectively.

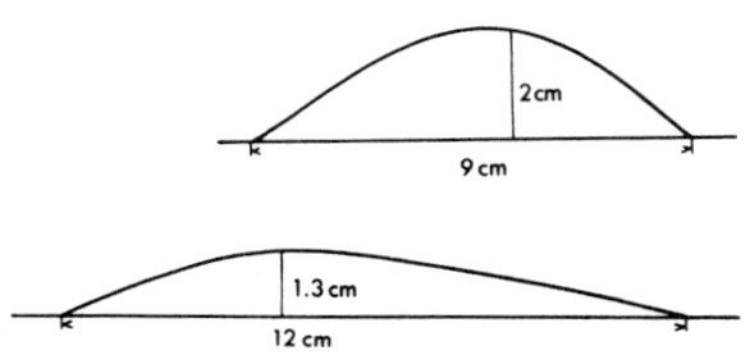

Figure 2.2. Shape and size of ridge of coloured grains before and after the experiment. At the top: before the experiment. At the bottom: after the experiment.

The spraying procedure has two disadvantages, the first being that the air stream from the spray may change the structure of the surface and the second being that the depth to which the spray penetrates differs from section to section. The sections were located successively, at 12.5 cm downwind of the original crest of coloured sand and then every 10 cm downwind until 122.5 cm; at 102.5 cm and 122.5 cm no glass tray samples were preserved. The locations for which both the central part and the glass tray parts were preserved are numbered 1 to 10, 1 being at 12.5 cm downwind and 10 at 112.5 cm downwind. The reason not to take more sand samples further downstream was that we thought (wrongly, as it turned out) that the amount of coloured grains would be too small for a reliable estimation.

Next, the spray was removed from the central part samples. This was done using acetone, which turned out not to affect the colouring of the grains. These samples were then separated into five fractions through a sieve system such that the grains have diameters between 0.063 mm - 0.125 mm, 0.125 mm - 0.177 mm, 0.177 mm - 0.250 mm, 0.250 mm - 0.354 mm, and 0.354 mm - 0.500 mm, respectively. For each size fraction the weight was recorded, and a sample from the fraction was spread on a piece of paper with millimeter division and put under a microscope. For four different square centimetres the number of coloured and uncoloured grains were counted.

Furthermore, counts of coloured and uncoloured grains, irrespective of grain size, in the uppermost surface layer were made from the glass tray samples, using a microscope having an ocular with a square grid.

3 ANALYSIS OF RESULTS AND MODEL FORMULATION

Table A.1 of the appendix shows the counts of coloured and uncoloured grains from the various size fractions and from the glass tray samples, and table A.2 gives the weights of the different size fractions.

As a preliminary step to the analysis of the data it is necessary to adjust the counts of uncoloured grains in each size fraction. In order to make the samples from different positions downwind comparable we must either know the surface area or the thickness of each of the samples, but as mentioned already in section 2, the spray method is somewhat arbitrary in these respects. For the sake of argument we shall assume that coloured grains are found only in the uppermost layer of grains, where the transport takes place. We are only interested in the fraction of coloured sand in this active surface layer, so that we have to estimate

δ_{ij} = fraction of coloured grains in the i'th size fraction of the active layer at location j, $j = 1,\ldots,10$, $i = 1,\ldots,5$ (1: 0.063 mm - 0.125 mm,..., 5: 0.354 mm - 0.500 mm).

The amount of coloured sand in the i'th size fraction is obtained directly from the fraction Δ_{ij} of coloured grains, as estimated from the counts of coloured and uncoloured grains, and the mass measurements M_{ij}. In order to calculate δ_{ij} we must estimate the amount of sand in the i'th size fraction in the active layer. If we know the size distribution in the active layer, all we need to evaluate is the total amount of sand in this layer. This quantity can be estimated by equating the number fraction of coloured grains in the active layer to the observed number fraction δ_j from the corresponding glass tray samples.

In order to determine the size distribution in the active layer we make the

Stationarity assumption: The particle size distribution is the same everywhere in the sample region and has not changed during the experiment.

This assumption is only valid because the experiment did not last more than two and a half minutes. If it had lasted considerably longer, longitudinal sorting in the active layer would have made the size distribution in this layer non-uniform and different from that of the layers below. One might fear that sorting during the half hour of blowing wind before the experiment did induce a size distribution in the active layer different from that in the layers below, but since the transport rate in this period was low, we ignore this possibility. The ridge might also have caused a longitudinal variation of the size distribution, but an examination of the empirical size distributions obtained from the weights in table A.2 revealed no systematic variation with the distance from the ridge (though the distribution at position 1 deviated somewhat from the other distributions).

Using the stationarity assumption we end up with the formula

$$\delta_{ij} = \Delta_{ij} \frac{\delta_j \sum_1 M_{1j}/\bar{d}_1^3}{\sum_1 \Delta_{1j} M_{1j}/\bar{d}_1^3} ,$$

where $\bar{d}_i$ is a typical diameter for the i'th size fraction, which we choose as $\bar{d}_i = \sqrt{d_i d_{i+1}}$, d_i being the mesh size of the i'th sieve.

In figure 3.1 we have made a plot of z_{ij} against position, where z_{ij} is the logistically transformed value of δ_{ij}, i.e. $z_{ij} = \ln\{\delta_{ij}/(1-\delta_{ij})\}$ = ln(number of coloured grains in the active layer/number of uncoloured grains in the active layer).

This figure indicates that z_{ij} may reasonably be described as a sum of two terms, one depending on i only, the other on j only. Furthermore, if we disregard the observations from the last position, the term depending on j appears to be reasonably approximated by a linear function of the distance from the ridge of coloured grains, although there is a large scatter around the linear trend. The observations from the last position do not accord with this linear behaviour. This may be due to errors in the sampling or analysing procedures, and these observations will therefore be ignored in the following considerations. We shall return to the observations from the last position in the discussion. The first hypothesis, of additivity, can be checked by the wellknown control plots for additivity; these plots showed no significant evidence against the hypothesis.

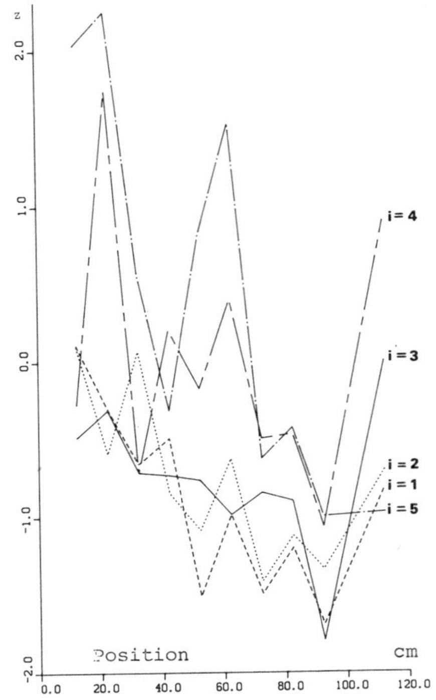

Figure 3.1. Estimated values z_{ij} of ln{number of coloured grains in the active layer/number of uncoloured grains in the active layer} plotted against location j = 1,...,10 for each size fraction i = 1,...,5.

The deviations from each of the five lines may be thought of as sums of many smaller errors, many of which are due to the fact that the experimental technique has not yet been refined. We will therefore suppose that the observed z_{ij}'s are stochastically independent and normally distributed with expectation $\beta_i - \alpha_i x_j$ and variance σ_i^2, where x_j is the distance from the ridge to position j and where α_i, β_i and σ_i^2 are parameters depending on size only. First, the hypothesis that the variances are independent of size is tested. This is a prerequisite for the next test but is also of independent interest. The hypothesis is accepted at the 2.5% significance level. Next, we test the hypothesis that the slopes of the lines are equal, i.e. $\alpha_1 = \ldots = \alpha_5 = \alpha$. This yields a test probability between 10% and 30%, which is acceptable. The estimates for α and β_i under the last hypothesis are given in table 3.1. (The hypothesis that the β's are equal is rejected by the ordinary F-test at the 0.1% significance level.)

Table 3.1. Estimated parameters of the model $z_{ij} \sim N(\beta_i - \alpha x_j, \sigma^2)$.

log-size = $\ln \bar{d}_i = s_i$	$\hat{\beta}_i$	$\varphi(s_i) = \exp(-\hat{\beta}_i)$
-0.8658	1.5814	0.2057
-1.2123	0.9509	0.3864
-1.5587	0.2069	0.8131
-1.9058	0.2817	0.7545
-2.4225	0.1302	0.8780

$\hat{\alpha} = 1.99\ m^{-1}$.

In this way we end up with the model

$$\zeta(s,x) = \beta(s) - \alpha x$$

where ζ is the theoretical value of z_{ij}, x is the distance to the ridge and s is the logarithm of the particle diameter. If we let $c(s,x)$ and $u(s,x)$ denote, respectively, the mass of coloured grains and the mass of uncoloured grains of size s per unit mass of the active layer at position x, the model may be written as

$$(3.1) \qquad \frac{c(s,x)}{u(s,x)} = e^{\beta(s)-\alpha x}.$$

The stationarity assumption implies that

$$(3.2) \qquad c(s,x) + u(s,x) = h(s),$$

where h is a function of size only. From (3.1) and (3.2) it follows that

$$(3.3) \qquad c(s,x) = \frac{h(s)}{1+\varphi(s)e^{\alpha x}},$$

where $\varphi(s) = \exp(-\beta(s))$.

In order to interpret (3.3), we now introduce

$\pi(x;s)$ = probability density function of the random variable whose value is the total distance travelled during the experiment by a coloured grain of size s, given that the grain actually left the ridge.

Since π can be interpreted as the fraction of the moved grains of size s that should ideally be found at position x, we conclude that π is of the form

$$\pi(x;s) = f(s)c(s,x),$$

where f is a function of s only, that is, from (3.3),

$$(3.4) \qquad \pi(x;s) = k(s)\frac{1}{1+\varphi(s)e^{\alpha x}},$$

where $k(s)$ is a norming constant, which is easily found to be

$$(3.5) \qquad k(s) = \alpha/\ln(1+\varphi(s)^{-1}).$$

A check of the model can be made by plotting $\ln \pi(x;s_i)$, as defined by (3.4) with the estimated values of α and $\varphi(s_i)$, together with $\ln \hat{f}(s_i) + \ln \hat{c}(s_i,x_j)$, where $\hat{c}$ denotes the observed equivalent of c and $\ln \hat{f}$ is found by minimizing the sum of squares of the residuals. The plots of $\ln \pi$ and $\ln \hat{f} + \ln \hat{c}$ are given in figure 3.2 and show a reasonable fit.

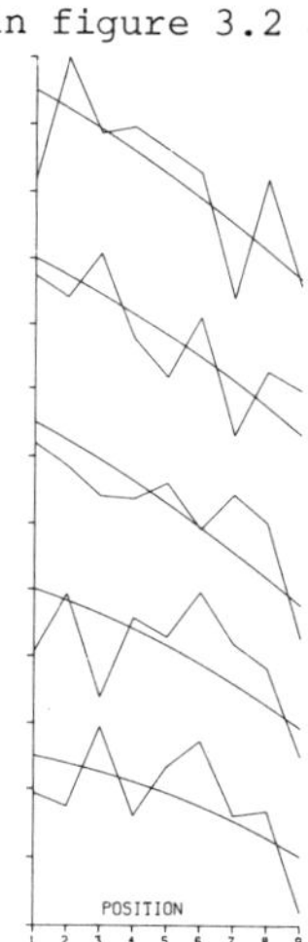

Figure 3.2. Check of model. For each size fraction $i = 1,\ldots,5$ the graph of $\ln \pi(x;s_i)$ (smooth curve) is shown together with the estimated values $\ln \hat{f}(s_i) + \ln \hat{c}(s_i,x_j)$ (connected by straight line segments).

4 ANALYSIS OF THE MODEL

The stochastic process describing the distance travelled in a time interval $[0,t]$ by a single sand particle of size s which enters the active layer at time 0, and the dependency of this process on s are the objects we are really interested in. We have, however, obtained the probability density function π of another random variable, viz. distance travelled in the time interval $[0,T]$, where T denotes the duration of the experiment, by a grain of size s that is only known to have entered the active layer sometime during $[0,T]$. It is therefore necessary to establish the connection between the two random phenomena.

Let $p(x;s,t)$ be the probability density function of the distance travelled by a particle of size s during a time period of length t, and, among the grains of size s that entered the active layer during the experiment, let $a(t;s)$ be the fraction of grains which did so at time t after the start of the experiment. If T is the duration of the experiment, $a(\cdot;s)$ is a probability density on the interval $[0,T]$. Let $\tau(s)$ and $\omega^2(s)$ denote the expectation and the variance of a $(= a(\cdot;s))$, and let $\mu(s,t)$ and $\sigma^2(s,t)$ be the expectation and variance of p $(= p(\cdot;s,t))$.

If we assume that the ridge of coloured sand does not affect the wind flow to an extent that the physical parameters determining the stochastic process are significantly changed along the test section, we have

$$\pi(x;s) = \int_0^T p(x;s,T-t)a(t;s)dt,$$

i.e. π is a time-mixture of p's with a as mixing distribution.

We now assume that the stochastic process of the distance travelled by a sand grain is additive. This essentially means that the increments of the process in disjoint time intervals of equal length are independent and identically distributed. The additivity assumption implies that

$$(4.1)\quad \mu(s,t) = t\mu(s,1) \qquad \sigma^2(s,t) = t\sigma^2(s,1).$$

The functions we are trying to determine are $\mu(s) = \mu(s,1)$ and $\sigma^2(s) = \sigma^2(s,1)$. The quantity $\mu(s,1)$ is the average velocity of sand grains of size s.

If $m(s)$ and $v(s)$ denote expectation and variance of π, respectively, we find from (4.1) by using conditional expectation that

$$(4.2)\qquad m(s) = \mu(s)(T-\tau(s))$$

and

$$(4.3)\quad v(s) = \sigma^2(s)(T-\tau(s)) + \mu(s)^2\omega^2(s).$$

Since $T-\tau(s)$ is the travel time of a typical grain of size s that did enter the active layer, (4.2) simply means that $m(s)$ is the distance travelled by a typical grain of size s during the experiment, whereas (4.3) shows that the variance $v(s)$ is larger than the variance for a typical grain. In the absence of any precise knowledge of τ a reasonable estimate is $T/2$.

Unfortunately, we cannot determine p explicitely from π and a, whereas it is easy to find $\mu(s)$ and $\sigma^2(s)$ from (4.2) and (4.3) if $m(s)$, $v(s)$, $\tau(s)$, and $\omega^2(s)$ are known. In the following we find $m(s)$ and $v(s)$ from (3.4). We could not observe $a(t;s)$ in our experiment, but it is likely that it is independent of s and that it does not vary too much with t. We can therefore assume that $a(t;s) \equiv 1/T$. This implies, in particular, that $\tau = T/2$, as suggested above. Under these assumptions (4.2) and (4.3) take a simpler form from which we obtain

$$(4.4)\qquad \mu(s) = 2m(s)/T,$$

$$\sigma^2(s) = 2(v(s) - \tfrac{1}{3}m(s)^2)/T.$$

Next, we calculate φ_π, the Laplace transform of π, m and v from (3.4) and (3.5). Obviously the domain of φ_π is $(-\infty,\alpha)$, and from formula (21), p.310, in Erdélyi et al. (1954) we find that

$$\varphi_\pi(\theta;s) = \frac{x}{\ln(1+\varphi(s)^{-1})}\int_0^\infty \frac{e^{\theta x}}{1+\varphi(s)e^{\alpha x}}dx$$

$$= \frac{1}{\ln(1+\varphi(s)^{-1})}\int_1^\infty \frac{u^{\theta/\alpha-1}}{1+\varphi(s)u}du$$

$$= \frac{F(1,1-\theta/\alpha;2-\theta/\alpha;-\varphi(s)^{-1})}{\varphi(s)(1-\theta/\alpha)\ln(1+\varphi(s)^{-1})},$$

where $F(a,b;c;z)$ is the hypergeometric function. We can find a simple series expansion of φ_π by the formulae

$$F(a,b;c;z) = (1-z)^{-a}F(a,c-b;c;z/(z-1)),$$

$c \notin \{0,-1,-2,\ldots\}$, (Olver, 1974, p.164) and

$$F(a,b;c;z) = \sum_{n=0}^{\infty} \frac{a(a+1)\ldots(a+n-1)b(b+1)\ldots(b+n-1)}{c(c+1)\ldots(c+n-1)n!}z^n,$$

$|z|<1$, (Olver, 1974, p.159). Using these formulae we obtain

$$\varphi_\pi(\theta;s)$$

$$= \frac{F(1,1;2-\theta/\alpha;(1+\varphi(s))^{-1})}{(1+\varphi(s))(1-\theta/\alpha)\ln(1+\varphi(s)^{-1})}$$

$$= \frac{\Gamma(1-\theta/\alpha)}{\ln(1+\varphi(s)^{-1})}\sum_{n=1}^{\infty}\frac{(n-1)!}{\Gamma(n+1-\theta/\alpha)}(1+\varphi(s))^{-n}.$$

The first and second moments of π can now be found by differentiation of φ_π. As the calculations are straightforward they are omitted. We find that the mean is

$$(4.5) \qquad m(s) = \frac{\sum_{n=1}^{\infty} \frac{a_n}{n} (1+\varphi(s))^{-n}}{\alpha \ln(1+\varphi(s)^{-1})}$$

and that the second moment around the origin is

$$(4.6) \qquad m^{(2)}(s) = \frac{\sum_{n=1}^{\infty} (a_n^2+b_n)\frac{1}{n}(1+\varphi(s))^{-n}}{\alpha^2 \ln(1+\varphi(s)^{-1})},$$

where $a_n = \sum_{k=1}^{n} k^{-1}$ and $b_n = \sum_{k=1}^{n} k^{-2}$.

Using the values of α and $\varphi(s_i)$ estimated from the data (cf. table 3.1), we can now make plots of $\mu(s)$ and $\sigma(s)$ as calculated from (4.4), (4.5) and (4.6). These are shown in figure 4.1.

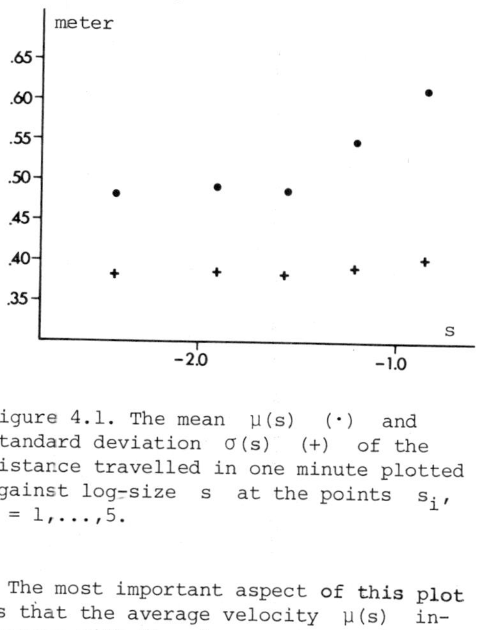

Figure 4.1. The mean $\mu(s)$ $(\cdot)$ and standard deviation $\sigma(s)$ (+) of the distance travelled in one minute plotted against log-size s at the points s_i, $i = 1,\ldots,5$.

The most important aspect of this plot is that the average velocity $\mu(s)$ increases with the size s. As mentioned in the introduction, this kind of dependency on size has been observed for sediment transport by water, but this, as far as we know, is the first time it has been reported for windblown sand. Note also that the standard deviation $\sigma(s)$ is nearly constant. As the average velocity $\mu(s)$ and the standard deviation $\sigma(s)$ of the distance travelled in a unit of time are of the same order of magnitude there is a considerable variation in the distance travelled. The departure from the trend of the other points shown by the third size fraction (which contains the largest amount of grains) will be discussed in Section 5.

5 DISCUSSION

This pilot experiment has shown that it is possible by the methods employed to obtain considerable information on the travel behaviour of wind-driven sand grains, as this depends on grain size. The large scatter of the data indicates that it is essential to investigate which parts of the experimental procedure give rise to the largest errors and hence to improve the procedure, for instance by repeating these parts.

The adjustment procedure, by means of the surface counts, of the number of uncoloured grains in the central part samples, which was carried out in order to make the samples comparable, is not very satisfactory. A check of how reasonable it is can be made by calculating the ratio of the mass of the original sample to the mass of the adjusted sample at each location and comparing these ratios. This reveals that the sample from the position closest to the ridge has been adjusted much less than the others, and as this sample is of average size compared to the others this must either be because the surface area of the sample was larger than that of the others, or because the active layer at the first position or at the other positions have been estimated incorrectly. If we believe in the density function π, an inspection of the residuals in figure 3.2 suggests that the second explanation is the appropriate one, since all the residuals at position 1 are negative. The adjustment procedure seems to have been reasonable at all the other locations, so the possible error in the adjustment of the sample from position 1 cannot have affected our results much. However, since the method of adjustment is disputable it would be convenient to avoid it. A possible way of doing this would be to take samples with a fixed surface area. Then the masses of coloured grains in the samples will be comparable, provided each sample is so thick that all coloured grains below the surface area of the sample are included in the sample. Since the quantity c of section 3 at any given location is proportional to the mass of sand in a sample that meets these requirements, with the same constant of proportionality at all positions, we can study the density function π directly by means of such samples. If we use this procedure we will not need the stationarity assumption or any equivalent assumption, and thus we do not have to care about sorting during the experiment, unless this sorting is so vigorous as to change the stochastic process we are studying either along the tunnel or in time. The stationarity assumption is needed, however, if we want to deduce a probability density function for the distance travelled during the experiment from the behaviour of the ratio of the numbers of coloured and uncoloured grains, in the way it is done in this paper.

The active layer is not needed in the analysis of comparable samples, but it is of independent interest to determine the thickness of this layer. One way of investigating it would be to place a zone of coloured grains somewhere in a bed of otherwise uncoloured sand in the windtunnel. Then after a period of transport the coloured grains in the active layer will have been replaced by uncoloured grains, and by fixating samples vertically through the bed in the originally coloured zone the thickness of the active layer can be determined. This thickness depends, of course, on various physical parameters, particularly the friction velocity.

The ridge must to some extent have affected the wind flow in the sampling region. Since no wind profiles were measured we cannot provide a precise description of this influence. We can only give some general physical considerations. There must have been a wake behind the ridge which may have affected the flow at the first sample position. Another consequence is that a turbulent boundary layer was reestablished downwind of the ridge. This may have affected the flow in most of the sample region, and there must have been some variation in the shear stress with the distance from the ridge. This picture may have been complicated by the fact that the ridge was placed so close to the intake of the windtunnel that the turbulent boundary layer was not fully developed at the position of the ridge. In future experiments the properties of the wind flow must be studied by measurement of the wind profiles. It would also be desirable to avoid the ridge. This can be done by replacing the sand in the active layer in a section of the wind tunnel by coloured sand, instead of introducing the coloured sand in the form of a ridge.

The sand sample taken furthest downwind from the ridge $(j = 10)$, which we have disregarded in the preceeding analysis, shows a continuing decrease in the amount of the largest grains but an increasing amount of small grains. This may be due to experimental errors, as already noted, but it may also be explained by assuming that sometime during the experiment the rate at which the small grains started from the ridge decreased, for instance because the size distribution of the sand at the surface of the ridge was changed in such a way that the large grains could offer the small grains a better shelter against the wind. If this is in fact what happened, the assumption that a is uniformly distributed in the interval $[0,T]$ is violated, and this could profoundly change our results. The question must be answered in future experiments by taking samples further downwind from the ridge and by checking the size distribution of the ridge before and after the experiment.

In future experiments we will also use a larger number of sieves in order better to determine the function $\varphi(s)$. This will also enable us to analyse the size distribution of the samples more accurately and thus to check the assumption of longitudinal stationarity.

The conclusion of the analysis presented in this paper is that once the experimental method has been refined, and the questions raised above have been investigated, this kind of experiment can, very likely, be used to determine how the first two moments of the stochastic process of distance travelled by a single grain depends on various physical parameters. Other aspects of the process can be investigated in a similar manner.

APPENDIX

This appendix contains some tables referred to in the text.

Table A.1. Counts of coloured and uncoloured grains, from each size fraction of the central part samples, and from the glass tray samples. c: coloured and u: uncoloured.

Position / Part of sample	1 c	1 u	2 c	2 u	3 c	3 u	4 c	4 u	5 c	5 u	6 c	6 u	7 c	7 u	8 c	8 u	9 c	9 u	10 c	10 u
0.500-0.354	35	33	4	25	4	33	5	32	8	21	6	52	18	132	3	43	4	41	1	39
	14	19	10	40	2	49	2	21	6	20	3	33	11	65	4	49	4	33	0	23
	17	19	7	39	4	39	0	33	3	15	8	54	8	64	6	42	0	40	2	35
	18	23	4	33	7	30	1	22	5	16	7	40	4	45	2	35	4	19	0	17
0.354-0.250	10	36	8	40	4	64	9	104	7	44	13	136	13	74	0	40	5	65	2	22
	11	32	6	42	0	43	11	103	6	37	8	93	14	103	1	35	4	38	4	41
	18	72	8	59	6	81	6	89	6	40	11	97	15	86	3	23	5	41	4	42
	18	50	8	36	2	31	12	87	7	22	13	150	9	84	10	63	3	55	3	30
0.250-0.177	24	98	8	86	7	234	12	232	6	59	9	227	13	72	5	70	3	88	4	63
	19	80	7	96	10	90	14	247	6	60	7	261	8	78	6	89	2	74	4	58
	42	150	7	73	7	87	14	196	10	58	17	232	7	85	3	76	6	67	11	141
	17	72	3	68	2	67	8	188	4	41	7	259	7	80	5	61	3	73	2	55
0.177-0.125	64	125	10	130	9	96	6	194	10	80	13	170	17	238	16	234	10	108	6	123
	58	165	7	122	12	103	11	219	6	73	14	300	16	233	8	203	9	103	7	100
	42	133	7	96	8	133	13	223	7	109	13	184	14	239	15	225	9	149	5	140
	37	94	5	101	10	106	13	198	9	84	8	261	16	191	9	235	5	121	4	89
0.125-0.063	64	158	18	179	10	140	9	129	19	244	7	138	21	317	10	203	10	183	9	157
	79	147	10	199	10	135	9	161	13	158	6	142	17	372	3	106	7	157	13	268
	64	191	17	200	12	240	19	242	11	157	7	233	31	369	11	151	10	152	3	158
	55	174	19	242	11	217	11	201	10	154	9	209	16	250	4	102	7	188	8	142
Glass tray	63	68	56	85	50	70	36	72	33	92	45	100	30	105	32	96	19	87	38	83

Table A.2. Weights in grams of the different size fractions.

Position / Size fraction	1	2	3	4	5	6	7	8	9	10
0.500-0.354	0.0304	0.0235	0.0502	0.0614	0.0582	0.0570	0.0610	0.0461	0.0566	0.0533
0.354-0.250	0.4764	0.2916	0.3823	0.4998	0.6256	0.6260	0.5218	0.3678	0.4828	0.4217
0.250-0.177	3.8353	2.5740	2.6072	3.5350	4.6171	4.1014	3.2787	2.3875	3.5613	3.2450
0.177-0.125	2.1729	2.3992	2.0125	2.8095	3.0524	3.2330	2.0198	1.9396	3.0300	2.8130
0.125-0.063	0.1889	0.4356	0.3233	0.4125	0.1708	0.5106	0.2575	0.3462	0.4025	0.3960
Total	6.7039	5.7239	5.3755	7.3182	8.5241	8.5280	6.1388	5.0872	7.5332	6.9240

REFERENCES

Bagnold, R.A. 1941. The physics of blown sands and desert dunes. London: Methuen.

Bagnold, R.A. and Barndorff-Nielsen, O. 1980. The pattern of natural size distributions. Sedimentology 27: 199-207.

Barndorff-Nielsen, O. 1977. Exponentially decreasing distributions for the logarithm of particle size. Proc.Roy.Soc.London A 353: 401-419.

Barndorff-Nielsen, O., Dalsgaard, K., Halgreen, C., Kuhlman, H., Møller, J.T., and Schou, G. 1982. Variation in particle size distribution over a small dune. Sedimentology 29: 53-65.

Barndorff-Nielsen, O. and Darroch, J.N. 1981. A stochastic model for sand sorting in a wind tunnel. Adv.Applied Prob. 13: 282-297.

Barndorff-Nielsen, O., Jensen, J.L., and Sørensen, M. 1981. The relation between size and distance travelled for wind-driven sand grains - results and discussion of a pilot experiment using coloured sand. Research Report 74, Dept. Theor.Statist., Aarhus University.

Byrne, R.J. 1965. Some effects of particle bed geometry in selective sediment sorting. Office Nav.Res.Contract Nonr-2121 (26), Technical Report No.3.

Einstein, H.A. 1937. Der Geschiebetrieb als Wahrscheinlichkeitsproblem. Zürich: Verlag Rascher and Co. Reprinted in English as appendix C in Shen, H.W. (ed.) 1972. Sedimentation symposium to honor Professor H.A.Einstein. Colorado: Fort Collins.

Erdélyi, A. et al. 1954. Tables on integral transforms, Vol. I. New York: McGraw-Hill.

Gilbert, K.G. 1914. The transportation of debris by running water. U.S.Geol.Survey, Prof. Paper 86.

Grigg, N.S. 1970. Motion of single particles in alluvial channels. J.Hydraul. Div., ASCE., 96, HY 12: 2501-2518.

Hubbell, D.W. and Sayre, W.W. 1964. Sand transport studies with radioactive tracers. J.Hydraul.Div., ASCE., 90, Hy 3: 37-68.

Hung, C.S. and Shen, M.W. 1979. Statistical analysis of sediment motions on dunes. J.Hydraul.Div., ASCE., 102, HY 3: 213-227.

Meland, N. and Norrman, J.O. 1969. Transport velocities of individual size fractions in heterogeneous bed load. Geografiska Annaler 51A, 3: 127-244.

Nevin, C. 1946. Competency of moving water to transport debris. Geol.Soc.Am.Bull. V.57.

Olver, F.W.J. 1974. Asymptotic and special functions. New York: Academic Press.

Yang, C.T. and Sayre, W.W. 1971. Stochastic model for sand dispersion. J.Hydraul.Div., ASCE., 97, HY 2: 265-288.

Euromech 156: Mechanics of Sediment Transport / Istanbul / 12-14 July 1982

On the mathematical modeling of aeolian saltation

JENS LEDET JENSEN & MICHAEL SØRENSEN
University of Aarhus, Denmark

ABSTRACT: It is argued that the development of a mathematical model for aeolian saltation is a promising way of obtaining further progress in the field of wind-blown sand. It is demonstrated how interesting quantities can be calculated from a model defined in general terms, and a specific model is defined and compared to previously published data on aeolian saltation. This comparison points out the necessity of discriminating between pure and real saltation.

1 INTRODUCTION

A number of empirically determined transport rate formulae for wind blown sand has appeared in the literature (see e.g. the review paper by Phillips and Willetts(1978)). Unfortunately these are not in mutual accordance, presumable due to different experimental conditions or errors induced by the sand traps. Such formulae can therefore only roughly give the order of magnitude of the transport rate, and in some contexts such as e.g. the study of sand sorting phenomena this is not sufficient. Furthermore, even careful measurements of transport rates of the total population of sand will rather yield complicated empirical formuelae than actual understanding of the physics of aeolian sand transport.

Further progress in the field of aeolian sand transport must in the authors opinion involve the development of a mathematical model of the working physical processes. Of course, parameters in the model must be determined by experiments, but since they are part of a model the possibility of interpreting the results and understanding discrepancies is much increased. Evaluation of parameters in the model can for instance be done by means of trap data as those obtained by Williams (1964), where sand is caught in various heights above the sand surface.

In section 2 a mathematical model for aeolian saltation is defined in general terms. The usefulness of such a model is demonstrated by the derivation of expressions for various quantities including the transport rate in terms of basic quantities. In section 3 it is discussed how to calculate the forces exerted on the grains during the saltation, and a specific way of doing so is chosen. This specific model is confronted with William's data in section 4. The comparison renders the definition of pure and real saltation necessary.

2 MODELLING OF SALTATION IN GENERAL TERMS

Large forces are exerted on a sand grain at the moment it starts a saltation jump. It may receive its initial momentum from a descending grain that shoots into the bed or from the turbulent velocity field, or on collision with the bed some of the forward momentum of a descending grain may be converted into upward momentum. We are not in the present paper interested in the nature or magnitude of the initiating forces. Our attention is confined to the result of these forces, namely the initial velocity of the saltating grain. When the grain has left the surface it is acted on by gravity, a drag force and a lift force.

We will assume that the motion of the saltating grains is two-dimensional and ignore the small change of the drag force and the lift force due to the neglected component of velocity. Furthermore, we will suppose that a saltating grain as soon as it has left the bed is only affected by the mean wind, such that we can disregard the turbulent eddies. Finally, we will assume that the sand surface is plane and horizontal, that it is defined by $y=0$, and that the wind is blowing in the direction of the x-axis.

Let $D(v)$ denote the drag on a sand grain at a relative velocity between the grain and the air of v, and let $L(v,y,\omega)$ be the lift force that, in addition to the relative velocity, may depend on the height above the surface y and the angular velocity of the grain ω. If we denote the mass of a sand grain by m and the wind speed in height y by $U(y)$ the equations of motion are

$$2.1\quad \ddot{x} = (D(v)/mv)(U(y)-\dot{x})+(L(v,y,\omega)/mv)\dot{y}$$
$$\ddot{y} = -(D(v)/mv)\dot{y}-g+(L(v,y,\omega)/mv)(U(y)-\dot{x})$$
$$I\dot{\omega} = M,$$

where I is the moment of inertia of the grain and M is the moment exerted on the grain.

Bagnold (1936) reports that rotation of the grains does not seem to be the rule, though occasionally a grain is seen to rotate once in ten diameters of path, whereas Chepil (1945) observed that between 50 and 70 per cent of the saltating grains are rotating at a speed of 200 to 1000 revolutions per second. White and Schulz (1977) tentatively included the effect of rotation in a model of saltation trajectories using an asymptotic expression for the Magnus force and moment on a sphere due to Rubinow and Keller (1961), valid for Reynold numbers in the Stokes region. They estimated the rotation by fitting the theoretical trajectories to photographically observed trajectories, and obtained initial spinning rates from 100 to 300 revolutions per second. Obviously there is considerable doubt about this procedure, and at least it can not in any way be accurate. Furthermore, knowledge of the initial spinning rate, which is difficult to obtain, is needed. Therefore, we will and can not use this procedure. Owen (1980) noted that it is very likely that a grain on collision with the ground gets a clockwise spin if it is moving from left to right, but that this spin will relatively quickly be damped by friction. In the light of this we will assume, as an approximation, that the momentum given to a grain by the Magnus force is included in the initial momentum of the grain and thus disregard the angular velocity.

If $U(y)$, $D(v)$ and $L(v,y)$ are specified, we can solve the first two equations of 2.1 numerically when the initial velocity is prescribed. In particular, we can calculate the time t_o, when the grain hits the surface again, the length of the jump $x(t_o)$, and the horizontal velocity component of the grain as it hits the surface. These quantities will turn up again in the following derivations.

We will assume that the sand population in question can be classified in classes of grains with the same average transport properties. The behaviour of each grain during the aeolian transport is supposed to vary randomly around an average behaviour due to slight differences between the grains as well as to the turbulence of the air stream and the random nature of the sand surface. In general, the transport properties of a grain depends on its size, shape and material, but often sand populations are sufficiently homogeneous with respect to shape and material that a classification can be based solely on the size of the grains. We will index the transport classes by the letter s, where s is varying in some appropriate set.

Let $\phi(s)$ be the mass of grains in class s that leaves a unit of area per second including grains starting from rest as well as rebounding grains. We assume that the saltation is stationary and uniform such that $\phi(s)$ does not depend on position and time and equals the mass of grains that hits the surface per second.

Since we are not here aiming at a detailed modelling of the initiation of the grains it is sufficient to assume that the initial velocity of a grain v_i is a random variable with density function $f_s(v_i)$. This density function will in addition to s at least depend on the friction velocity and the size distribution in the surface of the sand bed . The mass of grains in class s that leaves a unit of area with initial velocity v_i per second is

$$\psi(s,v_i) = \phi(s)f_s(v_i).$$

We denote by (u_1,v_1) respectively (u_2,v_2) the velocity vector of a grain when the vertical velocity component is positive respectively negative. It is easily verified that the time a sand grain spends between the heights y_1 and y_2 $(y_1<y_2)$ on its way up respectively down is

$$\int_{y_1}^{y_2} v_1(y)^{-1}dy \quad \text{respectively}$$

$$-\int_{y_1}^{y_2} v_2(y)^{-1}dy \,.$$

Since the saltation is uniform and stationary it follows that the concentration by mass in height y of grains of class s with initial velocity v_i on their way up respectively down is

$$2.2 \quad \psi(s,v_i)/v_1(y,s,v_i) \text{ respectively}$$

$$-\psi(s,v_i)/v_2(y,s,v_i).$$

These expressions can also be found by a continuity argument, as was shown by Owen (1964).

Now, let $y^*(s,v_i)$ denote the height of the trajectory traced out by a grain of class s that starts with initial velocity v_i. Then we can calculate the transport rate G(s) of grains of class s.

$$2.3 \quad G(s) = \int_0^{\infty}\int_0^{y^*(s,v_i)} \psi(s,v_i)\left\{\frac{u_1(y,s,v_i)}{v_1(y,s,v_i)} - \frac{u_2(y,s,v_i)}{v_2(y,s,v_i)}\right\} dy\, dv_i$$

$$= \phi(s)\int_0^{\infty}\int_0^{t_0(s,v_i)} \dot{x}(t,s,v_i)\,dt\, f_s(v_i)\,dv_i$$

$$= \phi(s)\ \bar{x}(t_o,s),$$

where $\bar{x}(t_o,s)$ is the mean of the jump length with respect to the distribution f_s, i.e. the conditional expectation of the jump length given the transport class. This formula is intuitively obvious. The total transport rate is

$$2.4 \qquad G_T = M\ \bar{\bar{x}}(t_o),$$

where M is the total mass of grains that leaves a unit of area per second and $\bar{\bar{x}}(t_o)$ denotes the mean of $\bar{x}(t_o,s)$ with respect to the size-distribution of the sand taking part in the saltation

$$\varphi(s) = \phi(s)/M$$

We assume that the shear stress is constantly equal to ρU_f^2 in the layer in which the saltation takes place. Here U_f denotes friction velocity, which is also well defined during saltation (see Bagnold (1936) and Owen (1964)), and ρ the density of the air. In the saltation layer the grains are, however, extracting momentum from the wind such that from the top of the layer to the surface of the sand bed an increasing part of the shear stress is carried by the grains The partition of the total shear stress in an air-borne shear stress τ and a grain-borne shear stress T was introduced in Bagnold (1954), while the idea was implicitly used in Bagnold (1941). Owen (1964) noted that if the mean wind and the sand transport are stationary and uniform it follows from the momentum equation that the gradient of the air-borne shear stress is given by

$$2.5 \quad \frac{d\tau}{dy} = \left\{\begin{array}{l}\text{the horizontal component of the}\\ \text{drag and lift force on all sal-}\\ \text{tating grains in the height y}\end{array}\right\}.$$

Since $T+\tau$ is constant it follows that $d\tau/dy = -dT/dy$. Now, let T_o and τ_o denote the values of T and at the surface. Using 2.1, 2.2 and 2.5 it follows that

$$2.6 \quad T_o = \rho U_f^2 - \tau.$$

$$= \int_0^{\infty} \frac{d\tau}{dy}\,dy$$

$$= \int_0^{\infty}\phi(s)\int_0^{\infty}\int_0^{t_o(s,v_i)} \ddot{x}(t,s,v_i)\,dt\, f_s(v_i)\,dv_i\, ds$$

$$= \int_0^{\infty}\phi(s)\ \bar{\dot{x}}(t_o,s)\,ds$$

$$= M\ \bar{\bar{\dot{x}}}(t_o),$$

where $\bar{\dot{x}}(t_o,s)$ denotes the mean with respect to the distribution f_s of the horizontal velocity of a grain that hits the surface, while $\bar{\bar{\dot{x}}}(t_o)$ is the mean of $\bar{\dot{x}}(t_o,s)$ with respect to the size-distribution $\varphi(s)$. We have disregarded the horizontal component of the initial velocity, since it is small compared to the final horizontal velocity component. The formula 2.6 states that T_o equals the total horizontal momentum of the grains that hit a unit of area per second. From the calculations it follows that the fraction $\theta(s)$ of T_o carried by the grains of transport class s is

$$2.7 \quad \theta(s) = \phi(s)\ \bar{\dot{x}}(t_o,s)/(\rho U_f^2 - \tau_o).$$

In preparation for evaluating the mathematical model by means of trap data we will next find the mass of grains of class s with initial velocity v_i that is caught in an ideal trap, which is one lenght unit wide and Δ units high, placed with its buttom y_o units above the surface of the sand bed:

$$\psi(s,v_i)\int_{y_o}^{y_o+\Delta}\left\{\frac{u_1(y,s,v_i)}{v_1(y,s,v_i)} - \frac{u_2(y,s,v_i)}{v_2(y,s,v_i)}\right\} dy$$

$$= \psi(s,v_i)\,[\Delta x_{up}(s,v_i) + \Delta x_{down}(s,v_i)],$$

where Δx_{up} respectively Δx_{down} is the horizontal distance travelled by a grain while its vertical position increases from y_o to $y_o + \Delta$ respectively decreases from $y_o + \Delta$ to y_o.

3 THE ACTUAL CALCULATION OF THE TRAJECTORIES

When it comes to actually calculating saltation trajectories the forces exerted on the grains, the wind profile and the initial conditions must be evaluated. In what follows these points are discussed and specific ways of calculating the necessary quantities are chosen. Similar specific models have been used by Bagnold (1936), Ford (1957). Owen (1964 and 1980),Ellwood et.al. (1975), White et.al. (1976) and White and Schulz (1977).

We have already neglected the Magnus effect, but there may still be a lift force due to the presence of the bed and the fact that the saltation takes place in a shear flow. The magnitude and direction of the lift force on a non-rotating sphere near a solid boundary and in a turbulent shear flow has been studied at Reynold numbers between 800 and 50000 (cf. Willetts and Murray (1981) and Bagnold (1974)). If the results can be extrapolated to saltating grains over a loose boundary and Reynold numbers of the order 10^2, the grains will experience a large lift force close to the bed. Remote from the bed this force is negligible compared to gravity and drag, so as for the Magnus effect we can, as a fair approximation, suppose the effect of the lift force to be included in the initial velocity.

The drag on a body moving relatively to a fluid is usually calculated by the formula

$$D(v) = \tfrac{1}{2}\rho c(R) v^2 A ,$$

where A is the area of the projection of the grain on a plane perpendicular to the relative velocity vector, and R is the Reynold number dv/ν. Here d is a typical linear dimension of the grain in question and ν is the kinematic viscosity of the air. The way the drag coefficient c depends on the Reynold number has to be determined empirically. Various such graphs has been published (see e.g. Engelund and Hansen (1972)) but they depend on the shape of the grains which makes them difficult to use. For a sphere the drag coefficient has been determined very accurately, and it is desirable to use this information to calculate an approximation to the drag on a sand grain.

A method for doing this was proposed and used by Bagnold (1936). When any lift force is disregarded the equations of motion 2.1 takes the simpler form

$$3.1 \quad \ddot{x} = (D(v)/mv)(U(y)-\dot{x})$$
$$\ddot{y} + g + (D(v)/mv)\dot{y} = 0.$$

These equations only involve drag through the ratio

$$\Sigma(v) = D(v)/mg,$$

which Bagnold called the susceptibility. Inspired by this fact, the drag on a sand grain can be approximated by the drag on a sphere with the same specific gravity as the grain and with the same susceptibility at a relative velocity typical of saltation. The only relative velocities at which it is easy to determine the drag on a sand grain is the terminal velocity of fall in a fluid. Bagnold chose the sphere made of the same material as the grain and with the same susceptibility at the terminal velocity of fall in air. It is easily seen that this is the sphere of the same terminal velocity of fall in air as the grain. This sphere is called the aerodynamically equivalent sphere , and its diameter is called the fall diameter. However, calculations using the model specified in this section reveal that the terminal velocity of fall in air is not a typical relative velocity during a saltation jump.The use of the the aerodynamically equivalent sphere in calculating the drag on a grain may still be useful, but this point needs clarification. In this connection it should be noted that the terminal velocity of fall in water can not be used.Willetts and Rice (1982) have demonstrated that methods based on the terminal velocity of fall in water is not capable of discriminating between even very different grain shapes for grains smaller than 300 µm, presumably because the Reynold numbers corresponding to the terminal velocity of fall in water for such small grains are in a range of small Reynold number, where the drag does not depend much on shape. These Reynold numbers are considerably smaller than those characteristic of aeolian saltation.

A measure of grain shape which enables the calculation of the drag on a sand grain is highly desirable , but perhaps unattainable. In the absence of anything better we will use Bagnold's aerodynamically equivalent sphere in our calculations, and calculate the drag coefficient for a sphere by the formula given by Schiller and Nauman (1933)

$$3.2 \quad c(R) = (24/R)(1 + 0.15\, R^{0.687}).$$

This equation is useful for R less than 700, which is appropriate for aeolian saltation. However, since the drag coefficient is known to increase as the grain concentration in the air is increased, the correct value of c will be larger than that found from 3.2. The clarification of this point must wait until a precise determination of the grain concentration in various heights has been made. Presumably the error induced by using 3.2 will only be of any significance close to the surface. Here the presence of the bed will also render 3.2 unreliable.

From 3.2 it follows that

$$\Sigma = \frac{18\rho\nu}{\sigma}(1 + 0.15\ R^{0.687})d^{-2},$$

where d is the fall diameter, $R = dv/\nu$ and σ is the density of the grains.

The authors are not aware of any reliable measurements of the wind profile in the saltation layer close to the ground. In the whole saltation layer Bagnold(1936) found that a logarithmic profile

$$U(y) = 2.5U_f\ \ln(y/k')+V$$

agreed closely with his measurements.

Here k' is the height of a focus point at which the velocity should be the same, viz. V, irrespective of the friction velocity. There has been some discussion as to what is the appropriate value of von Kármáns constant in the presence of saltating grains in the air (see Zingg (1953). We will not join this discussion.

A fixed focus point was not present in Zingg's (1953) measurements, and a dimensionless plot of the wind data obtained by Bagnold, Chepil (1945)and Zingg made by Owen (1964, figure 3) indicates that the logarithmic profile is only valid outside the saltation layer. The velocities in the saltation layer are higher than would be expected from the logarithmic profile. Owen (1980) therefore assumed that inside a saltation layer of a thickness of approximately $U_f^2/2g$ the mean wind velocity is constant, the argument being that the effect of the saltating grains on the air stream is similar to the effect of a perforated screen. This is, of course, not exactly true. Based on simple assumptions Owen (1964) derived an asymptotic expansion for the wind profile in the saltation layer. The first term of this expansion is a constant proportional to U_f.

The departure from the logarithmic velocit profile close to the surface of the sand bed may or may not be an artefact caused by a break down of the measurement techniques in the vicinity of the surface. But until this has been investigated further, we will use a logarithmic profile above the height where the wind speed equals 6 U_f and below this point assume the wind speed to equal 6 U_f. The speed of 6 U_f was assessed from figure 3 in Owen (1964).

Bagnold (1941) and Chepil (1945) concluded from their photographical registrations of saltation trajectories that saltating grains start almost vertically. In contrast to this White and Schulz (1977) found by similar methods that the mean lift off angle for saltating spheres is approximately 50°. Since the more refined method used by White and Schulz enabled them to record trajectories as low as 0.5-0.75 cm, the results obtained by Bagnold and Chepil may be due to a systematic sampling error, since these authors presumable observed only relatively high trajectories. It is likely that the initial momentum of saltating grains usually have a horizontal component, but this component is small compared to the forward momentum that the grains receive from the wind during a jump. What really matters is how far into the air stream the grain reaches, because this is what determines how far the grain jumps and how much momentum it will receive. Also, when we consider the horizontal velocity of the grain when it hits the surface again we can ignore the initial horizontal velocity component. We will therefore, considering the uncertainty concerning the magnitude of the initial horizontal velocity, simply assume that the grains start vertically.

4 EVALUATION OF THE MODEL

The model defined in section 2 and 3 can be evaluated by comparing it to the data on aeolian saltation obtained by Williams (1964). He caught saltating grains in five heights (0-1, 2-3, 5-6, 9-10 and 15-16 cm) above the sand bed by means of a trap with five gabs. The experiment was carried out for variuos wind speeds, grain shapes and initial size-distributions of the bed material.

For each combination of wind speed, size-distribution and grain shape three runs were made. In each run the wind was turned on for a period of 1-10 minutes depending on the wind speed, and afterwards the total amount of sand that had entered each of the gabs was weighed and a size analysis was made by means of sieving. The sand was divided into five size classes. The results from the three runs was combined in order to eliminate sampling errors.

For naturally worn sand particles the transport rate at height y was found to be given by the equation

$$q = q_o e^{-By},$$

where q_o and B are constants. Unfortunately there is a discrepancy between the values of the total transport for glass spheres given in figure 15 and in table II of Williams' paper, so we cannot safely use these data. The transport rate in each size class is not decreasing exponentially with height, but in order to estimate the rate of transport for particles of size s at heights between the five gabs, we assume that between two successive gabs the transport rate of each size class does in fact decrease exponentially. Thus the transport rate at any given height is found by simple interpolation between the two neighbouring gabs.

During each run wind velocity measurements were taken at heights 16.5, 10.7, 7.1 and 4.0 cm above the bed. For the heighest wind speed the nethermost measurement was omitted because of lack of time. The results are given in terms of the friction velocity U_f and the wind speed in height 16.5, $U_{16.5}$. Williams claims that at the position of the trap , which was 8.84 m from the intake of the wind tunnel, the logarithmic profile layer extended as far as 16.5 cm above the surface. This is contrary to our experience, and it is likely that the velocity measurements in height 16.5 cm, and presumably 10.7 cm as well, are lower than would be expected from a logarithmic profile. Therefore, the calculated friction velocities must be expected to be too small, particularly for the fast wind speed. In fact, the wind data reported by Williams are not in accordance with the previously mentioned non-dimensional plot in Owen (1964, figure 3). We have tried to overcome this problem by using in the model a wind profile of the form

$$4.1 \quad U(y)=\begin{cases} 2.5\, U_f \ln(2y/33)+U_{16.5} & \text{if } y \geq y_o \\ 6\, U_f & \text{if } y < y_o \end{cases}$$

where $y_o = 16.5 \exp(0.4(6-U_{16.5}/U_f))$.

The calculation of the drag on the grains is complicated by the fact that Williams uses a geometric shape description, viz. sphericity, which does not tell how to calculate the drag. As pointed out in section 3 a shape description based on the terminal velocity of fall in air would be more informative. We have, however, simply estimated the fall diameter for naturally worn sand by 75 per cent of the volume diameters reported by Williams. The volume diameter is the diameter of a sphere with the same volume as the grain in question. Bagnold (1941) found that the fall diameter is 75 per cent of the sieve diameter for wind-blown desert sand, but Williams does not give sieve diameters, and since the 75 per cent is a rough estimate anyhow, it is not worth-while taking the difference between volume diameter and sieve diameter into account.

We will now present a method of estimating $\psi(s,v_i)$, as defined in section 2,from Williams' data and thus evaluate the mathematical model of the saltation trajectories. The procedure is based on the assumption that the trajectory of any grain caught in the trap is reasonable well described by the mathematical model of section 2 and 3, such that we can calculate the quantities needed by means of the model. We divide the layer from the bed to 16 cm above it into 16 zones each 1 cm high and label these 1 to 16, the first being closest to the bed. In what follows only one size class is considered at a time, so we can suppress the index s.

The estimation of $\psi(s,v_i)$ is done by a two-step iterative procedure. Let the initial velocities v_{0i}, i=1,...,16, be chosen such that their corresponding trajectories will have the maximal height equal to the middle of the i'th zone. In the first step we assume that the distribution of the initial velocity is concentrated on the sixteen values v_{0i} with point probabilities p_i, i = 1,...,16. Then we can find $\psi(v_{0i}) = \phi p_i$ from the equations

$$4.2 \quad q_j = \sum_{i=j}^{16} \phi p_i (\Delta x^j_{up}(v_{0i}) + \Delta x^j_{down}(v_{0i})),$$

where q_j denotes the transport rate through zone j for the size class considered, and $\Delta x^j_{up}(v_{0i})$ respectively $\Delta x^j_{down}(v_{0i})$ denotes the horizontal distance travelled by a grain from the size class with initial velocity v_{0i} on its way up respectively down through the j'th zone. The equations 4.2 are obtained from a result in section 2. The essence of 4.2 is that we adjust the observed mass in each zone for the grains caught in the zone on their way to or from another zone above the zone in question. First we find ϕp_{16}, then ϕp_{15} and so on. If we encounter a negative number it is replaced by zero and the process is continued. A negative number will, of course, indicate that the saltation model is not adequate for some of the grains caught in one or more of the zones above the zone with the negative number. Finally ϕ and the p_i's are determined by the equation $\sum_{i=1}^{16} p_i = 1$.

Now choose initial velocities $\tilde{v}_i$, i=1,...,16 such that the corresponding trajectory has maximal height equal to the top of the i'th zone, and define for each zone the ten initial velocities

$$\tilde{v}_{0ij}=j(\tilde{v}_i-\tilde{v}_{i-1})/10+\tilde{v}_{i-1} \quad j=1,...,10 \quad i=1,...,16$$

where $\tilde{v}_0 = 0$.

In the second step we attach weights $\tilde{p}_{ij}$, with $\sum_{j=1}^{10} \tilde{p}_{ij}=1$, such that the $\ln \tilde{p}_{ij}$'s $j=1,...,10$ are equidistant and increases or decreases linearly with the slope determined from the two points $(v_{0i}, \ln p_i)$ and $(v_{0i+1}, \ln p_{i+1})$ obtained in step one. Now we assume the distribution of the initial velocity v_0 to be concentrated on the values v_{0ij} with point probabilities $\tilde{p}_i\tilde{p}_{ij}$ and as in step one we can calculate $\tilde{p}_i$, $i=1,...,16$ from the equations

$$q_K=\sum_{i=k}^{16} \sum_{j=1}^{10} \phi\tilde{p}_i\tilde{p}_{ij}(\Delta x^j_{up}(\tilde{v}_{0ij})+\Delta x^j_{down}(v_{0ij}))$$

and

$$\sum_{i=1}^{16} \tilde{p}_i=1.$$

Step two is carried out in order to make sure that the fact that we have assumed the distribution of v_0 to be discrete does not affect the calculated mean values seriously. It turned out that step two almost did not change these quantities from the values obtained in step one.

When applied to Williams' data for naturally worn sand grains the estimation procedure gave a host of negative values for the three finest size fractions, particularly for the fast wind. First one might think that the drag is calculated incorrectly, but application of other methods such as the drag coefficient curve for sand grains published by Engelund and Hansen (1972) does not change the results appreciably. We are therefore led to conclude that the motion of some of the grains caught in the trap are not well described by the mathematical model of section 2 and 3.

We will call the motion described by this model pure saltation. Correspondingly, we define pure suspension as the motion of a grain, which when it has left the bed is prevented from settling again by the upward turbulent currents. To the whole spectrum of the transport modes between the two extreems we attach the name real saltation. This transport mode is to some extend affected by the turbulent eddies.Most of the grains presumably has a transport mode close to pure saltation, but the nature and magnitude of the departures from this transport mode still has to be investigated.

In the light of Chepil's (1945) investigations of the importance of pure suspension in various heights it is most likely that the explanation of the negative values is that much of the finer sand caught in the upper zones is transported in a way that is closer to pure suspension than to pure saltation. Naturally, we shall not adjust in the lower zones for suspended grains caught in the upper zones, since they will not pass the lower zones. It is not obvious how the grains with a transport mode between suspension and saltation are moving. If the turbulent currents will somehow make them pass the lower zones more rapidly than it is expected from the pure saltation model, this would also explain the negative values. It is possible that the inclusion of a lift-force in the pure saltation model would have this effect too.

In the absence of any knowledge about the motion of grains with an intermediate transport mode, we have to divide the caught grains into saltating grains and suspended grains. We do this by supposing that the grains caught below a height H are saltating and those caught above H are suspended. This is a rather crude way of dividing the observed grains, but the fact that Chepil (1945) found that pure suspension is mainly found in the upper zones supports it. The assumption underlying the method is that the dominant transport mode changes from close to pure saltation to pure suspension as we move away from the bed.

The height H is found by trial and error. The estimation procedure was applied to the data of sand caught below different heights, and the largest height for which no negative values occurred, and for which the values of p_i did not occillate too vigorously was chosen. A vigorous occillation indicates that we are still adjusting too much in the lower zones. The values of H determined in this way are given in Table 1.

We find that an H value below 5 cm is unreasonable. Therefore the procedure failed in the case of the finest size class of the uniform sand for the fast wind speed. The variation in Table 1 is in accordance with what would be expected if suspension is the reason why the saltation model is not in accordance with the whole data set. This gives further support to our division procedure.

Table 1. The estimated height of the layer of pure saltation (H).

Size class	Symmetric Size-Distribution		
	Fast	Interm.	Slow
0.161	5	6	7
0.215	7	10	10
0.295	10	16	16
0.414	16	16	16
0.550	16	16	16

Size class	Uniform Size-Distribution		
	Fast	Interm.	Slow
0.161	-	6	7
0.215	7	10	10
0.295	10	16	16
0.414	16	16	16
0.550	16	16	16

Size class	Truncated Size-Distribution		
	Fast	Interm.	Slow
0.161	5	6	7
0.215	7	10	10
0.295	10	16	16
0.414	16	16	16
0.550	16	16	16

We conclude this section by mentioning that we have some objections to the construction of Williams' trap which we will not touch in this paper. See, however, Jones and Willetts (1979). Further details of the model and more results of the analysis of Williams' data can be found in Jensen and Sørensen (1982).

5 CONCLUSION

It has been demonstrated that a mathematical model of aeolian saltation is a powerful tool for obtaining further progress in the field of wind-blown sand, but that there is still a fair amount of work to be done in the development of such a model. In particular, the model defined in the present paper is not capable of describing the motion of grains with a transport mode far from pure saltation.

REFERENCES

Bagnold, R.A. (1936). The movement of desert sand. Proc.Roy.Soc. London Ser. A, 157, 594-620.

Bagnold, R.A. (1941). The Physics of Blown Sand and Desert Dunes. Methuen, London.

Bagnold, R.A. (1954). Experiments on a gravity-free dispersion of large solid spheres in a Newtonian fluid under shear. Proc.Roy.Soc. London Ser. A, 225, 49-63.

Bagnold, R.A. (1974). Fluid forces on a body in shear-flow; Experimental use of 'stationary flow'. Proc.Roy.Soc. London Ser. A, 340, 147-171.

Chepil, W.S. (1945). Dynamics of wind erosion: I. Nature of movement of soil by wind. Soil Sci., 60, 305-320.

Ellwood, J.M., Evans, P.D., and Wilson, I.G. (1975). Small scale aeolian bedforms. Journal of Sedimentary Petrology, 45, 554-561.

Engelund, F.A. and Hansen, E. (1972). A Monograph on Sediment Transport in Alluvial Streams. Teknisk Forlag. København.

Ford, J. (1957). The transport of sand by wind. Transaction, American Geophysical Union, 38, No.2, 171-174.

Jensen, J.L. and Sørensen, M. (1982). A model for saltation trajectories - evaluated by means of trap data. Research Report 85, Dept. Theor. Statist, Aarhus University.

Jones, J.R. and Willetts, B.B. (1979). Errors in measuring uniform aeolian sand flow by means of an adjustable trap. Sedimentology, 26, 463-468.

Owen, P.R. (1964). Saltation of uniform grains in air. J. Fluid.Mech., 20, Part 2, 225-242.

Owen, P.R. (1980). The physics of sand movement. Lecture Notes, Workshop on Physics of Desertification, Trieste, 10-28 Nov., 1980.

Phillips, C.J. and Willetts, B.B. (1978). A review of selected literature on sand stabilisation. Coastal Engineering, 2, 133-147.

Rubinow, S.I. and Keller, J.B. (1961). The transverse force on a spinning sphere moving in a viscous fluid. J.Fluid.Mech., 11, 447-459.

Schiller, L. and Nauman, A. (1933). Z.Ver. Dent. Ing., 77, 318.

White, B.R., Greeley, R., Iversen, J.D. and Pollack, J.B. (1976). Estimated grain saltation in a Martian atmosphere. J. Geophys. Res., 81, 5643-5650.

White, B.R. and Schulz, J.C. (1977). Magnus effect in saltation. J.Fluid.Mech., 81, Part 3, 497-512.

Willetts, B.B. and Murray, C.G. (1981). Lift exerted on stationary spheres in turbulent flow. J.Fluid.Mech., 105, 487-505.

Willetts, B.B. and Rice, M.A. (1982). Practical representation of characteristic grain shape of sand. Draft manuscript.

Williams, G. (1964). Some aspects of the eolian saltation load. Sedimentology, 3, 257-287.

Zingg, A.W. (1953). Wind-tunnel studies of the movement of sedimentary material. Iowa Univ., Studies Eng., Bull. 34, 111-135.

Euromech 156: Mechanics of Sediment Transport / Istanbul / 12-14 July 1982

Forces on a single sediment grain and their dependence on the surrounding flow field*

P.HILLE, E.MAGENS & W.TESSMER
University of Kiel, Germany

ABSTRACT: The lift force on a 5 mm sediment grain was studied in dynamical flume experiments with a force sensor with sensitivity in the micro Newton range. The force was sampled simultaneously with the mainstream velocity and the vertical velocity in realtime. The velocities were measured with a two component Laser Doppler Anemometer in a crossbeam setup. Probability densities and RMS values of the lift in dependence of grain Reynolds number were determined. Reynolds shear and other turbulence characterizing parameters were measured for comparison.

1 INTRODUCTION

The interaction between the turbulent flow in the boundary layer region and the bed-load is a topic of interest. The knowledge of lift and drag forces on sediment grains is necessary for studies in fields such as sediment movement, entrainment and suspension, and for the determination of bedload rates.

First measurements of the lift and interpretations of the obtained data were made by H. A. Einstein and El-Samni (1949). They deduced the lift coefficient from pressure difference measurements on half spheres in a waterflow. Chephil (1958) used hemispheres to study the lift in an airflow. A review of lift and drag was given by Benedict and Christensen (1972). In his book Raudkivi refers to the thesis of Apperley (1968) who used a strain gauge balance for lift and drag measurements on a 6 mm sphere. Some experiments on moving and rotating cylinders were made by Bagnold (1974).

In addition to lift force experiments, the structure of the turbulent boundary layer was studied in the last years under the aspect of so-called "coherent structures". Kim and Kline (1967, 1971) found that most of the turbulent energy produced in the boundary layer region is transported into the wake region by short bursts. Sumer and Deigaard (1981) studied particle motion under the aspect of the bursting process. It is useful to study this probably more deterministic rather than statistical process by correlation methods, flow visualisation and conditional sampling methods.

In this paper the lift force fluctuations on a single sediment particle are studied in a 10 cm x 10 cm x 2 m open water channel (small channel) and a 0.56 m x 0.73 m x 20 m flume (big channel).

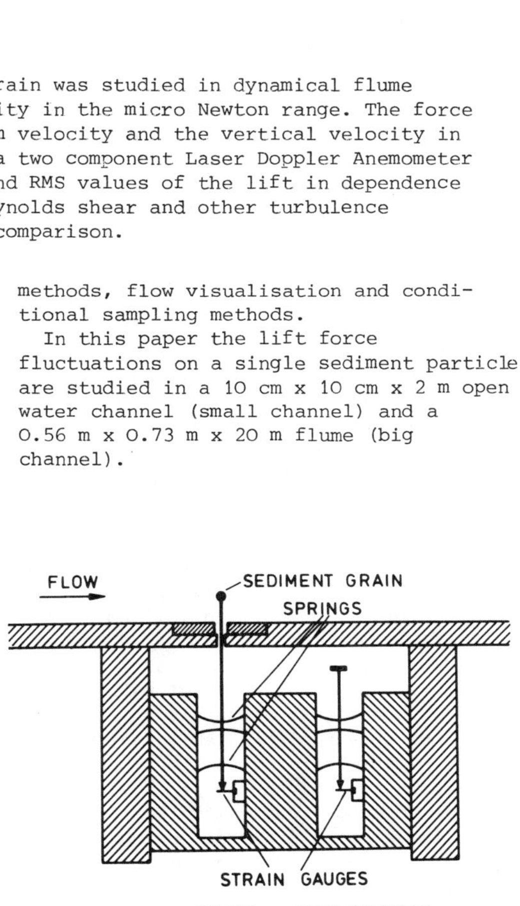

Fig. 1
Lift force sensor with compensation device

* Publication No. 397 of the "Joint Research Programme", Kiel University (Sonderforschungsbereich 95 der Deutschen Forschungsgemeinschaft).

2 EXPERIMENTAL SETUP

The sediment grain had a diameter of 5 mm and its center was positioned between 3 mm and 6 mm above ground. The lift force was measured with a device designed especially for this investigation. This sensor operates up to 1 kHz within a range from 1 to 10000 μNewton. The Laser Doppler Anemometer allows the measurement of cartesian velocity components. We used it for determination of the mainstream velocity (x) and the vertical (y) velocity. Force and velocity may be sampled up to 1 kHz and are processed by a Nova mini computer.

The force sensor, shown in Fig. 1, operates with a piezoresistive bending beam semiconductor strain gauge. This device responds to static and to dynamic forces as well. The lift is coupled by a rod from the sediment grain through the bottom of the flume onto the strain gauge. The rod is mounted on three springs to prevent a cross talk from the drag to the lift force. For sensitivity in the micro Newton range the system has to be "hard", this means it must have a high resonance frequency (1 kHz). The resonance is damped to get a linear frequency response. The lift force is fed through a narrow gap sealed by highly viscous silicone grease through the bottom of the flume. At the feed point, the sensor has an effective area of (0.08 cm^2) and is weakly sensitive to pressure (60 dyn/1 cm H_2O). To make sure that the measured force is caused by the lift of the grain and not by pressure, we made experiments with and without the grain (fig. 2) and found the influence of pressure, fluctuations small compared with the lift as long as no high waves with long wavelength are travelling along the channel.

Another essential problem is the sensitivity of the force sensor to shock and vibrations. We solved the problem by using a compensation system. The compensation lowers the influence of shock and vibrations from the flume pump by a factor 10. The largest fluctuations due to vibrations have a magnitude of 30 μNewton at high velocities (V^* = 2.5 cm/sec). This includes all types of fluctuations. In view of the construction of the sensor, disturbances by drag force fluctuations are not measured directly. Static measurements are more difficult, because there are temperature and drift

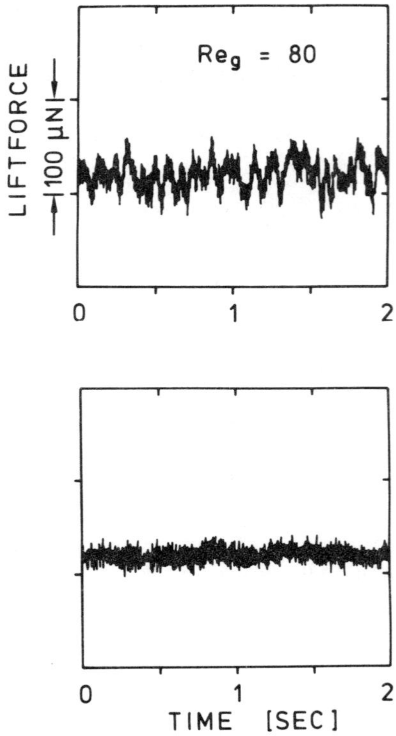

Fig. 2
Lift force, measured with 5 mm grain (above) and without grain (below) on the 0.56 m x 20 m flume

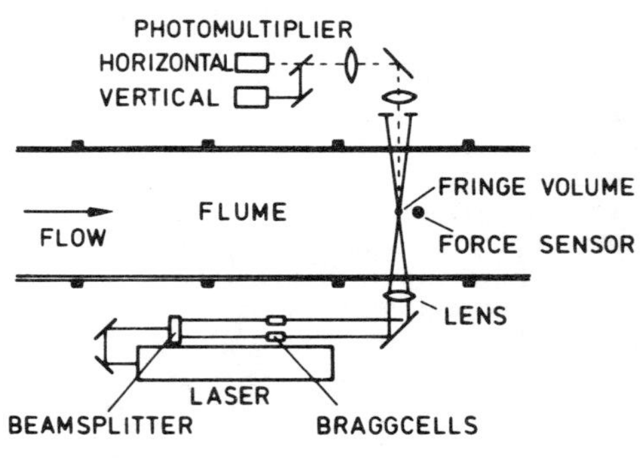

Fig. 3
Laser Doppler Anemometer setup at a flume. The light path of the horizontal mainstream component is shown

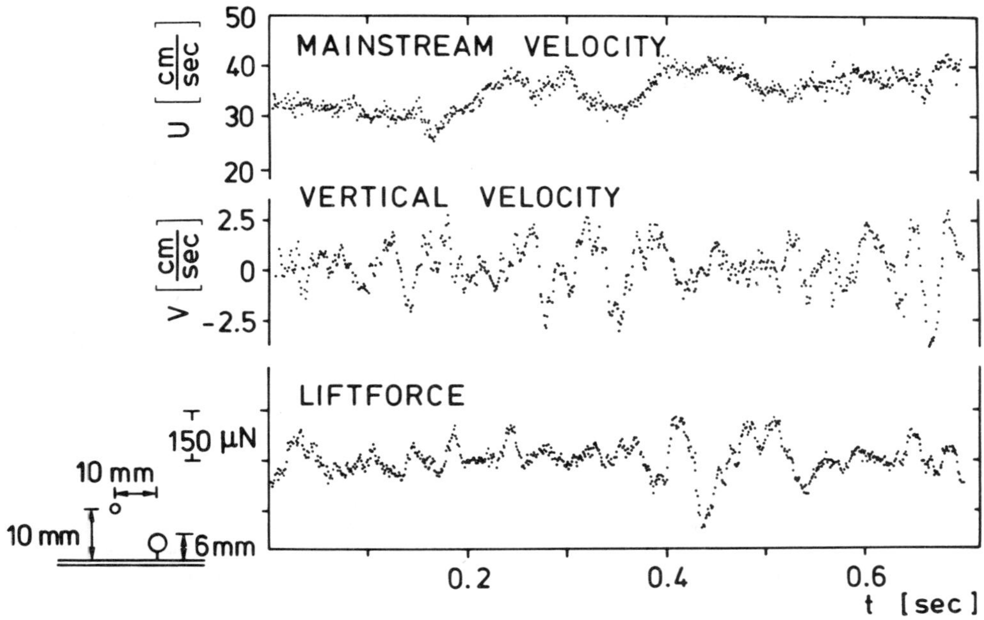

Fig. 4
Realtime sampling of velocity and force, $V^* = 2.2$ cm/sec $Re_g = 110$, big flume

problems and a finite influence of static pressure (water level).

For velocity measurements we use a modified commercial system OEI type LDO-A-212. The system is sketched in Fig. 3. It features a two colour optic using the green and blue emission of an argon ion laser. Four bragg cells are used for frequency shift of the laser beams, so we are able to distinguish negative and positive values. The beams intersect in a 1 mm^2 x 5 mm fringe volume. The scattered light is separated into the green laser doppler bursts from the vertical (y) and the blue bursts from the horizontal (x) component.

An advantage of a Laser Doppler system is the strict linearity of velocity determination compared with hotfilm/hotwire probes and non-intrusive measurement inside a well defined measuring volume. Nevertheless it is planned to install hotfilm probes to get more information on the flow, especially space-time correlations and conditional sampling measurements. With LDA it is possible to determine onepoint statistics, but no spacial correlations. Realtime measurements of the velocity up to sampletimes of 1 msec are done with signal processing by a

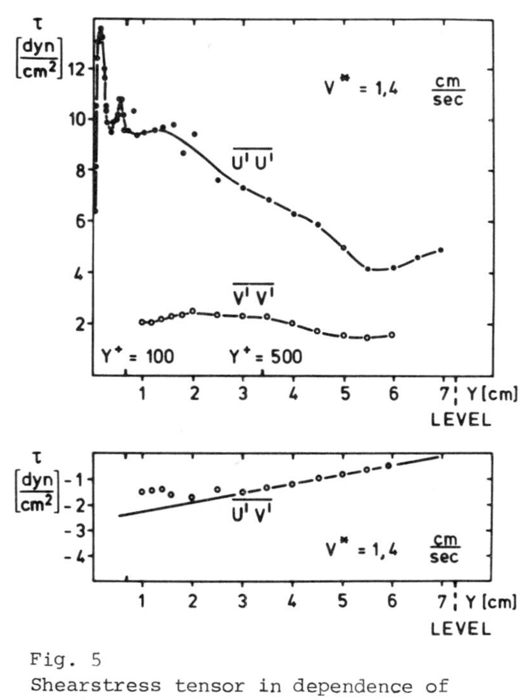

Fig. 5
Shearstress tensor in dependence of vertical position, big flume

frequency tracker. This is a demodulating circuit for LDA photomultiplier signals on the basis of a phase locked loop.

3 DATA PROCESSING

The realtime data are sampled by a microprocessor on Magtape and are later analyzed by a Nova C4 mini computer. In a first step we obtain the statistics of the individual components, in a second step we obtain Reynolds shear, shear velocity, and autocorrelation functions and spectra of these components. In a further step crosscorrelation functions and conditional sampling methods will be tried in order to get an agreement with current theories about the sedimentation process. Still further methods of data evaluation have to await the development of suitable software.

4 MEASUREMENTS

Our first experiments were done on the small open channel. The sphere was 6 mm above hydraulically smooth ground. Further experiments were done in the big flume, also with hydraulically smooth walls. In these cases, the grain itself produces a strong wake region, and vortex shedding occurs at a grain Reynolds number around 120. ($Re_g = V^* \cdot d/\nu$). It is difficult to obtain velocity signals from the grain's wake region, because the turbulent fluctuations are rapid behind the grain. This turbulent separation may cause the relatively strong lift forces.

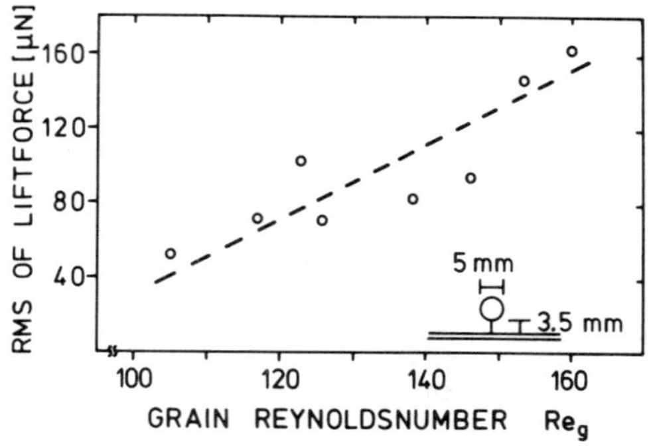

Fig. 7
Mean square force on the small test flume. The waterlevel was about 2.5 cm, V^* was computed from channel's slope

5 RESULTS

A first analysis of our data was made by comparison of their time dependence. In Fig. 4 the velocities measured 1 cm above ground and the lift force are shown. These measurements belong to a relative high shear velocity of 2.2 cm/sec and were made at the big flume. The center of the 5 mm diameter grain was positioned 3.5 mm above ground. From these measurements no direct relation between these components is evident. Also crosscorrelations between single components did not lead to clear indications up to now. From our previous experiments a large

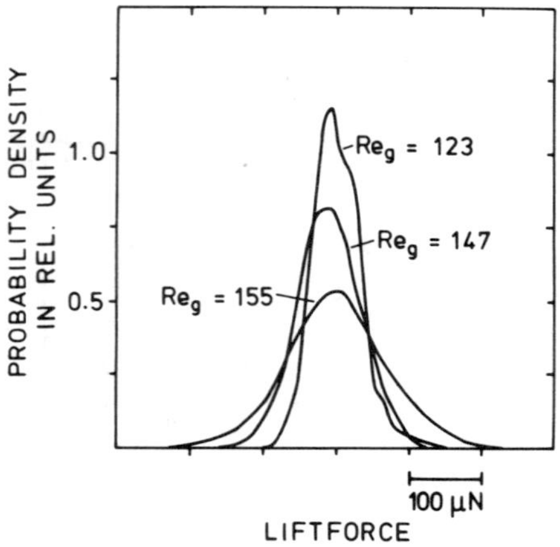

Fig. 6
Lift fluctuations, small test flume

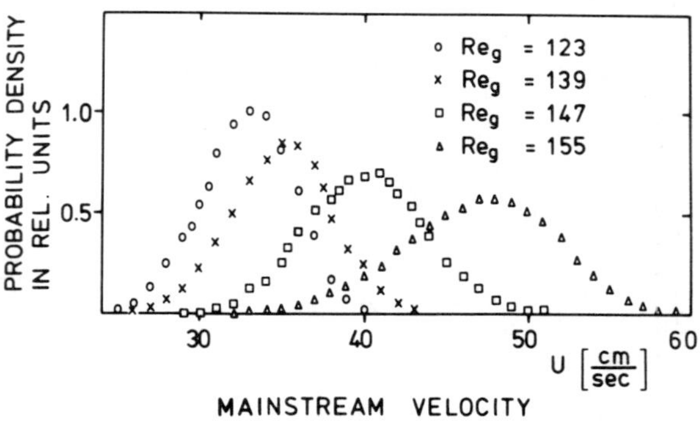

Fig. 8
Fluctuations of the horizontal velocity component. The fringe volume was positioned 1 mm above grain. Small test flume

dependence of the lift amplitudes and frequencies on the position of the grain is expected.

The shear velocity was determined in the big flume from logarithmic profiles, in the small test flume it was determined from the channel slope, and it turns out to be higher here. Fig. 5 shows profiles of some shear stress tensor components; the results are in good agreement with the theory. It could be that a determination of the third component of the velocity, which is associated with secondary velocities, may also be necessary for a correct interpretation of lift force measurements.

Some other results were obtained at the testflume. Fig. 6 shows the probability densities of the lift. In this case the velocity measuring volume was 1 mm above the grain, the center of the 5 mm grain was 6 mm above ground, the level was 2 cm. The calculated RMS values are shown as functions of the grain Reynolds number in Fig. 7. Here a grain Reynolds number of 100 seems comparable with grain Reynolds numbers of 80 in the big flume. Fig. 8 shows the probability of the mainstream velocity, measured above the grain.

In our experiments we found high lift force fluctuations, compared to the static results of other authors. In future work we will install a rough bed and measure simultaneously static and dynamic lift forces. This seems to be possible with our present measurement apparatus.

We thank Prof. Dr. Schulz-DuBois, physicist and Prof. Dr. E. Walger, geologist, for many helpful discussions and we thank the SFB 95/Deutsche Forschungsgemeinschaft for financial support.

6 REFERENCES

Einstein, H. A., El-Samni, E. A., 1949.
Hydrodynamic forces on a rough wall.
Rev. Mod. Phys. 21 No. 3 pp. 520.

Chephil, W. S., 1958.
The use of evenly spaced hemispheres to evaluate aerodynamic forces on soil surfaces.
Trans. Am. Geophys. Union, 39, pp. 397.

Benedict, B. A., Christensen, B. A., 1972.
Hydrodynamic lift on a stream bed.
In: Sedimentation; Symposium to honor of Prof. H. A. Einstein.

Apperley, L. W., 1968
The effect of turbulence on sediment entrainment.
Thesis at School of Engineering, University of Auckland, New Zealand, Rep. No. 45.

Bagnold, R. A., 1974.
Fluid forces on a body in shearflow.
Proc. Roy. Soc. London A 340, pp. 47.

Kim, H. T., Kline, S. J., Reynolds, W. C. 1971.
The production of wall region in turbulent flow.
J. Fluid Mech. 50, pp. 133.

Sumer, B. M., Deigaard, R., 1981.
Particle motions near the bottom in turbulent flow.Part 2.
J. Fluid Mech. 109, pp. 311

Grass, A. J., 1971.
Structural features of turbulent flow over smooth and rough boundaries.
J. Fluid Mech. 50, pp. 233

Durst, F., Melling, A., Whitelaw, J. H., 1976.
Principles and practice of laser doppler anemometry.
Academic Press 1976.

Raudkivi, A. J., 1976.
Loose boundary hydraulics.
Pergamon Press.

Hille, P., Koepp, N., Vehrenkamp, R., 1982.
Dynamical force measurements in the μNewton range.
Conference Sensor'82 in Essen Germany.

Euromech 156: Mechanics of Sediment Transport / Istanbul / 12-14 July 1982

Migration of spherical particles suspended in shear flows

F.FEUILLEBOIS, A.LASEK & S.SPITZ
Laboratoire d'Aérothermique du C.N.R.S., Meudon, France

SYNOPSIS: The motion of a small spherical particle is considered, when suspended in an incompressible viscous fluid undergoing; 1) a simple shear flow, 2) a boundary layer flow. The particle is small, so that the Reynolds number of the relative fluid flow is small. The particle inertia is fundamental to the problem. The relative importance of the different forces acting on the particle is discussed: The Stokes force, the unsteady forces ("added mass" and Boussinesq-Basset), the force due to a pressure gradient (Faxen), corrections to the friction coefficient due to wall effects, and transverse migration forces due to fluid inertia (Saffman, Harper and Chang).

Some trajectories are calculated, assuming the Stokes force and the transverse migration forces to be the important ones. Possible implications of the results on sediment transport are given.

INTRODUCTION

Consider a small, spherical particle suspended in a viscous incompressible fluid. The purpose of this paper is to calculate the trajectories of such a particle moving in a shear flow. This problem is of importance in certain natural phenomena and in chemical engineering processes. The notations are given in Figure 1 for a sphere moving in a boundary layer flow parallel to the solid wall y = 0, with velocity u_p along the wall (x direction) and velocity v_p perpendicular to the wall (y direction). Let $\underset{\sim}{v}_p$ be the translational velocity vector and $\underset{\sim}{\omega}_p$ be the rotational velocity. Let u_f be the fluid flow velocity in the absence of the sphere ("unperturbed" velocity), at the distance y_p from the wall, i.e. at the location of the sphere centre. The velocity gradient at the location of the sphere centre is denoted by k. These notations are also valid for a sphere moving in a pure shear flow (for which k is then constant). Only for the boundary layer flow, the velocity profile has a curvature, say k_2.

A number of different expressions for the force acting on a particle were derived under the assumption that the "relative" Reynolds number (based on the radius of the sphere and on the velocity of the fluid relative to the sphere) is small.

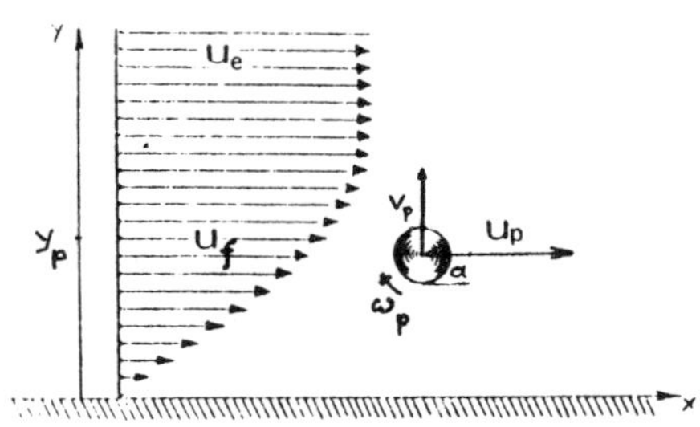

Figure 1

Notations for the sphere moving in a boundary layer flow

Actually several different "relative" Reynolds numbers may be defined, depending upon which relative fluid velocity is chosen:

$$Re = \frac{\sup|\underset{\sim}{v}_f - \underset{\sim}{v}_p|a}{\nu}$$

$$Re_k = \frac{ka^2}{\nu}$$

$$Re_{k_2} = \frac{k_2 a^3}{\nu}$$

$$Re_\omega = \frac{\sup|\underset{\sim}{\omega}_p|a^2}{\nu}$$

All these Reynolds numbers are assumed to be smaller than unity. The inertia of the fluid motion relative to the particle is then neglected in the first approximation. The Stokes "creeping flow" equations apply. The original Stokes' result for the drag force on a sphere moving steadily in an indefinite fluid at rest has to be modified when different factors are taken into account, i.e.:

- the unsteadiness of the fluid motion relative to the sphere;
- the unperturbed flow field in which the sphere is moving;
- the influence of the wall.

In this approximation of negligible inertia of fluid motion relative to the particle, and with the geometry shown on figure 1, the drag force on the sphere is parallel to the relative fluid velocity $\underset{\sim}{v}_f-\underset{\sim}{v}_p$. Now if the next approximation in the small Reynolds numbers is considered, the inertia of the fluid provokes a small migration force perpendicular to the relative fluid velocity $\underset{\sim}{v}_f-\underset{\sim}{v}_p$, if the particle is moving in a shear flow and/or if it is rotating.

A considerable literature exists concerning each separate factor enumerated above, much less concerning two coupled factors, and practically none concerning the coupling of more than two factors. Therefore, in the practical application of these results, care should be taken that the factors which are not taken into account be actually negligible.

We will evaluate the orders of magnitude of the different forces which may influence the motion of a sphere embedded in

- a simple shear flow, in presence of a wall;
- a boundary layer flow.

We will present first three types of forces that are actually neglected in our calculations of trajectories, i.e. the unsteady forces, the force due to a pressure gradient, and the influence of the wall.

We then present the migration forces that are used in the calculation of trajectories.

UNSTEADY FORCES

The unsteady motion of the fluid induces a "hereditary" force on a sphere moving in a fluid at rest (Boussinesq 1885, Basset 1888). A "hereditary" couple on a rotating sphere was calculated as well (Feuillebois, Lasek 1978). For a large ratio of the sphere density to the density of the fluid:

$$\frac{\rho_p}{\rho} >> 1$$

these unsteady effects contribute terms of order $O(\rho/\rho_p)^{1/2}$ to the force, and $O(\rho/\rho_p)$ to the couple (Feuillebois 1980). We assume that these correction terms can be neglected, which is particularly justified for solid particles in air.

Now, for a sphere moving in a non uniform flow field, the force and couple have more complicated expressions as derived by Gatignol (1982/1983), the unsteady part of which are also small for large ρ_p/ρ.

FAXEN FORCE

Consider a spherical particle embedded in a boundary layer over a flat plate. From boundary layer theory it is well known that over a flat plate the pressure gradient is negligible:

$$\frac{\partial p}{\partial x} \simeq 0$$

on the reference scale ℓ on the plate. But on the scale a, radius of the sphere, such that,

$$a << \ell$$

the pressure gradient may become important. This pressure gradient is directly proportional to the curvature of the unperturbed velocity profile in the vicinity of the sphere. For, if we develop in Taylor series the boundary layer velocity profile u(y) in the vicinity of the sphere centre:

$$u(y) \simeq u_f + k(y-y_p) + k_2(y-y_p)^2$$

all terms of higher order have to be neglected so as to satisfy the Stokes equation, which writes then:

$$\rho \frac{\partial p}{\partial x} = 2 \mu k_2$$

Such a pressure gradient in Stokes flow gives, after Faxen (1924) a force on the sphere:

$$F = 2 \pi a^3 \mu k_2$$

To see whether Faxen's force is actually negligible, we take a standard velocity profile in the boundary layer and evaluate some orders of magnitude. Let u_e be the velocity outside the boundary layer and

$$\theta = \frac{u_p-u_f}{u_e}$$

be the local slip ratio, and

$$Re_\ell = \frac{u_e \ell}{\nu}$$

be the large Reynolds number based on the length ℓ of the flat plate. Let also

$$x^* = \frac{x}{\ell} \; ; \quad y^* = \frac{y}{\ell}$$

The assumption of a small relative Reynolds number writes

$$\theta \, Re_\ell \frac{a}{\ell} << 1$$

which is generally true for a very small particle.

The Faxen force can be written as

$$F = - 6\pi a\mu u_f (\frac{a}{\ell})^2 Re_\ell B$$

where B is a function of the coordinates x*, y*, given by Figure 2. It is seen that B is the largest close to the leading edge of the boundary layer, where the curvature of the boundary layer velocity profile is relatively more important. Outside this limited region, the Faxen force can be neglected.

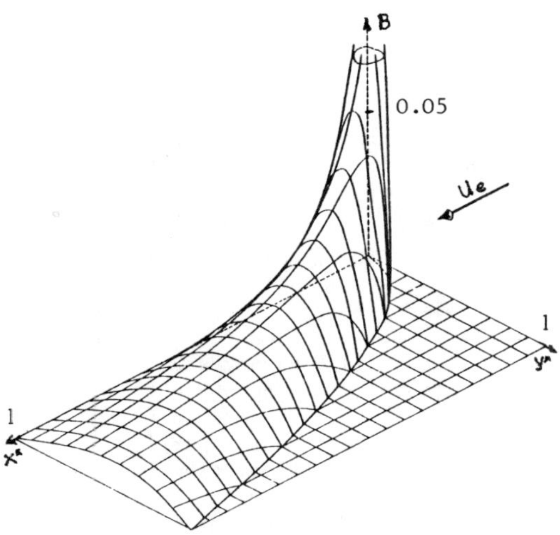

Figure 2

Relative importance of the Faxen force on the sphere, in the boundary layer.

For a pure shear flow, the Faxen force is identically zero.

WALL EFFECTS

Concerning the wall effects, results exist for the force acting on a sphere embedded in different flow fields.

We will be interested here in,

- a sphere in motion parallel to a plane wall in a fluid at rest (Goldman, Cox, Brenner, 1967a);
- a sphere in motion perpendicular to a plane wall in a fluid at rest (Brenner, 1961);
- a sphere at rest in a Couette flow parallel to a solid wall (Goldman, Cox, Brenner, 1967b);
- a sphere at rest in a parabolic flow parallel to a solid wall (Goren, O'Neill, 1971).

These flow results may be added up by virtue of the linearity of the Stokes equations of fluid motion.

As a result, we get the force acting on a sphere moving in a flow field with velocity,

$$u(y) = \lambda y + \lambda_2 y^2$$

The components of the force along x, y, respectively are :

$$F_x = -6\pi a\mu(u_p f_x - \lambda y_p f_1 - \lambda_2 y_p^2 f_2)$$

$$F_y = -6\pi a\mu \; v_p \; f_y$$

y_p being the distance from the sphere centre to the solid wall. Limiting values of f_x, f_1, f_2, f_y for large values of y_p/a may be used to evaluate the errors in neglecting the wall effects:

$$f_x = 1 + \frac{9}{16} \frac{a}{y_p}$$

$$f_y = 1 + \frac{9}{8} \frac{a}{y_p}$$

(Lorentz, 1897)

$\frac{y_p}{a}$ =	5	10	21	51	∞
f_1 =	1.123	1.059	1.028	1.011	1
f_2 =	1.135	1.062	1.028	1.011	1

(f_1 after Goldman, Cox, Brenner, 1967b)
(f_2 after Goren, O'Neill, 1971).

Note, that the effect of the wall on the motion of the sphere along y is relatively more important. However, the velocity of the sphere in this direction is generally small. Thus, we may accept that the wall effects are negligible within a 3 % error for a particle 10 diameters away from the wall.

MIGRATION FORCES

The inertia of the fluid represents a small effect for a very viscous fluid. it gives a correction to the Stokes'force plus, eventually, some lateral "migration" force for a particle in a shear flow. This "migration" may have important physical implications, as we will see. Saffman (1965) calculated the force on a sphere moving with a velocity relative to a shear flow, u = ky, and parallel to it (u_p along x), under the assumptions that,

$$Re = \frac{Sup \left| u_f - u_p \right| a}{\nu} << 1$$

$$Re_k = \frac{k\ a^2}{\nu} << 1$$

$$\sqrt{Re_k} >> Re$$

and assuming the fluid to be of infinite extent.

Harper and Chang (1968) generalized Saffman's calculation to the case where the sphere is moving in any direction. Their results for the components of the force in the x and y directions are :

$$F_x = -6\pi a\mu\ [(1+0,944\ \sqrt{Re_k}(u_p-u_f) + 0,620\sqrt{Re_k}\ v_p]$$

$$F_y = -6\pi a\mu\ [(1+0,326\sqrt{Re_k})v_p + 0,343\sqrt{Re_k}(u_p-u_f)]$$

These equations were used for a sphere in a simple shear flow. Now, for a sphere in a boundary layer flow, with the velocity parallel to a flat plate (y=0)

$$u_p = u_p(y)$$

the curvature of this velocity profile should be considered. But if the first order Faxen force is neglected, so all higher order effects due to the curvature of the velocity profile have to be neglected too.

To see whether Harper and Chang force is applicable in the boundary layer, we take a standard velocity profile and evaluate some orders of magnitude, as done for the Faxen force.

The assumption $\sqrt{Re_k}$ >>Re gives an upper bound for the quantity $Re_\ell\ \theta^4$. It is seen that outside a region close to the stagnation point of the flat plate, this upper bound is too small to be compatible with a large Re_ℓ, unless the slip ratio is small (figure 3).

Thus Harper and Chang's result is valid only in some limited region of the flow field.

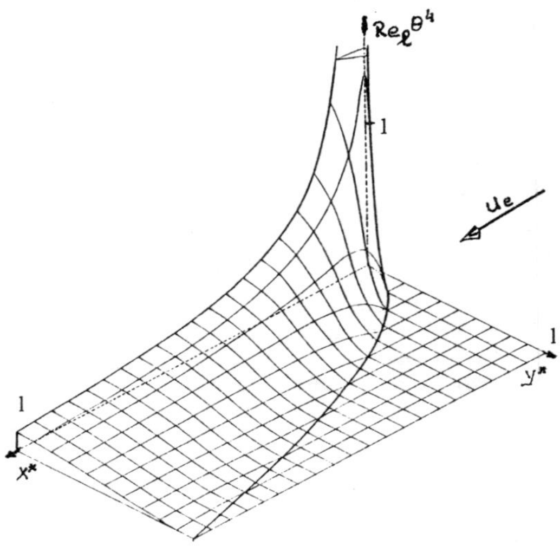

Figure 3

Region of validity of Saffman's and Harper and Chang's forces on the sphere, in the boundary layer.

A NEARLY VERTICAL FLOW FIELD

Nevertheless, in the absence of more information, if we assume (arbitrarily), that Saffman's and Harper and Chang's forces are valid in the whole flow field, we are able to obtain some interesting results for the motion of a particle entrained in a nearly vertical flow field.

The particle is entering the flow field with a uniform velocity which is u_e.

For a pure shear flow, u_e is defined as the fluid velocity at a distance d from the wall. Now for the boundary layer, u_e is the fluid velocity at its edge (figure 1).

The equation of motion of the sphere is,

$$\frac{4}{3}\pi a^3\ \rho_p \frac{d\underset{\sim}{v}_p}{dt} = \underset{\sim}{F} + \frac{4}{3}\pi a^3(\rho_p-\rho)\underset{\sim}{g}$$

where $\underset{\sim}{g}$ is the gravity and $\underset{\sim}{F}$ is Harper and Chang's force.

It was integrated analytically for the case of a pure shear flow. Some examples of calculated trajectories of the sphere, depending upon the introduction point and the inclination of the plate, are shown on figure 4. The cases where the plate is tilted to the right and to the left are grouped on the same sketch.

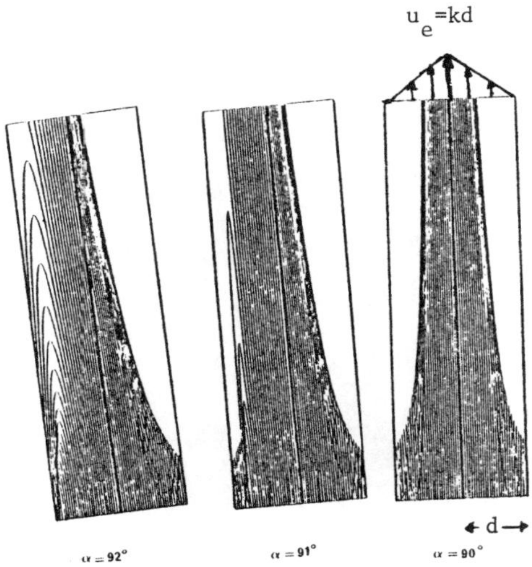

Figure 4

Trajectories of a sphere in a nearly vertical shear flow.

For the case of a boundary layer flow, the velocity profile in the boundary layer was taken as a polynomial, and the equation of motion of the sphere was integrated numerically by a standard Runge-Kutta procedure. Figure 5 shows the entrainment by a boundary layer flow. The remarkable feature is that a small tilt in the direction of the main flow may induce drastic changes in the particle trajectories. In the boundary layer, some regions even appear, where particles seem to accumulate. This suggests that sediment transport in vertical channels may produce regions of high concentration.

Only a few remarkable trajectories have been shown here. A number of other cases are available from the authors.

CONCLUSION

In despite of the limited theoretical validity of the migration forces which could be used, the results obtained should be taken as a warning, that drastic changes may occur in vertical sediment transport.

Future research in this field is certainly needed.

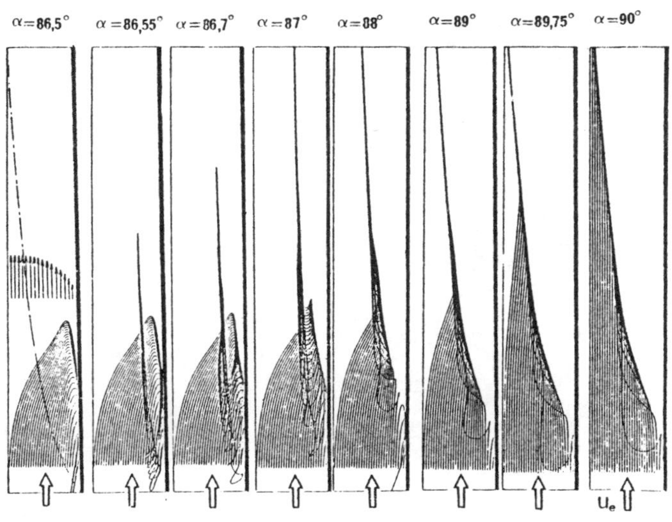

Figure 5

Trajectories of a sphere in a nearly vertical boundary layer flow.

REFERENCES

Basset, A.B. 1888, A treatise on hydrodynamics, Cambridge.

Boussinesq, J. 1885, Sur la résistance qu'oppose un fluide... C.R. Ac. Sc., Paris, 100, p. 935.

Brenner, H. 1961, The slow motion of a sphere through a viscous fluid. Chem. Eng. Sc., 16, p. 242.

Faxen, H. 1924, Der Widerstand gegen die Bewegung einer starren Kugel. Arkiv. Mat. Astron. Fys., 18, 29.

Feuillebois, F., Lasek, A. 1978, On the rotational historic term in non-stationary Stokes flow. Quart. J. Mech. Appl. Maths., 31, 4, p. 435.

Feuillebois, F. 1980, Certains problèmes d'écoulements mixtes fluide - particules solides. Thèse de Doctorat d'Etat, Univ. P. & M. Curie, Paris.

Gatignol, R. 1982 or 1983, The Faxen formulae for a rigid particle in an unsteady non-uniform Stokes flow. To be published in Journal de Mécanique Théorique et Appliquée.

Goldman, A.J., Cox, R.G., Brenner, H. 1967(a), Slow viscous motion of a sphere parallel to a plane wall, I. Chem. Eng. Sc. 22, p. 637.

Goldman, A.J., Cox, R.G., Brenner, H. 1967(b), Slow viscous motion of a sphere parallel to a plane wall, II. Chem. Eng. Sc. 22, p. 653.

Goren, S.L., O'Neill, M.E. 1971, On the hydrodynamic resistance to a particle of a dilute suspension when in the neighbourhood of a large obstacle. Chem. Eng. Sc., 26, p. 325.

Harper, E.Y., Chang, I.D. 1968, Maximum dissipation resulting from lift in a slow viscous shear flow. J.F.M., 32, 2, p. 209.

Lorentz, H.A. 1897, A general theorem concerning the motion of a viscous fluid and a few consequences derived from it. Versl. Kon. Acad. Wet. Amst., 5, p. 168.

Saffman, P.G. 1965, The lift on a small sphere in a slow shear flow. J.F.M., 22, 2, p. 385 (corrigendum 1968 : J.F.M., 31, 3, p. 624).

Euromech 156: Mechanics of Sediment Transport / Istanbul / 12-14 July 1982

Particle dynamics equations for turbulent suspensions

PETER J.MURPHY
University of Massachusetts, Amherst, USA

ABSTRACT: A new equation is developed for the fluid force on a low Reynolds number particle in a turbulent flow. Some of the effects of the spatial non-uniformity of the flow are examined.

1 INTRODUCTION

In this paper the equations which determine the force on a small sphere that is moving through a turbulent flow field will be examined. The equations in common use today will be shown to have omitted a number of non-uniform flow effects, and some of these effects will be included in equations developed here.

The development of these current fluid force equations is based on the quasi-steady creeping flow equations. Oseen (1) used the low Reynolds number approximation to find the force on a sphere moving through static fluid. Tchen (2) applied Oseen's results to a sphere moving through an unsteady but uniform flow as an approximation of a turbulent flow field. Corrsin and Lumley (3) attempted to include the effects of the non-uniformity of the turbulent flow, but were only partially successful.

This paper will examine the fluid force by starting with the Navier-Stokes and continuity equations for a constant density fluid and making a weaker approximation than that of the quasi-static creeping flows. Only the normal advection term of the Navier-Stokes equations will be omitted. This approximation will lead to a modified form of the solution procedure by Oseen (1) for his low Reynolds number equations. That modified procedure will be then applied to unsteady non-uniform flow fields to obtain a new equation for the fluid force on a sphere in turbulent flow.

1.1 Low Reynolds number analysis

The fluid force on a sphere is a combination of the pressure and shear stress force given by

$$F_i = \iint_A \left(-Pn_i + \mu n_k \left(\frac{\partial u_i}{\partial x_k} + \frac{\partial u_k}{\partial x_i}\right)\right) dA. \qquad (1)$$

where u_i is the fluid velocity, μ is the viscosity, P is the pressure, and n_i is the unit normal to the surface area A of the sphere. The pressure and shear stress are solutions of the Navier-Stokes and continuity equations,

$$\frac{\partial u_i}{\partial t} + u_j \frac{\partial u_i}{\partial x_j} = -\frac{1}{\rho}\frac{\partial P}{\partial x_i} - g\frac{\partial z}{\partial x_i} + \nu \frac{\partial u_i}{\partial x_j \partial x_j}, \qquad (2)$$

and

$$\frac{\partial u_j}{\partial x_j} = 0. \qquad (3)$$

In order to develop the quasi-static creeping flow approximation let $x_i = x_i'a$, $t = T_o t'$, $u_i = U_o u_i'$, and $P = \frac{\mu U_o}{a} P'$. Inserting the dimensionless primed variables in the fluid flow equations gives

$$\left(\frac{a^2}{\nu T}\right)\frac{\partial u_i'}{\partial t'} + \left(\frac{U_o a}{\nu}\right) u_j' \frac{\partial u_i'}{\partial x_j'} = -\frac{\partial '}{\partial x_i'} - \left(\frac{a^2 g}{\nu U_o}\right)\frac{\partial z'}{\partial x_i'} + \frac{\partial^2 u_i'}{\partial x_j' \partial x_j'} \qquad (4)$$

and

$$\frac{\partial u'_j}{\partial x'_j} = 0. \qquad (5)$$

The advective Reynolds number $\frac{U_o a}{\nu}$ is assumed to be so low that the advection term can be omitted and the quasi-static creeping flow equations are the result. Returning to dimentional variables, these linearized fluid flow equations are

$$\frac{\partial u_i}{\partial t} = -\frac{1}{\rho}\frac{\partial}{\partial x_i}(P + \rho g z) + \nu \frac{\partial^2 u_i}{\partial x_j \partial x_j} \qquad (6)$$

and

$$\frac{\partial u_j}{\partial x_j} = 0. \qquad (7)$$

The Oseen (1) solution procedure for these equations begins with the separation of the velocity vector into its potential flow and rotational flow components,

$$u_i = -\frac{\partial \Phi}{\partial x_i} + V_{ri}. \qquad (8)$$

The velocity potential Φ must satisfy Laplace's quation, $\nabla^2\Phi = 0$, while the rotational component V_{ri} is given by

$$V_{ri} = \frac{\partial G_i}{\partial t} - \nu \frac{\partial^2 G_j}{\partial x_i \partial x_j}, \qquad (9)$$

where G_i is the solution of the equation,

$$\frac{\partial G_i}{\partial t} - \nu \frac{\partial^2 G_i}{\partial x_j \partial x_j} = 0. \qquad (10)$$

Once the general solution for Φ and G_i are found, the boundary conditions are applied. First the normal velocity boundary conditions are used to determine ϕ_1, one of two parts of the potential flow Φ. Then the tangential velocity boundary conditions are used to find G_i and finally both ϕ_1 and G_i are used to determine the second potential flow component ϕ_2. The resulting solution is

$$u_i = -\frac{\partial}{\partial x_i}(\phi_1 + \phi_2) + \frac{\partial G_i}{\partial t} - \nu \frac{\partial}{\partial x_i}\left(\frac{\partial G_j}{\partial x_j}\right) \qquad (11)$$

$$P = \rho \frac{\partial}{\partial t}(\phi_1 + \phi_2) \qquad (12)$$

Applying this procedure to the problem of a sphere of radius a moving with velocity U_i through static fluid, Oseen determined the fluid force on the sphere to be

$$F_i = \tfrac{1}{2}\left(\frac{4\pi a^3}{3}\right)\rho \frac{dU_i}{dt} - 6\pi\mu a U_i$$

$$- \frac{6\pi\mu a^2}{\sqrt{\pi\nu}} \int_o^t \frac{\frac{dU_i}{dt'}\,dt'}{(t-t')^{\frac{1}{2}}}. \qquad (13)$$

The components of this force are the added mass, the Stokes drag, and the Basset drag terms respectively. The buoyancy effect was taken for granted.

Tchen (2) applied the Oseen procedure to the problem of a sphere of radius a moving with velocity U_i through an unsteady but spatially uniform flow described by the velocity $U_{\infty i}$ and pressure P_∞. He obtained the force

$$F_i = \left(\frac{4\pi a^3}{3}\right)\rho\frac{dU_{\infty i}}{dt} - \tfrac{1}{2}\left(\frac{4\ a}{3}\right)\rho\left(\frac{dU_i}{dt} - \frac{dU_{\infty i}}{dt}\right)$$

$$- 6\pi\mu a(U_i - U_{\infty i}) - \frac{6\pi\mu a^2}{\sqrt{\pi\nu}} \int_{-\infty}^t \frac{\left(\frac{dU_i}{dt'} - \frac{dU_{\infty i}}{dt'}\right)dt'}{(t - t')^{\frac{1}{2}}}. \qquad (14)$$

His result adds the unsteady but uniform pressure gradient effect to the already obtained effects of added mass, steady and unsteady drag, and buoyancy.

Tchen (2) and later Corrsin and Lumley (3) also considered the effect of the spatial variation of the far field velocity $U_{\infty i}$ by introducing terms containing $\partial U_{\infty i}/\partial x_j$ in the equation for the fluid force. However the addition of these terms was not based on the Oseen procedure for the solution of the quasi-static creeping flow equations. Some of these new terms are related to the advective terms of the Navier-Stokes equations and so the low Reynolds number approximation will be re-examined to see how some of the effects of the spatial variation of $U_{\infty i}$ can be properly introduced in the equation for the fluid force.

1.2 Modified Oseen procedure

The advective term of the Navier-Stokes equations represents both the spatial acceleration along a streamline and the centripetal acceleration towards the center of curvature of the streamline. Thus the equations can be written as

$$\frac{\partial u_i}{\partial t} + \frac{\partial}{\partial x_i}(\tfrac{1}{2}u_j u_j) + [\frac{u_j u_j}{R} - n_j \frac{\partial}{\partial x_j}(\tfrac{1}{2}u_k u_k)]n_i$$

$$= -\frac{1}{\rho}\frac{\partial}{\partial}(P + \rho g z) + \nu \frac{\partial^2 u_i}{\partial x_j \partial x_j} \qquad (15)$$

where R is the radius of curvature of the streamline and n_i is the unit normal directed towards the center of curvature. The gradient part of the advective term can be separated from the other acceleration terms by introducing the Bernoulli function, B, which is given by

$$B = P + \rho g z + \tfrac{1}{2}\rho u_j u_j. \qquad (16)$$

Thus equation 15 becomes

$$\frac{\partial u_i}{\partial t} - \nu \frac{\partial^2 u_i}{\partial x_j \partial x_j} + \frac{1}{\rho}\frac{\partial B}{\partial x_i} =$$

$$-[\frac{u_j u_j}{R} - n_j \frac{\partial}{\partial x_j}(\tfrac{1}{2}u_k u_k)]n_i. \qquad (17)$$

Now, instead of using the low Reynolds number approximation to obtain the quasi-static creeping flow equations, the approximation will be made that the normal acceleration is unimportant and its effects will be neglected. This approximation gives the linear equations

$$\frac{\partial u_i}{\partial t} = -\frac{1}{\rho}\frac{B}{x_i} + \nu \frac{\partial^2 u_i}{\partial x_j \partial x_j} \qquad (18)$$

and

$$\frac{\partial u_j}{\partial x_j} = 0 \qquad (19)$$

which are identical to the quasi-static creeping flow equations with the exception that B replaces $(P + \rho g z)$.

The solution to these equations can be found by using the Oseen procedure for u_i and B. The resulting solution is

$$u_i = -\frac{\partial}{\partial x_i}(\phi_1 + \phi_2) + \frac{\partial G_i}{\partial t} - \nu \frac{\partial}{\partial x_i}(\frac{\partial G_j}{\partial x_j}) \qquad (20)$$

and

$$B = \rho \frac{\partial}{\partial t}(\phi_1 + \phi_2), \qquad (21)$$

where the equation for B can be rewritten as

$$P = \rho \frac{\partial}{\partial t}(\phi_1 + \phi_2) - \rho g z - \tfrac{1}{2}\rho u_j u_j. \qquad (22)$$

This solution includes the advective effects along a streamline but omits the effects of acceleration across the streamlines.

Since the modified Oseen procedure is linear in ϕ_1, ϕ_2, G_i, and B, solutions due to different physical phenomena may be added. The force due to shear stress is also linear and so the shear forces due to different phenomena may be summed. All of the components of the pressure force except that due to $\tfrac{1}{2}\rho u_j u_j$ may also be added. However, the kinetic energy component of the pressure force is a quadratic function of ϕ_1, ϕ_2, and G_i and so this component of the forces may not be included in the linear superposition procedure.

The flow around a moving sphere will next be examined by considering the individual phenomena that sum to the fluid force on the moving sphere. The phenomena are described in the boundary conditions of the modified Oseen procedure.

1.3 Physical phenomena analysis

The uniform flow forces of buoyancy, added mass, Stokes and Basset drag, and the unsteady pressure gradient were given by equation 14. They are also included in the forces that result from the modified Oseen procedure.

The first non-uniform flow effect to be examined is the force on a sphere moving with constant velocity $(U_o, 0, 0)$ through a steady potential flow, $U_{\infty i}$, which has a pressure gradient along the path of a sphere. Let

$U_{\infty i} = (\frac{U_b x}{b}, -\frac{U_b y}{2b}, -\frac{U_b z}{2b})$ and assume

the sphere moves along the x axis so that $x_o = U_o t$. Then the modified Oseen procedure gives the velocity potential

$$\phi_1 = -\frac{U_b}{4b}[2(x_o + x')^2 - y'^2 - z'^2] + U_o x'$$

$$-\frac{U_b a^5}{6b\, r^3}(\frac{3x'^2}{r^2} - 1) - \tfrac{1}{2}(\frac{U_b x_o}{b} - U_o)\frac{a^3 x'}{r^3}, \qquad (23)$$

where (x', y', z') is the position vector relative to the center of sphere, $r^2 = x'^2 + y'^2 + z'^2$, and $x = x_o + x'$. ϕ_2 and G_i are identically zero. The force due to this flow is

$$F_x = \frac{3}{2}\rho(\frac{4\pi a^3}{3})(U_{cx} - U_o)\frac{\partial U_{cx}}{\partial x}$$

$$+\frac{3}{2}\rho(\frac{4\pi a^3}{3})U_o\frac{\partial U_{cx}}{\partial x}, \quad (24)$$

where $U_{cx} = U_{\infty x}$ evaluated at the center of the sphere and where the force has been separated into the component due to the non-uniform distribution of $\frac{1}{2}\rho u_j u_j$ and the component due to the variation of U_{cx} with time. The first of these parts is associated with the unsteady pressure force component of equation 14 by the $\frac{\partial \phi_1}{\partial t}$ term of equation 22.

A second non-uniform flow effect is the force on a sphere moving with velocity $(0, V_o, 0)$ through a steady, inviscid shear flow where $U_{\infty i} = (U_b y/b, V_b, 0)$. Assume that the sphere moves along the y axis with its center at $Y_o = V_o t$. Then the modified Oseen procedure gives

$$\phi_1 = -\frac{U_b Y_o}{b}(x' - \frac{1}{2}\frac{a^3 x'}{r^3})$$

$$-\frac{U_b}{2b}(x'y' - \frac{2}{3}\frac{a^5 x'y'}{r^5})$$

$$+(V_o - V_b)(y' + \frac{1}{2}\frac{a^3 y'}{r^3}), \quad (25)$$

while ϕ_2 and G_i are zero since $\nu=0$. The force due to this flow is $F_i=(F_x,F_y,0)$ where

$$Fx = \frac{3}{4}\rho(\frac{4\pi a^3}{3})(V_b - V_o)\frac{\partial U_{cx}}{\partial y}$$

$$+\frac{3}{4}\rho(\frac{4\pi a^3}{3})V_o\frac{\partial U_{cx}}{\partial y}, \quad (26a)$$

and

$$F_y = \frac{3}{4}\rho(\frac{4\pi a^3}{3})U_{cx}\frac{\partial U_{cx}}{\partial x}. \quad (26b)$$

The force component F_x has again been separated into the component due to the non-uniform distribution of $\frac{1}{2}\rho u_j u_j$ and the component due to the variation of U_{cx} with time. The F_y component is a lift force due to the non-uniform distribution of $\frac{1}{2}\rho u_j u_j$. As with the previous flow the two force components due to $\frac{1}{2}\rho u_j u_j$ are associated with the steady shear flow and the second term in equation 24a is related to the unsteady pressure term of equation 14 by the $\frac{\partial \phi_1}{\partial t}$ term of equation 22.

These two inviscid flow examples indicate that $\frac{1}{2}\rho u_j u_j$ and $\frac{\partial \phi_1}{\partial t}$ produce a pressure force on the sphere where there is spatial variation of $U_{\infty i}$ either along or across the streamlines. The forces due to $\frac{\partial \phi_1}{\partial t}$ depend on the velocity $U_{\infty i}$ linearly, and so those pressure forces may be added to the first two terms of equation 14 giving

$$F_i = \frac{3}{2}\rho(\frac{4\pi a^3}{3})(\frac{\partial U_{\infty i}}{\partial t} + U_j\frac{\partial U_{\infty i}}{\partial x_j})$$

$$-\frac{1}{2}\rho(\frac{4\pi a^3}{3})\frac{dU_i}{dt}. \quad (27)$$

These are the unsteady or added mass forces due to the acceleration of the fluid and the sphere respectively. The forces due to $\frac{1}{2}\rho u_j u_j$ may not be simply added to those already obtained, since these pressure forces do not depend linearly on the velocity $U_{\infty i}$. The forces which result from the non-uniform distribution of kinetic energy must be calculated separately for each different case of spatial variation of $U_{\infty i}$.

A third effect of the spatial variation $U_{\infty i}$ is a steady viscous force. Faxen's (4) laws indicate that there is an additional viscous drag force on a sphere in a steady, low Reynolds number, non-uniform flow. That force is

$$F_i = \pi\mu a^3\nabla^2 U_{ci}. \quad (28)$$

Since this is a linear viscous force, it too may be added to those already determined.

The last effect of the spatial variation of $U_{\infty i}$ that will be discussed is the unsteady viscous force. Corrsin and Lumley (3) indicated that the Basset force for flows with spatial variation is

$$F_i = -\frac{6\pi\mu a^2}{\sqrt{\pi\nu}}\int_{-\infty}^{t}\frac{(\frac{dU_i}{dt'} - \frac{\partial U_{\infty i}}{\partial t'} - U_j\frac{\partial U_{\infty i}}{\partial x_j})dt'}{(t-t')^{\frac{1}{2}}}. \quad (29)$$

Again the part of this force due to the spatial variation of $U_{\infty i}$ is linear and may be added.

Although the examples just given do not include all of the effects of the spatial variation of a turbulent flow, they are a beginning. Thus the force on a sphere moving with velocity U_i through a turbulent flow represented by $U_{\infty i}$, P_∞ is

$$F_i = (\frac{4\pi a^3}{3})\rho g\frac{\partial z}{\partial x_i} - \frac{1}{2}(\frac{4\pi a^3}{3})\rho\frac{dU_i}{dt}$$

$$- 6\pi\mu a(U_i - U_{ci})$$

$$+ \pi\mu a^3\nabla^2 U_{ci} + \tfrac{3}{2}\left(\frac{4\pi a^3}{3}\right)\rho\left(\frac{\partial U_{ci}}{\partial t} + U_j \frac{\partial U_{ci}}{\partial x_j}\right)$$

$$- \frac{6\pi\mu a^2}{\sqrt{\pi\nu}} \int_{-\infty}^{t} \frac{\left(\frac{dU_i}{dt'} - \frac{\partial U_{ci}}{\partial t'} - U_j \frac{\partial U_{ci}}{\partial x_j}\right) dt'}{(t - t')^{\frac{1}{2}}}$$

$$+ \iint_A \tfrac{1}{2}\rho u_j u_j n_i \, dA + \ldots \quad . \qquad (30)$$

1.4 Discussion

Tchen's equation 14 for the force on a sphere in a low Reynolds number, unsteady, uniform flow is a reasonable approximation only if the sphere is much smaller than the shortest length scale of the turbulence. For larger particles the spatial variation of the turbulent velocity and pressure are important. The modified Oseen procedure presented here permits the examination of some of this spatial variation and some advective effects. It also relaxes the low Reynolds number condition. However it does this at the expense of the centripetal acceleration, and so the fluid force normal to the streamlines is not expected to be correct when the modified Oseen procedure is used.

A number of additional forces on the sphere resulted from the study of the spatial and advective effects. These were all fairly well known, having been developed previously by others who used either the high or low Reynolds number approximation. With the modified Oseen procedure they can come together.

REFERENCES

Oseen, C.W., Hydrodynamik, Akademische Verlagsgesellschaft M.B.H., Leipzig, 1927.

Tchen, C.M., "Mean values and correlation problems connected with the motion of small particles suspended in a turbulent fluid," thesis presented at the University of Netherlands in Delft, 1947, in partial fulfullment of the requirements for the Degree of Doctor of Philosophy.

Corrsin, S. and Lumley, J., "On the equation of motion for a particle in turbulent fluid," Applied Scientific Research, Vol. 6, 1956, pp. 114-116.

Happel, J. and Brenner, H., Low Reynolds Number Hydrodynamics, Prentice-Hall, Inc., Englewood Cliffs, NJ, 1965.

3 Initiation, formation and behaviour of ripples and dunes

Euromech 156: Mechanics of Sediment Transport / Istanbul / 12-14 July 1982

Shape and dimensions of ripples and dunes*

JØRGEN FREDSØE
Technical University of Denmark, Lyngby

ABSTRACT: The governing equations leading to prediction of the shape of bed forms in steady flow are discussed. Further on, methods to find the exact dimensions of dunes are outlined. Finally, the behaviour of dunes in unsteady flow is slightly considered.

1 APPLICATION OF STABILITY ANALYSIS

The work concerning bed forms formed by flowing water has been rather comprehensive during the last decades.

Especial attention has been paid to why these bed forms develop, and the main tool for these investigations has been the stability analysis. In stability analysis, an original plane bed is given an (infinitely) small disturbance. The influence from this disturbance on the flow pattern is then calculated, and in combination with a sediment transport formulae and the equation of continuity for sediment it is possible to predict whether the bed remains stable or breaks up into different kind of bedforms. Hereby, it has been succeeded to explain:

(i) formation of current ripples (rough bed: (Richards 1980), smooth bed: (Sumer et al. 1982))

(ii) formation of dunes (for a review, see (Kennedy (1969), (Engelund and Fredsøe 1982) (also (iii) and (iv)))

(iii) transition from dunes to plane bed

(iv) formation of antidunes.

The reason why stability analyses are applied rather much is that the calculations are simplified, because the disturbance is assumed infinitely small, so the flow equations for the perturbed flow become linear.

However, stability analyses are only able to predict whether bed forms develop or not and can say nearly nothing about the final dimension (height, length) and shape of the bed forms.

2 SHAPE OF BED FORMS IN STEADY FLOW

In order to determine the shape of bed forms a much more complex problem has to be solved:

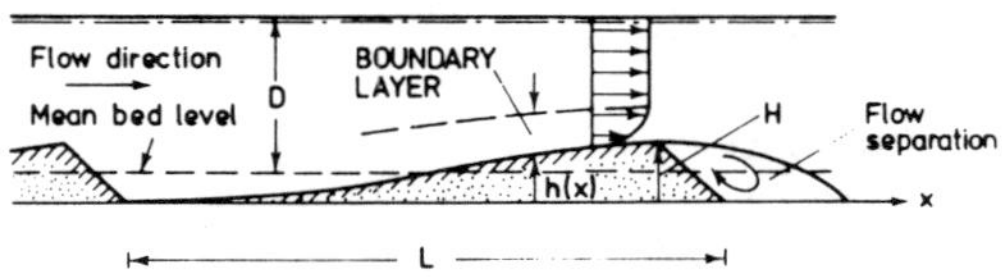

Figure 1. Definition sketch of migrating dunes.

If we consider figure 1, which shows a schematic representation of a sand dune in a river bed, it is seen that the flow is highly non-uniform. Just behind the dune top we have a strong adverse pressure gradient (leading to flow separation) while along the mild positive slope of the dune a slightly favourable pressure gradient is present, which forms a kind of boundary layer along the bed.

This flow in fact has to be solved in order to describe the shape of dunes in the right way, as seen from the following reasoning:

the continuity equation for sediment is given by

$$\frac{\partial q}{\partial x} = -(1-n)\frac{\partial h}{\partial t} \qquad (1)$$

q = sediment transport rate, n = porosity and t = time.

In case of STEADY FLOW the bed forms can be modelled to migrate downstreams with a velocity a without changing form

$$h = h(x-at) \qquad (2)$$

so Eq. (1) gives

* This paper was presented by invitation at the colloquium as a survey lecture.

$$q = q_0 + a(1-n)h \quad (3)$$

In case of pure bed load, the dimensionless sediment transport $\Phi = q/\sqrt{(s-1)gd^3}$ (s = relative density, g = acc. of gravity and d = grain diameter) is only a function of the local shields parameter $\theta = \frac{\tau}{\rho(s-1)gd}$ (τ = local bed shear stress) and the local bed slope:

$$\Phi = \Phi(\theta, \frac{dh}{dx}) \quad (4)$$

The migration velocity a is given by

$$a = \frac{q_{top}}{(1-n)H} \quad (5)$$

q_{top} being the sediment transport at the top of the bed form, H the height. Further, q_0 in Eq (3) is zero (the sediment transport in the trough). Hereby we obtain

$$h = \Phi(\theta, \frac{dh}{dx}) \frac{H}{\Phi_{top}} \quad (6)$$

which is the governing differential equation to describe the bed form in steady flow. If the relationship Eq (4) is known (for instance by applying a Meyer-Peter formulae) Eq (6) can be solved if θ is known. The variation of θ along the dune is a function of the shape itself, however:

$$\theta = \theta(h(x)) \quad (7)$$

Hence, if the flow (and hence the bed shear stress) above an arbitrary bed h(x) can be calculated, the shape of the bed form h = h(x) can be calculated from Eq (6) if H and Φ_{top} are given in advance.

The remarks until now are quite general and not restricted to dunes, but also valid for instance for ripples.

The key problem is, as mentioned earlier, to determine the relationship Eq (7).

More or less advanced models have been introduced to solve this problem, but common for all of them is that they utilize the fact that the shape of the bed forms is comparable with flow over a negative step (for instance (Sündermann, Vollmers & Puls 1980), (Klaassen 1979)). That several scientists are aware of this is confirmed by an increasing number of experimental papers, studying in detail the flow picture behind a negative step (after the basis work of Bradshaw and Wong, 1972 for instance Etheridge and Kemp 1978, 1979). This also seems to be a fruitful way to go.

If we for instance look at the bed shear stress measurements by Bradshaw and Wong, as depicted in figure 2, this shows the variation in θ in the flow direction under the condition that the bed is flat down-

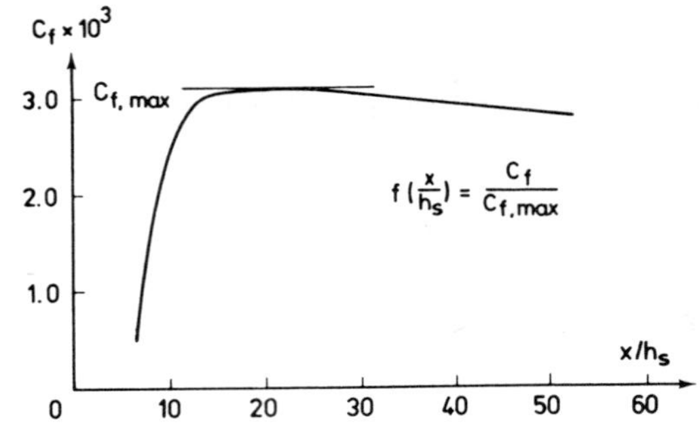

Figure 2. Variation in bed shear stress downstream a rearward-facing step. After Bradshaw and Wong.

stream the step. It is seen that reattachment occurs about 6 times the step height h_s downstream the step, after which the bed shear increases rapidly until it becomes maximum about 16 - 18 times H_s downstream the step. Further downstream the bed shear stress decreases again, due to growth in boundary layer thickness.

If we, as a starting point, assume that Eq. (7) is nearly identical to the form shown in figure 2 - which means that the presence of the mild sloping part of the following dune is disregarded - we obtain a picture of the bed form as shown in figure 3, obtained by use of Eq. (6).

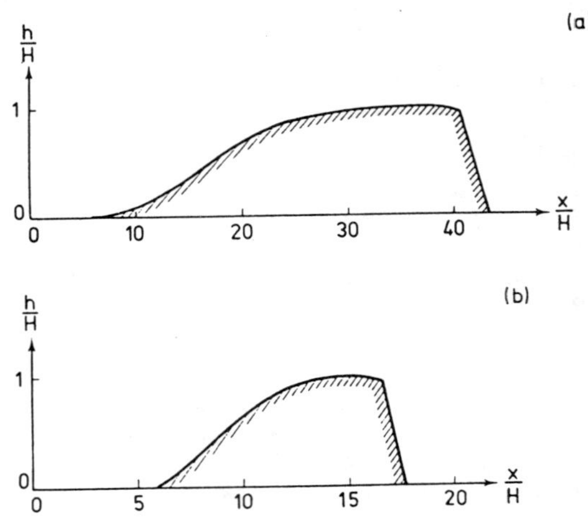

Figure 3. Examples on calculated dune profiles.
(a): θ_{top} = 0.057, H/D = 0.04, θ_c = 0.05
(b): θ_{top} = 0.30, H/D = 0.27, θ_c = 0.05

Just downstream the maximum dune-height, the flow diverges, which will lead to flow separation so the bed form ends.

If we concentrate on dunes, these are

normally very large - several times the water depth - and in this case the presence of the bed form downstream the negative step can be approximated by assuming the effect on the bed shear distribution, figure 2, is according to hydrostatic pressure.

Hence, the variation in local dimensionless bed shear stress θ^* past a dune is suggested to be given as

$$\frac{\theta^*}{\theta^*_{top}} = \frac{(1-\frac{H}{2D})^2}{(1+\frac{H}{2D}-\frac{h}{D})^2} \; f(\frac{x}{H}) \qquad (8)$$

where θ^*_{top} is the bed shear stress at the dune top. D is the depth, defined as the distance from the mean bed (about $\frac{H}{2}$ above the dune trough) to the water surface.

The first factor on the right hand side of Eq (8) takes into account the changes in cross-section, while the second factor f is the non-dimensional function given in figure 2.

Introducing the correction Eq (8) into the calculation of the bed forms shown in figure 3 gives only small changes of the shape. However, the expression given by Eq. (8) is important, as demonstrated later.

Concerning ripples, Eq (8) can not be valid. The length of ripples is small compared to water depth (and independent of water depth), so the water depth does not play the same role. Instead it might be possible to find another vertical length parameter, which may replace the water depth in Eq (8). This can be the length away from the bed where the ripples disturbe the flow significantly, so this length is a function of the ripple length (suggested by F. Engelund).

If we make a comparison with the much more developed field of water waves, dunes are similar to long waves, where hydrostatic pressure can be assumed to be valid along a vertical. Ripples on the other hand must be compared with short waves, which form and velocity are independent of the water depth.

However, as mentioned with regard to dunes, the shape of the bed form is only affected very little by the modification, Eq. (8). This is consistent with the fact that the shape (but not dimensions!) of ripples and dunes is quite similar.

3 DIMENSIONS OF DUNES

The shape of bed forms depicted in figure 3 can in fact be calculated for an arbitrary value of H, for which a solution can be obtained. That means that Eq (6) and (8) can not predict the absolute value of the dune height.

As the length of the bedforms becomes a function of the height (through the calculations of the shape) this can not be predicted immediately either.

One way to solve this problem has been demonstrated by Fredsøe 1982, who demonstrated that the calculated shape, as shown in figure 3, is only stable to small disturbances $\tilde{h}$, as indicated in figure 4 for specific value of H/D. The idea of the analysis is to introduce the small distur-

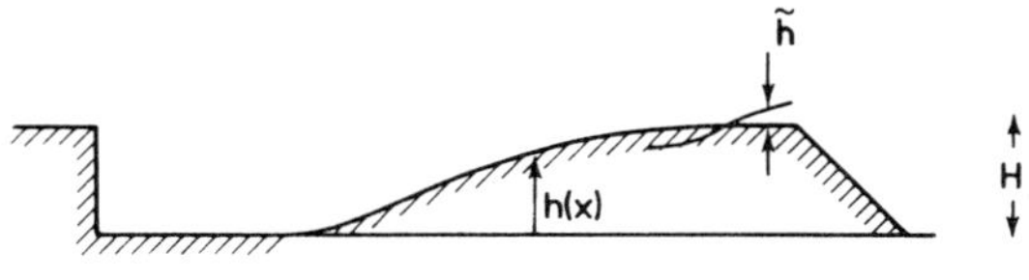

Figure 4. Definition sketch of the perturbation of the dune form, presented in figure 3.

bance $\tilde{h}$ in the equation of continuity, Eq (1), and assume Eq (8) to be valid for the perturbed flow also. Hereby, the change in $\tilde{h}$ with time can be calculated. If H/D is below a certain value, $\tilde{h}$ will increase with time, so the dune height becomes larger until it reaches a sufficient large value where the shape becomes stable. The result of this analysis is that H/D is determined by

$$\frac{H}{D}\Big/(1-\frac{H}{2D}) = \Phi\Big/(2\theta\frac{d\Phi}{d\theta}) \qquad (9)$$

as long as the sediment transport is bed load only.

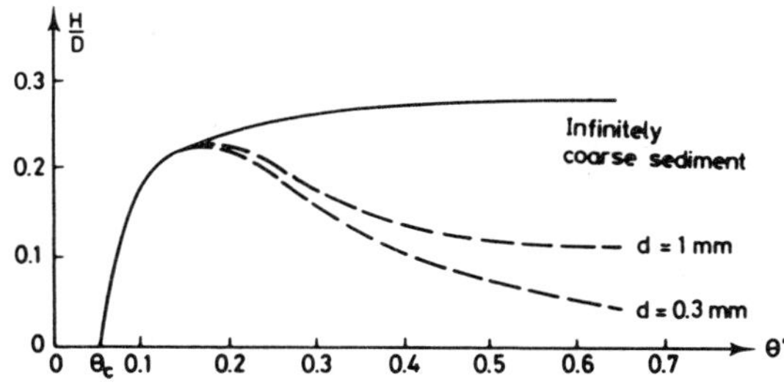

Figure 5. Variation in dune height - water depth ratio with bed shear stress for different grain sizes. θ_c is the critical shields parameter.

These results are shown in figure 5 with the full drawn line. The dotted lines show the results from a similar analysis where the effect of suspended sediment is taken into account: if part of the sediment is carried into suspension, this part will normally not settle on the front of the dune, when it passes the crest. Hence, in the expression for the dune migration velocity Eq (5), q only shall stand for q_b (bed load transport). This leads to a slightly changed version of Eq (9):

$$\frac{H}{D}\Big/\left(1-\frac{H}{D}\right) = \Phi_b\Big/\left(2\theta\left(\frac{d\Phi_b}{d\theta}+\frac{d\Phi_s}{d\theta}\right)\right) \qquad (10)$$

This leads to a decrease in H/D at large values of θ (transition to plane bed).

If we return to the wavelength, the ratio H/L is shown in figure 6. At small values

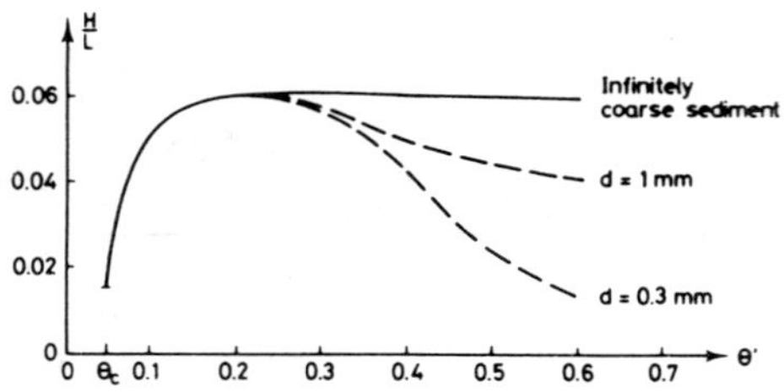

Figure 6. Variation in length-height ratio with bed shear stress for different grain sizes.

of θ, the effect of gravity (sloping bed) on the bed load is important as it involves small ratios of H/L. At larger values of θ, this effect becomes less pronounced and H/L ~ 1/16, which is accordance with the position of the maximum in bed shear downstream a negative step, cf. figure 2. At even larger values of θ, the ratio H/L increases again, because of the effect of suspended load, which removes the point of maximum transport downstream the point of maximum bed shear stress. As the position of the point of maximum sediment transport and the point of maximum bed level is the same (Eq (6)), this means that the length of the dunes increases.

4. UNSTEADY FLOW

If we shortly consider unsteady flow in alluvial rivers, the theoretical work is even more sparse, due to the complexity of the problem.

As described in the previous, the dimensions of the bed forms - at least the dunes - are a function of the bed shear stress. Hence, if we for instance change the water-discharge in the river from one value to another, the dimensions of dunes will change. How this change occurs is a rather complex problem: it is obvious that the changes in the dunes must take some time if we consider the theoretical example that the water-discharge suddenly change from one value Q_1 to another value Q_2. Allen 1976a, 1976b was one of the first to consider this problem and suggested that the dunes have to travel some times their own length in order to adjust their shape and dimensions to the new flow conditions. Further on, he argued that an individual dune has a restricted life-span, after which it is destroyed and replaced by new dunes which are more consistent with the new flow conditions. Very little is known about this life-span, which is connected to the stochastic behaviour of the dunes in river beds.

Fredsøe 1979, 1981 has made some preliminary theoretical investigations from a deterministic point of view. Idealizing the shape of the dunes to be triangular, it is possible, by an analysis very similar to the perturbation analysis described in the former section, to predict whether the dune-height will grow or decrease with time, and further on, how fast this process occurs. An indirect test of the theoretical findings is shown in figure 7, where the changes in water-depth, due to change in dune-dimensions and hence change in flow-resistance, are given as function of time. The measurements are carried out by Gee, 1973, and the solid curves are the theoretical findings.

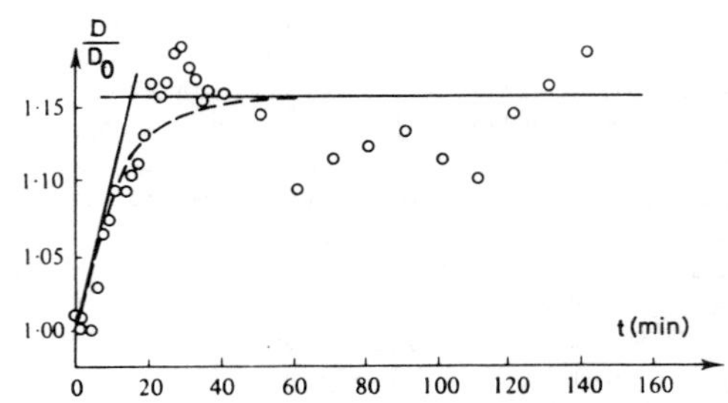

Figure 7. Changes in water depth with time after a sudden change in water discharge. The circles are the data obtained by Gee, 1973.

More accurate analyses of this problem need to be carried out by taking the correct shape of the dunes into account and by introducing the stochastic nature in the process.

REFERENCES

Allen, J.R.L. 1976a. Computation models for dune time-lag: General ideas, difficulties and early results. Sediment. Geol. 15, 1-53.

Bradshaw, P. and Wong, F.Y.F. 1972. The re-attachment and relaxation of a turbulent shear layer. J. Fluid Mech. Vol. 52, pp 113-135.
Engelund, F. and Fredsøe, J. 1982: Sediment Ripples and Dunes. Ann. Rev. Fluid Mech. 14, 13-37.
Etheridge, D.W. and Kemp, P.H. 1978. Measurements of flow downstream a rearward-facing step. J. Fluid Mech. 86, 3.
Etheridge, D.W. and Kemp, P.H. 1979: Velocity measurements downstream of rearward-facing steps, with special reference to bed instability. J. Hyd. Res. 17, 2.
Fredsøe, J. 1979: Unsteady flow in straight alluvial streams: modification of individual dunes. J. Fluid Mech. 91, 497-512.
Fredsøe, J. 1981: Unsteady flow in straight alluvial streams. Part 2. Transition from dunes to plane bed. J. Fluid Mech. 102, 431-53.
Fredsøe, J. 1982: Shape and dimensions of stationary dunes in rivers. J. Hyd. Div., ASCE 108, HY 8.
Gee, D.M. 1973: Sediment transport in non-steady flow. Univ. Calif. Berkeley, Rep. HEL 22-3.
Kennedy, J.F. 1969. The formation of sediment ripples, dunes and antidunes. Ann. Rev. Fluid Mech. 1, 147-68.
Klaassen, G.J. 1979. Sediment transport and hydraulic roughness in relation to bed forms. Publ. 213, Delft hydraulics laboratory.
Richards, K.J. 1980. The formation of ripples and dunes on an erodible bed. J. Fluid Mech. 99, 597-618.
Sumer,B.M., Bakioğlu, M. and Bulutoğlu,A.1982 Ripple formation on a bed of fine cohesionless granular sediment. In B.M.Sumer and A.Müller (eds.), Euromech 156: The mechanics of sediment transport, Rotterdam, Balkema.
Sündermann, J., Volmers, H.J. and Puls, W. 1980. A numerical model for dune dynamics. Coast. Eng. Conf. 17, 1584-1598, Sydney.

Euromech 156: Mechanics of Sediment Transport / Istanbul / 12-14 July 1982

Ripple formation on a bed of fine, cohesionless, granular sediment

B.MUTLU SUMER, MEHMET BAKIOĞLU & ARGUN BULUTOĞLU
Technical University of Istanbul, Turkey

ABSTRACT : Making use of an analysis of flow over a small-amplitude smooth, wavy wall, a bedform instability analysis is worked out. The bed grains are such that the bed behaves as an hydraulically smooth boundary. The effect of gravity is included in the instability work through the local inclination of the bed. It was found that the bed is unstable for a particular range of wavenumbers. The range of the most unstable wavenumbers gives a range of wavelengths at which one would expect a bed disturbance to grow. Agreement with available limited ripple data has been observed.

1 INTRODUCTION

The most recent information about the sediment ripples and dunes can be found in the comprehensive review by Engelund and Fredsøe (1982).

The present study deals with the ripple formation on a bed which behaves as an hydraulically smooth boundary. Mantz(1978) observed that ripple bedforms develope with sediment of very small size (as small as 15 μm). In Mantz's experiments, the initial flat bed stress was such that $du_*/\nu \sim 0.1$ for the 15 μm-grade tests and $du_*/\nu \sim 0.9$ for the 66 μm-grade tests he conducted. The latter means that the bed prior to the formation of the bedforms acted as an hydraulically smooth boundary, since $du_*/\nu < 4$.

Richards (1980) presented a bedform instability analysis. Richards' theory successfully predicted the occurrence of two separate modes of instability, with wavelengths related to the roughness of the bed and the depth of the flow. It was postulated that the former mode corresponds to the formation of ripples and the latter to that of dunes. Richards, in his formulation of the problem, included the roughness length of the bed y_o, which in turn was related to the grain size d by a relation like $y_o = \alpha d$. He finally ended up with the result that the ripple-mode instability occurs with a wavelength of $203d < L < 4050d$, which is comparable to that of observed ripples with coarse sediment.

In the case of fine, granular sediment, the bed behaves as an hydraulically smooth boundary (as has been the case in Mantz's (1978) experiments); hence, the roughness length should drop from the formulation, and the viscosity should be taken into consideration instead. This leads to that Richards' theory does not apply to the problem of ripple formation on a fine, granular sediment bed.

In the context of formation of ripples, Engelund and Fredsøe (1982), in their review, follow the same line of thinking as in the preceding paragraph and suggest that it would be of interest to carry out an analysis similar to that of Richards for an hydraulically smooth bed, pointing out that ripples, usually, are associated with a smooth bed.

This paper is an attempt to work out a bedform instability analysis for the above mentioned case. Briefly, a) making use of Thorsness et al. 's (1978) study, the problem of flow over a small-amplitude smooth, wavy wall is analysed; b) the latter is incorporated into a bedform instability analysis in which the effect of the local bed slope on the sediment transport is taken into account; and c) the results are compared with Mantz's observations. The present work does not investigate the final dimensions of ripples, nor does it study their final shapes. A detailed account of the shape and dimensions of ripples and dunes has been given in the survey article by Fredsøe (1982).

2 FLOW OVER A SMALL-AMPLITUDE WAVY WALL

It is assumed that the time scale of the flow development over the bedform is small compared with the time scale of the bedform development--an assumption the validity of which is tested in §4. The latter suggests that the flow over an erodible bed can be assumed to be steady for times small compared with the development time of the bedforms.

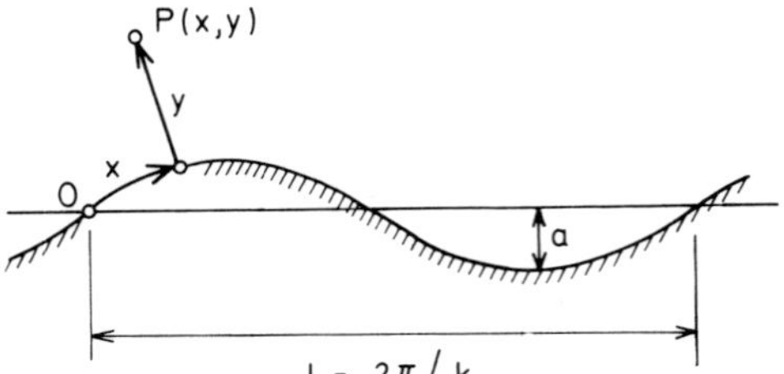

Fig.1. Definition sketch

Thorsness et al. (1978) studied the flow over a smooth, small-amplitude, solid wavy wall. For convenience, Thorsness et al.'s linear theory will be summarized herein briefly.

The equation of motion (Reynolds equation) :

$$\underset{\sim}{u} \cdot \nabla \underset{\sim}{u} = -\frac{1}{\rho} \nabla p + \nu \nabla^2 \underset{\sim}{u} + \frac{1}{\rho} \nabla \cdot \underset{\sim}{r} \qquad (1)$$

where $\underset{\sim}{u}$ = the velocity, p = the pressure, $\underset{\sim}{r}$ = the Reynolds stress tensor, ρ = the density, ν = the kinematic viscosity.

Taking curl of the preceding equation, we have the vorticity equation :

$$-\nabla \times (\underset{\sim}{u} \times \underset{\sim}{\omega}) + \nu \nabla \times (\nabla \times \underset{\sim}{\omega}) = \frac{1}{\rho} \nabla \times (\nabla \cdot \underset{\sim}{r}) \quad (2)$$

where ω = the vorticity vector.

For convenience, we use the boundary-layer coordinate system shown in Figure 1, the linearized metrical coefficients of which are

$$h_x = 1 + ak^2 \, y e^{ikx} , \quad h_y = 1 \qquad (3)$$

where a = the amplitude of the wavy wall, k = the wavenumber, and a is the perturbation parameter of the problem.

The vorticity equation (eq.2) with respect to this coordinate system becomes

$$\frac{\partial(\psi,\nabla^2\psi)}{\partial(x,y)} + \nu \, h_x \, \nabla^4\psi =$$

$$= \frac{1}{\rho}\Big[-\frac{1}{h_x^2}\frac{\partial h_x \partial h_x}{\partial x \, \partial y} r_{yy} - \frac{1}{h_x^2}\frac{\partial h_x}{\partial x}\frac{\partial r_{xy}}{\partial x} +$$

$$+ \frac{1}{h_x^2}\frac{\partial h_x}{\partial x}\frac{\partial h_x}{\partial y} r_{xx} + \frac{1}{h_x}\frac{\partial^2 h_x}{\partial x \, \partial y} r_{yy} +$$

$$+ \frac{1}{h_x}\frac{\partial h_x}{\partial y}\frac{\partial r_{yy}}{\partial x} + \frac{\partial^2 r_{yy}}{\partial x \, \partial y} + \frac{1}{h_x}\frac{\partial^2 r_{xy}}{\partial x^2} -$$

$$- \frac{1}{h_x}\frac{\partial h_x}{\partial y}\frac{\partial r_{xx}}{\partial x} - \frac{1}{h_x}\frac{\partial^2 h_x}{\partial x \, \partial y} r_{xx} -$$

$$- \frac{\partial^2 r_{xx}}{\partial x \, \partial y} - 3\frac{\partial h_x}{\partial y}\frac{\partial r_{yx}}{\partial y} - h_x \frac{\partial^2 r_{yx}}{\partial y^2}\Big] \qquad (4)$$

where ψ = the stream function defined by

$$u = \frac{1}{h_y}\frac{\partial\psi}{\partial y} = \frac{\partial\psi}{\partial y} , \quad v = -\frac{1}{h_x}\frac{\partial\psi}{\partial x} \qquad (5)$$

The stream function ψ is written as

$$\psi = \int_o^y \bar{U}(y) \, dy + aF(y) \, e^{ikx} \qquad (6)$$

The first term on the right-hand-side of the preceding equation corresponds to the unperturbed flow and the last term to the wave-induced part.

Substituting eq.6 into the vorticity equation (eq.4) gives (to the first order in a) the following equation for F :

$$ik\left[\bar{U}\,(F''-k^2F) - \bar{U}''F + k^2\bar{U}^2\right] =$$

$$= F^{IV} - 2k^2F'' + k^4F + 2k^2\bar{U}'' - k^4\bar{U} + R \qquad (7)$$

where

$$R = ik^3\bar{r}_{xx} - ik^3\,\bar{r}_{yy} + 3k^2\,\bar{r}'_{xy} +$$

$$+ ik\,(\hat{r}'_{xx} - \hat{r}'_{yy}) + k^2\,\hat{r}_{xy} + \hat{r}''_{xy} \qquad (8)$$

Herein $r_{ij} = \bar{r}_{ij} + a\hat{r}_{ij}\, e^{ikx}$ where the term with ^ is the wave-induced part of the Reynolds stress.

The boundary conditions are

$$F = 0 \;,\quad F' = 0 \quad \text{at} \quad y = 0 \qquad (9)$$

$$F = \bar{U} \;,\quad F' = \bar{U}' \quad \text{at large } y \qquad (10)$$

The wall shear stress can be obtained from the function F through the following equation :

$$\tau_w = \bar{\tau}_w + a\, F''(0)\, e^{ikx} \qquad (11)$$

and, similarly, the wall pressure :

$$p_w = -\frac{ia}{k}\left[F'''(0) + k^2\, \bar{U}'(0) \right] e^{ikx} \qquad (12)$$

To bring a solution to eq.7, one needs to model the Reynolds stresses. Thorness et al. adopted the following scheme with this purpose :

$$r_{xy} = \frac{\nu_T}{\nu}\,(2\, e_{xy}) \qquad (13)$$

$$\frac{\nu_T}{\nu} = \ell_m^2\, |2\, e_{xy}| \qquad (14)$$

$$\ell_m = \kappa y \left[1 - \exp\left(- \frac{y\, \tau_w^{1/2}}{A} \right) \right] \qquad (15)$$

$$A = \bar{A} + \frac{a\, k_1 \bar{A} ik\, \hat{p}_w}{1 + ik\, k_{LP}}\, e^{ikx} \qquad (16)$$

where

$$2e_{xy} = \bar{U}' + a(k^2 F - k^2 \bar{U} + F'')e^{ikx} \qquad (17)$$

Herein ν_T = the turbulence kinematic viscosity, ℓ_m = the mixing length, κ = the Karman constant, A = the van Driest's damping factor, and k_1 and k_{LP} = two parameters of the model. Note that equations 13-17 are in non-dimensional forms, non-dimensionalization being achieved by the wall parameters $u_*(=\sqrt{\bar{\tau}_w/\rho})$ and ν . In the calculations, the effect of the normal stresses r_{xx}, r_{yy} were neglected, and a van Driest mixing length equation was used to determine the unperturbed velocity profile:

$$\bar{U} = 2 \int_o^y \frac{dy}{1 + \{1 + 4\kappa^2 y^2 \left[1 - \exp(-\frac{y}{\bar{A}})\right]^2\}^{1/2}} \qquad (18)$$

3 SEDIMENT TRANSPORT AND INSTABILITY ANALYSIS

In the present context, we consider the sediment transport as bed load, neglecting any suspended load on the ground that only the bed-load transport mode which is involded in ripple formation (see, for example, Kennedy 1980).

3.1 Sediment transport formula

Bagnold's (1956) bed-load formula is adopted for the present purpose (with some modifications, however). It relates the rate of work done in moving the bed load along the bed against total resistance to the available energy in the flow where the bed-load transport takes place. This gives

$$q_b = \frac{B e_b}{(s-1) g \rho^{3/2}}\; \frac{(\tau_w - \tau_{cr})\, \tau_w^{1/2}}{(c\, \tan\phi + \tan\alpha)\cos\alpha} \qquad (19)$$

where q_b = the bed-load discharge, α = the local inclination of the bed to the horizontal, ϕ = the friction angle, e_b = an efficiency factor of order 0.1, s = the specific gravity of grains, g = the acceleration due to gravity, τ_{cr} = the critical shear stress. What B and c stand for now follows.

In Bagnold's original derivation, the product Bu_* was defined as the velocity which the fluid would have at the effective plane of its thrust on the grains; B was therefore taken to be 8.5, assuming the latter plane to lie at $y \simeq d$ and employing the logarithmic velocity distribution $\bar{U} = 2.5 u_* \log(30y/k_s)$ where k_s = the equivalent sand roughness, $k_s \sim d$. In the present case, the bed behaves as an hydraulically smooth boundary. Since the bed-load transport takes place at locations $y \simeq d$, and also $du_*/\nu < 4$, the velocity Bu_* in the present case should be determined from the linear velocity profile in the viscous sublayer :

$$Bu_* = \left. \frac{u_*^2\, y}{\nu} \right|_{y \simeq d} \qquad (20)$$

which gives B as

$$B = du_* / \nu \qquad (21)$$

This is the first modification made in Bagnold's bed-load formula.

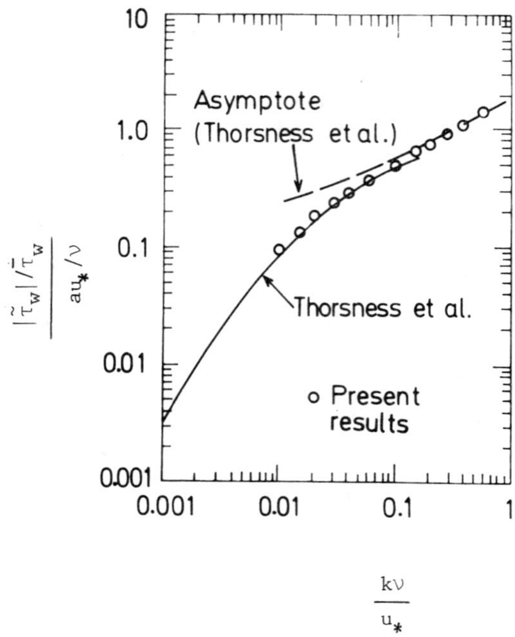

Fig.2. Amplitude of the wave-induced part of the wall shear stress, defined by $\tilde{\tau}_w = |\tilde{\tau}_w| e^{i\theta}$.

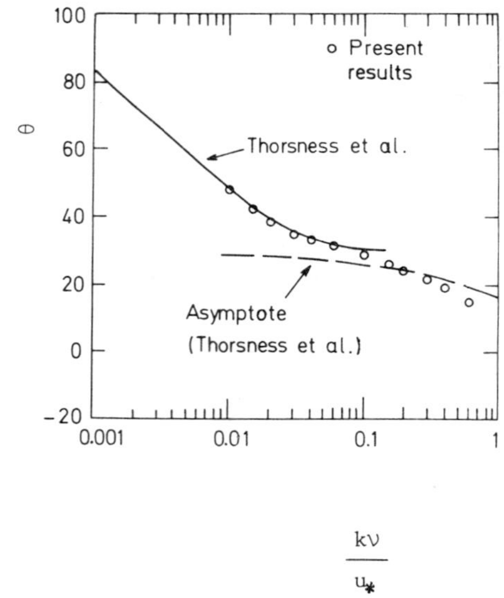

Fig.3. Phase angle of the wave-induced part of the wall shear stress. See caption of Fig.2 for the definition of θ.

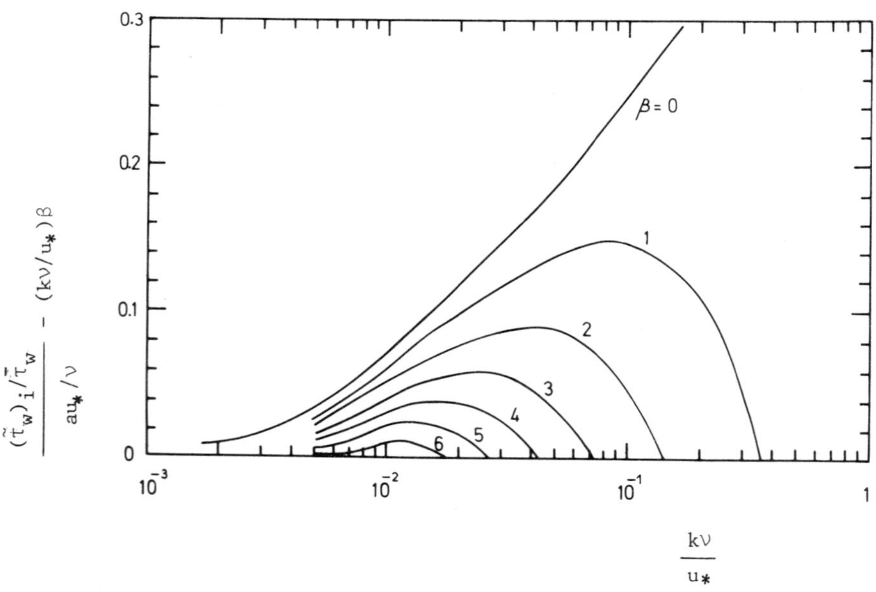

Fig.4. Bedform instability curves.

The second modification concerns the factor c appearing in eq.19. By definition, the bed-load motion takes place in the so-called bed-layer immediately above the bed, the thickness of which is proportional to the grain size : $a' = \xi d$. For a bed of local inclination α, the resisting force opposing the bed-load motion is the sum of a) the gravity force and b) the friction force, and plus c) other resisting forces which are of minor importance. Let C_b denote the bed concentration by weight. The gravity-induced resisting force per unit area of the bed is then $C_b\, a' \sin\alpha$. As to the friction force, it is provided by the normal force exerted by the bed-load particles across a surface of the non-moving bed. However, some part of the bed-load particles move in short jumps. This, on an average, will cause the normal load on the non-moving bed to be reduced: this means that an average reduction in the friction force should be expected. Thus the normal force per unit area of the bed surface is $c\, C_b\, a' \cos\alpha$ and then the friction force $c C_b a' \cos\alpha \tan\phi$ where the factor c is included to compensate for the latter effect. This factor would be unity if there had been no "jumping" mode in the bed-load transport while it would tend to zero if all the particles of the bed-load had been in "jumping" mode. Therefore c should be $0 < c < 1$. To provide an estimate for c, we can assume that the friction force is provided by only those part of the bed-load particles which are in contact with the non-moving bed, and write the friction force to be $C_b\, d \cos\alpha \tan\phi$, which implies that $c = 0.5$, taking the proportionality constant ξ in $a' = \xi d$ as 2. Although the friction force is expected to be a little higher than the above value owing to the possible interactions between the particles moving in contact with the bed and those moving in "jumping" mode, the above reasoning implies that the factor c is of order 0.5. Incorparating these ideas into Bagnold's formula gives eq.19 with c as the multiplier of $\tan\phi$.

3.2 Instability analysis

The sediment continuity equation is

$$\frac{\partial q_b}{\partial x} = -(1-n)\frac{\partial h}{\partial t} \tag{22}$$

where h = the bed height, n = the porosity of the bed, t = the time.

The bed height is given by

$$h = a\, e^{i(kx-\sigma t)} \tag{23}$$

where a = the amplitude of the bed wave (the perturbation parameter), k = the wavenumber and σ = the growth rate.

Substituting eq.23 and the linearized form of eq.19 into eq.22, we obtain, to the first order in a,

$$\sigma = \frac{B e_b \beta}{(s-1) g \rho^{3/2}} \frac{k}{1-n} \frac{3\bar{\tau}_w - \tau_{cr}}{2} \bar{\tau}_w^{1/2} \times$$

$$\times \left(\frac{\tilde{\tau}_w / \bar{\tau}_w}{a} - \beta k i \right) \tag{24}$$

where $\tilde{\tau}_w$ is the wave-induced part of the wall stress defined by

$$\tau_w = \bar{\tau}_w + \tilde{\tau}_w\, e^{i(kx-\sigma t)}, \tag{25}$$

and β is the parameter representing the effect of the combined action of gravity and friction:

$$\beta = \frac{1}{c \tan\phi} \tag{26}$$

The bed is unstable when $\sigma_i > 0$ where σ_i is the imaginary part of σ. This leads to the condition that the bed is unstable when

$$\frac{(\tilde{\tau}_w)_i / \bar{\tau}_w}{a} - \beta k > 0 \tag{27}$$

where $(\tilde{\tau}_w)_i$ is the imaginary part of $\tilde{\tau}_w$.

4 RESULTS AND DISCUSSION

4.1 Wall shear stress

Employing Thorsness et al.'s (1978) linear formulation, numerical solutions to eq.7 were obtained through a finite-difference technique. The fourth-order differential equation and also boundary conditions are approximated to the finite-difference equations, so that derivatives are expressed up to fifth or sixth order central differences at the pivotal points. Also difference correction terms, beginning with fifth or sixth order central differences, were taken into account to improve the solution (Fox 1957).

In the calculations, the parameters k_1 and k_{LP} were taken as -60 and 3000, respectively, as recommended by Thorsness et al. Also, κ was taken as 0.41 and $\bar{A}$ as 25. Present numerical results are plotted in Figures 2 and 3 where the variation of the amplitude and the phase angle of the wall shear stress are illustrated in the wavenumber range of interest of this study. In Figures 2 and 3, the "asymptote" corresponds to the limiting solution of large k where it is assumed that the wave-induced flow field is contained in a region of $yu_*/\nu < 5$ for which $\bar{U}=u_*^2 y/\nu$ and for which the influence of turbulence on the wave-induced flow can be ignored (Thorsness et al. 1978). As to the formation of bedforms, one of the most significant consequences of these findings is that the wall shear stress appears to follow the bedform with a phase lag of 25°-50° in the wavenumber range of interest. One other consequence is that the wave-induced part of the wall stress can take fairly high values. For example, for a ripple bedform with a wavelength of $Lu_*/\nu = 400$ and a height-to-length ratio of $2a/L=0.06$, the stress maximum occurs at location upstream of the bedform crest, with a phase lag of 41°, its magnitude being 2.2 times of the average stress.

4.2 Ripple formation

Employing Thorsness et al.'s (1978) linear theory, the growth rate (eq.24) was calculated; the bedform instability curves are given in Figure 4 for various values of the parameter β.

In Figure 5 are plotted the particular wavenumbers at which the growth rate is a positive maximum. The variation of the

TABLE 1. Mantz's (1978) ripple data

Run	Sed. size (μm)	Time after start of run	Stress ratio $\bar{\tau}_w/\bar{\tau}_{cr}$	Dimensionless applied stress $\bar{\theta}_w = \frac{\bar{\tau}_w/\rho}{g(s-1)d}$	Shear veloc. $u_* = \sqrt{(\bar{\tau}_w/\rho)}$ (cm/s)	Bedform	Measured ripple wavelengths: L (cm)	Measured ripple wavelengths: $\frac{Lu_*}{\nu}$
(1)	(2)	(3)	(4)	(5)	(6)	(7)	(8)	(9)
Run 1; Table 1 and Fig.1	15	0	1.0	0.23	0.88	Flat bed	-	-
		2 h	1.8	0.41		Many small ripples	1.8	158
		6 h	1.3	0.30	0.85	{Isolated ripples on otherwise flat bed}	4.1	349
		22 h	1.4	0.33	0.90	"	7.5	675
Run 2; p.92 and Fig.2(d)	15	0	2	0.46	1.06	Flat bed	-	-
		"several mins."				{Transverse ripples}	4.5	477
Run 2; Table 2 and Fig.4	66	0	1.5	0.19	1.42	{Few isolated ripples on otherwise flat bed}	-	-
		1 h	1.5	0.19	1.42	{Isolated ripples on otherwise flat bed}	~6.4	~908
		2 h	2.2	0.28	1.73	"	~7.5	~1297

latter as a function of β can be approximated by the following empirical equation :

$$\frac{k\nu}{u_*} = 0.09 \quad \beta^{-1.18} \qquad (28)$$

The wavenumber at which the growth rate attains a positive maximum value (and thus the growth is the fastest) gives a preferred wavelength at which one would expect a bed disturbance to grow. For the reported range of the friction angle quoted in Richards (1980) $0.32 < \tan\phi < 0.75$ and using eq.26 (where c is taken as 0.5), eq.28 gives the following wavenumber range

$$0.010 < \frac{k\nu}{u_*} < 0.028 \qquad (29)$$

at which the growth rate is a positive maximum. This corresponds to the wavelength range

$$220 < \frac{Lu_*}{\nu} < 610 \qquad (30)$$

The instability results are strongly dependent on β. The parameter β is expected to be a function of particle properties (such as particle size, shape, density and its surface texture) as well as near-bed flow conditions; but we are at present short of any such information. Thus the wavelength range given in eq.30 should be considered speculative owing to the uncertainties involved in β values.

To the authors' knowledge, Mantz's (1978) study is the only one which provides experimental information on ripple formation when the bed behaves as an hydraulically smooth boundary. In fact, in Mantz's experiments, $du_*/\nu \sim 0.1$ for the 15 μm-grade tests and ~ 0.9 for the 66 μm-grade tests. In Table 1 are given Mantz's (1978) ripple data for comparison. Note that Table 1 does not include (a) Run 2 of the 15 μm-grade test and (b) Run 1 of the 66 μm-grade test of Mantz's experiments. (a) The former was discarded on the ground that the bed, already having a ripple bedform on it, had been further eroded (with an increase in discharge) to generate new ripple forms--a situation which is not consistent with the linear theory outlined in the present paper. (b) On the other

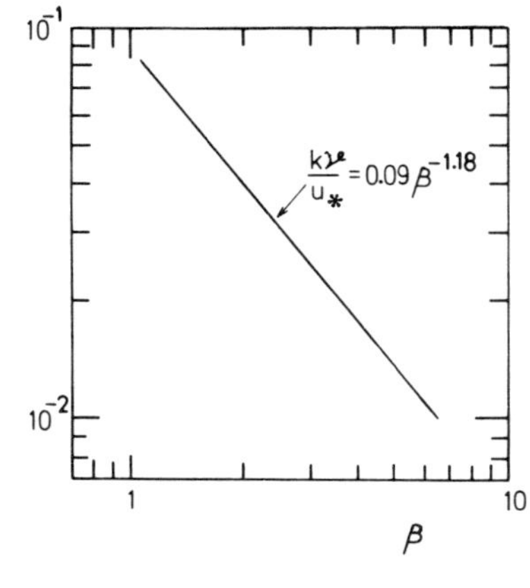

Fig.5. Wavenumbers at which the growth rate of bed disturbance is a positive maximum

hand, data corresponding to Run 1 of the 66 μm-grade test were discarded because it was difficult to specify a definite wave length in the photographes owing to the very small number of isolated ripples which developed.

Comparison of the observed wavelengths of Mantz experiments and the range in eq. 30 indicates that the theory predicts wavelengths which are in agreement with the observations. Although the observed wavelengths corresponding to 66 μm-grade are still comparable with those predicted by the theory, this latter data should be considered with some reservation for similar reasons in (a) in the preceding paragraph.

The time scale for the ripple development is $T_b \sim 1/\sigma_i$. From eq.(24), T_b is estimated for the 15 μm-grade data given in Table 1. Taking $L = 1.8$ cm and also $\tan\phi = 0.63$ as an average value(Bagnold 1973), the development time is found to be $T_b \sim 53$ min, which compares well with 2 h duration. For the data with the stress ratio 2.0, taking $L = 4.5$ cm and $\tan\phi = 0.63$ we find $T_b \sim 22$ min,which is comparable to the reported time duration "several minutes". On the other hand, the flow development time for a flow depth y can be estimated by $T_f \sim y^2/\nu_T$. Herein, ν_T is taken to be $0.5 \kappa y u_*$ (as an average value).If the depth of the flow affected by the bed wave is assumed to be $y \sim L$, then the flow development time is $T_f \sim 5 L/u_*$. This gives $T_f \sim 20$ s for the experiments Mantz conducted; thus the assumption made in §2 is confirmed, since $T_f << T_b$.

As an implication of the present study, one might think, yet falsely, that ripples may form in laminar flows as well, since the near-bed flow conditions of the present case resemble those in laminar flows as $du_*/\nu < 4$. Observations show the contrary, however. In fact Yalin and Karahan (1979) reports that "... in the case of a laminar flow, the grains of the uppermost layer (of the bed) are dragged (all together) in the form of layer (grain carpet)..." and Karahan (personal communication) pointed out that no ripple formation has been observed. This implies that even though the grains are well embedded in the viscous sublayer, as have been in the case of Mantz's 15 μm-grade tests where $du_*/\nu \sim 0.1$, it is the turbulence which seems to be the primary mechanism in the initiation of ripple bedforms. Recent observations in relation to the so-called bursting process (see for ex. Hinze, 1975) show that the influence of the wall turbulence can penetrate deeply into the lowermost part of the viscous sublayer; this provides further support for the above hyhothesis.

Furthermore, there are observational evidences showing that initial deformations of a flat bed are caused by the bursting process (Williams and Kemp 1971 and Grass 1982), where grains are swept into lowspeed wall streaks, causing piles up of grains by a mechanism similar to that explained in Sumer et al. (1978, 1981). However, one can pose the following question: How do these three-dimensional disturbances change into two-dimensional, well defined ripple forms? A composite picture of what is happening right from the beginning of the formation of ripples to the fully developed stage remains an object of speculation, unless carefully designed three-dimensional observations (in the line of Williams and Kemp) are made.

5 CONCLUSIONS

The present study gives a bedform instability analysis for a bed of fine, cohesionless, granular sediment where the bed behaves as an hydraulically smooth boundary.

It was found that the wavenumber of the bedform at which the growth rate of a bedform disturbance attains a positive maximum satisfies the following empirical equation:

$$\frac{k\nu}{u_*} = 0.09 \quad \beta^{-1.18}$$

where β is a parameter representing the effect of the combined action of gravity and friction. ($\beta = 1/(c \tan\phi)$; c = a factor of order 0.5 and ϕ = the friction angle). The latter equation gives a wavelength range of ripples as $220 < Lu_*/\nu < 610$. Owing to the uncertainties involved in the values of the parameter β, the preceding wavelength range should be considered speculative, although it compares favourably well with Mantz's (1978) experimental findings.

As the results are strongly dependent on β, and yet we are at present short of satisfactory information about this latter parameter, investigation of this parameter and its influence on the bedform development seems to be significant in the context of future sediment-transport research.

For the case when the grain Reynolds number is in the range c.a. $4 < du_*/\nu <$ c.a. 20 in which almost all the ripple data have been reported (see, for example, Vanoni 1975), it seems to be worth extending the present work so as to cover the above-mentioned interval, by properly modelling the flow near a bed of transitional-boundary character.

The authors would like to thank to Dr. E. Karahan for his comments on the first draft of the paper.

REFERENCES

Bagnold, R.A. 1956. Flow of cohesionless grains in fluids. Phil.Trans R.Soc. 249 A, 964: 235-297.

Bagnold, R.A. 1973. The nature of saltation and of "bed-load" transport in water. Proc. R.Soc. Lond. A.332: 473-504.

Engelund, F.A. and J.Fredsøe 1982. Sediment ripples and dunes. Ann.Rev.Fluid Mech. 14 : 13-37.

Fox, L. 1957. The numerical solution of two-point boundary problems in ordinary differential equations. Oxford at Clarendon Press.

Fredsøe, J. 1982. Shape and dimensions of ripples and dunes. In B.M.Sumer and A. Müller (eds.), Euromech 156: The mechanics of sediment transport, Rotterdam, Balkema.

Grass, A.J. 1982. The influence of boundary layer turbulence on the mechanics of sediment transport. In

B.M.Sumer and A.Müller (eds.), Euromech 156 : The mechanics of sediment transport, Rotterdam, Balkema.

Hinze, J.O. 1975. Turbulence McGraw-Hill.

Kennedy, J.F. 1980. Bed forms in alluvial streams : some views on current understanding and identification of unresolved problems. In H.W.Shen and H.Kikkawa (eds.), Application of stochastic process in sediment transport, p. 6a-1--6a-13. Water Resources Publications, Colorado, U.S.A.

Mantz, P.A. 1978. Bedforms produced by fine, cohesionless, granular and flakey sediments under subcritical water flows. Sedimentology. 25 : 83-103.

Richards, K. 1980. The formation of ripples and dunes on an erodible bed, J.Fluid Mech. 99 : 597-618.

Sumer, B.M. and B.Oğuz 1978. Particle motions near the bottom in turbulent flow in an open channel. J.Fluid Mech. 86 : 109-127.

Sumer, B.M. and R. Deigaard 1981. Particle motions near the bottom in turbulent flow in an open channel - Part 2. J.Fluid Mech. 109 : 311-337.

Thorsness, C.B., P.E. Morrisroe and T.J. Hanratty 1978. A comparison of linear theory with measurements of the variation of shear stress along a solid wave. Chem. Engrg. Science. 33 : 579-592.

Vanoni, V. 1975. Factors determining bed forms of alluvial streams.Closure.Proc. ASCE 101 No.HY11 : 1435-1440.

Williams, P.B. and P.H. Kemp 1971. Initiation of ripples on flat sediment beds. Proc. ASCE 97 No.HY4 : 505-522.

Yalin, M.S. and E.Karahan 1979. Inception of sediment transport. Proc.ASCE 105 No.HY11 : 1434-1443.

Euromech 156: Mechanics of Sediment Transport / Istanbul / 12-14 July 1982

Sand wave formation due to irregular bed load motion

TETSURO TSUJIMOTO & HIROJI NAKAGAWA
Kyoto University, Japan

ABSTRACT: Initiation and early stage of sand wave development are intimately related to the essential characteristics of bed load motion. In this paper, stochastic model for bed load transport process based on an Eulerian interpretation and sand bed instability analysis of Fourier transformed version are coupled to explain the spectral evolution of sand waves in early stage of sand wave development. Theoretical treatments are also followed by some experimental investigations.

1 INTRODUCTION

Small scale sand wave formation, particularly at least in early stage of dune or ripple formation by bed load motion, must depend upon irregular motion of individual bed material particles. Bed load motion is essentially irregular and discrete, and necessarily causes a non-equilibrium situation, and it cannot be represented without inspecting the internal structure of bed load transport process. In this paper, a stochastic model for bed load transport process of which constituent elements are sediment pick-up rate and step length and its Eulerian interpretation will be developed, and the essential characteristics of bed load motion will be investigated. Next, based on the properties of bed load motion preserved in this model, incipient process of bed surface irregularity, namely initiation of sand waves will be explained by stochastic techniques. It may depend upon bed instability mechanism whether such an initial bed disturbance grows up to sand waves or not. Sand bed instability analyses have been conducted by many researchers to date since Kennedy (1963). In these studies, the phase lag as a cause of bed instability due to flow properties has been minutely investigated, but the lag distance as another cause due to the characteristics of non-equilibrium sediment motion has been hardly investigated. Such an imperfection would be made up by using the transport model explained in this paper (Nakagawa & Tsujimoto 1980b), by which the so-called lag distance can be clearly evaluated (Nakagawa & Tsujimoto 1981). Moreover, in the previous instability analyses, validity of the theory has never been verified by experiments, and initiation of sand waves has not been discussed since initial bed disturbance has been assumed. In this paper, incipient process will be described in consideration of stochastic behavior of bed materials, and consequently initial bed disturbance becomes quite random. Hence, the instability analysis for Fourier transformed sand bed must be significant, where the spectral evolution process due to instability mechanism can be explained and it may be comparatively easily verified by experiments. Instability analysis of Fourier transformed version has been first proposed by Jain & Kennedy (1974), but their analysis involves some ambiguous assumptions concerned with the lag distance and the theoretical result has not been sufficiently compared with experimental data. Therefore, sand bed instability analysis of Fourier transformed version and stochastic model for bed load transport process will be coupled here, and the theoretical treatment and its results will be experimentally verified. For this purpose, some experimental investigations will be also conducted as for sand wave development from an initially flattened bed.

2 FUNDAMENTAL EQUATIONS FOR NON-EQUILIBRIUM BED LOAD TRANSPORT

Although most of bed load transport formulae were derived under equilibrium conditions, even a slight change of the situation is apt to generate non-equilibrium state of bed load transport. Sand wave initiation and early stage of its development must depend upon such non-equilibrium as seen in many alluvial processes such as armor coat formation, scouring and so on. Hence, it is necessary to establish non-equilibrium bed load transport formula. In case of bed load motion, the phenomenon is quite probabilistic and discrete, and the moving period of one step is negligibly short compared with the rest period. According to these characteristics, the formulation of bed load transport process can be obtained as follows (see Fig.1):

$$q_B(x)=(A_3d/A_2)\int_{x_0}^{x} p_s(x')\int_{x-x'}^{\infty} f_X(\xi)\,d\xi\,dx' \quad (1)$$

$$\partial y(x)/\partial t = A_1 d[p_d(x)-p_s(x)] \quad (2)$$

$$p_d(x) = \int_0^{x-x_0} p_s(x-\xi)f_X(\xi)\,d\xi \quad (3)$$

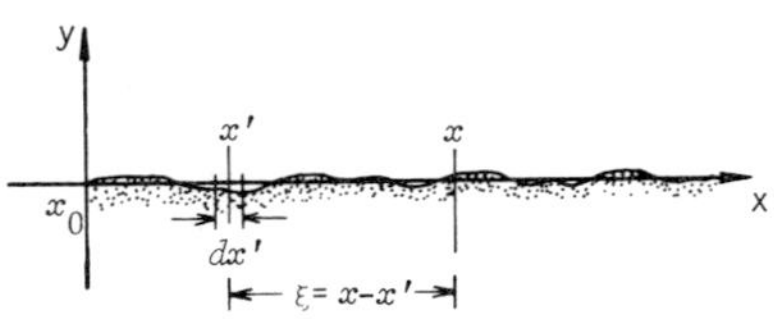

Fig.1. Definition sketch for modelling.

where q_B: bed load transport rate, d: sand diameter, p_s: sediment pick-up rate, p_d: deposit rate, $f_X(\xi)$: probability density function of step length, y: height of sand bed, and A_1, A_2, A_3: 1-, 2- and 3-dimensional geometrical coefficients of sand. The step length is defined here as a distance for a particle to travel from its incipient motion to the next definite stop. According to the above equations, sediment pick-up rate and the distribution of step length are important elements of bed load transport, and these equations represent well the essential characteristics of bed load motion. For example, although the pick-up rate is uniquely determined by local bed shear stress, the local sediment transport rate is influenced by the sediment transport states in upstream regions by action of step length, and this fact brings about the so-called lag distance as a cause of sand bed instability.

By the way, the above equations are of Eulerian type of the stochastic model for bed load transport. Though a stochastic model for bed load transport first derived by Einstein (1937) was certainly faithful to the actual phenomenon, only a Lagrangian model has been developed for describing the bed load dispersion and for predicting the subsequent transport rate under equilibrium condition.

Under flat bed condition or conditions slightly shifted from flat bed, pick-up rate and step length are correlated with the scales of sediment particle itself (diameter and fall velocity), while these are affected by the scales of bed geometry when bed deformation becomes appreciable. Initiation and early stage of sand wave formation must correspond to the former case, and its growth process after some time elapsed to the latter.

In this paper, only early stage of sand wave formation, at which bed load shows nearly same motion as that on flat bed, is investigated. Under a simple case of non-equilibrium situation of bed load transport which is slightly shifted from equilibrium flat bed condition, the applicability of the above equations has been already verified experimentally (Nakagawa & Tsujimoto 1980a, b).

3 CHARACTERISTICS OF BED LOAD MOTION

According to the authors' study on incipient motion of bed materials (1976a), dimensionless sediment pick-up rate can be estimated against the dimensionless bed shear stress as

$$p_{s*} \equiv p_s\sqrt{d/(\sigma/\rho-1)g}$$

$$= 0.03\tau_*[1-(0.035/\tau_*)] \quad (4)$$

$$\tau_* \equiv \tau/[(\sigma-\rho)gd] \quad (5)$$

where τ: bed shear stress, σ: density of sand, ρ: density of fluid, g: gravity acceleration (see Fig.2).

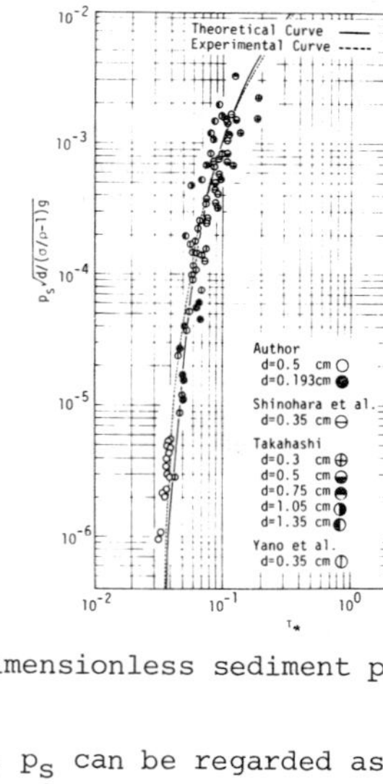

Fig.2. Dimensionless sediment pick-up rate.

Moreover, p_s can be regarded as the probability density per unit time for a particle to be dislodged, and also as the reciprocal of the mean rest period. When the rest period is represented by $\{T\}$, it is a random variable and $p_s=1/E[T]$ ($E[\cdot]$ represents the mean). The distribution of rest period expressed by the probability of exceedence is shown in Fig.3, which has been obtained by tracer tests in research programme of stochastic model for bed load transport. It is recognized from this figure that the rest period follows approximately an exponential distribution and the process of particle dislodgement can be almost regarded as a Poisson process. The probability density function of rest period can be approximated as

$$f_T(\tau) = p_s \exp(-p_s \tau) \tag{6}$$

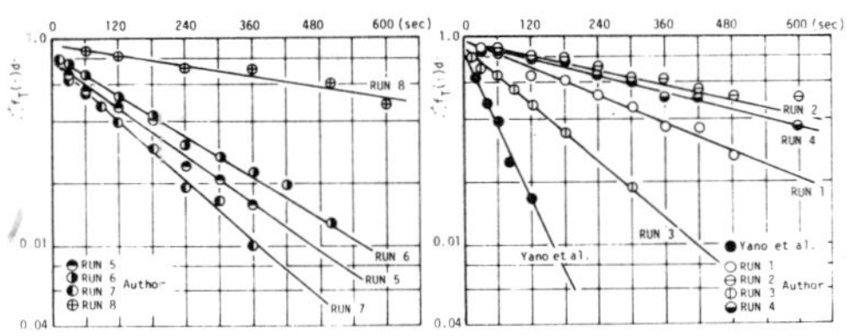

Fig.3. Distribution of rest period.

On the other hand, the distribution of step length has been also inspected by tracer tests and some of the results are shown in Fig.4, where they are expressed by the probability of exceedence. According to this figure, it is recognized that the step length on a flat bed condition follows an exponential distribution, and

$$f_X(\xi) = (1/\Lambda)\cdot\exp(-\xi/\Lambda) \tag{7}$$

where Λ is the mean step length. The relationship between dimensionless mean step length ($\lambda\equiv\Lambda/d$) and dimensionless bed shear stress τ_* is shown in Fig.5, where the experimental data directly observed and those indirectly obtained by using the stochastic dispersion or transport model are shown separately. Putting $x\to\infty$ and $p_s(x)=p_s$ (constant) in dimensionless expression of Eq. (1) (equilibrium situation), the mean step

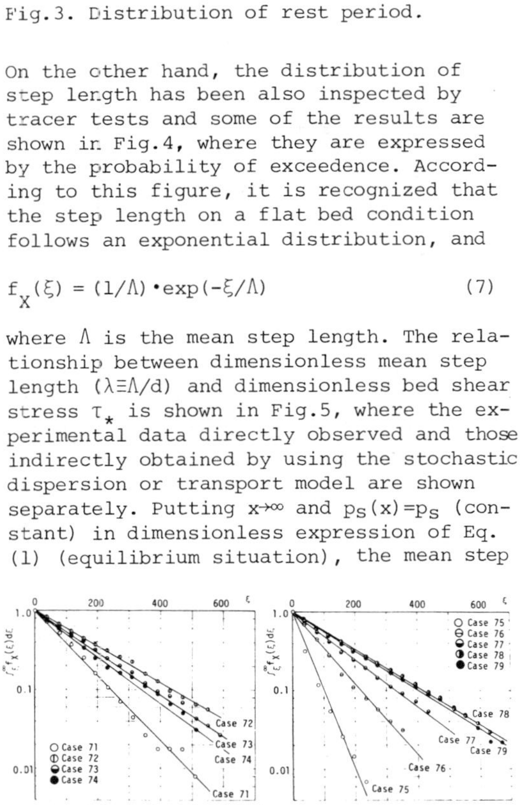

Fig.4. Distribution of step length.

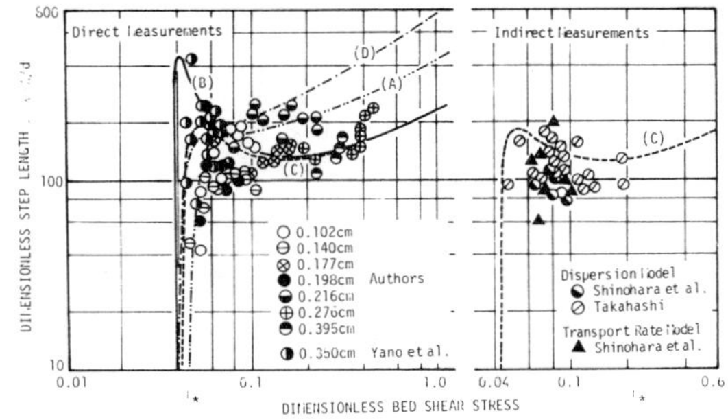

Fig.5. Mean step length of bed load.

length can be given by the following equation.

$$\lambda \equiv \Lambda/d = (A_2/A_3)\cdot q_{Be*}/p_{s*} \tag{8}$$

where q_{Be} is bed load transport rate under equilibrium flat bed condition and q_{B*} is dimensionless expression of q_B ($q_{B*}\equiv q_B/\sqrt{(\sigma/\rho-1)gd^3}$). Since q_{Be*} is generally connected to τ_* by transport formula, the relation between λ and τ_* can be obtained from Eq.(8). Four curves indicated in Fig.5 have been obtained by assuming the following formulae as the relationship between q_{Be*} and τ_*: (A) Meyer-Peter & Müller's (1948); (B) and (C) Bagnold's (1956) for $\tau_{*c}=0.04$ and 0.045 (τ_{*c}: dimensionless critical tractive force), respectively; and (D) Ashida & Michiue's (1972).

The above mentioned characteristics of bed load motion can be applied to the early stage of sand wave formation, where bed deformation is not so appreciable that individual bed load motion is not affected by bed forms. As mentioned in the preceding chapter, sediment motion over appreciably undulated bed is strongly influenced by bed geometry, and as for such a stage of sand wave formation, behaviors of individual sand waves should be noticed rather than those of individual sand particles (Nakagawa & Tsujimoto 1976b).

4 INCIPIENT PROCESS OF SAND BED IRREGULARITY

When $\{N_s(x,t)\}$ represents the number of particles dislodged from the bed surface occupied by one sand particle at x during the time interval (0, t), the following equation can be obtained by the similar consideration to Eqs.(2) and (3).

$$y(x,t)=[\int_0^\infty N_s(x-\xi,t)f_X(\xi)d\xi - N_s(x,t)]A_1 d \tag{9}$$

And, the spectrum of sand bed can be given by

$$S(\kappa, t) = \overline{F[y(x,t)] \cdot F^*[y(x,t)]} \qquad (10)$$

where $S(\kappa)$: wave number spectral density function of sand bed, and κ: angular wave number. $F[\cdot]$ indicates the Fourier transformation ($x \to \kappa$), $\cdot^*$ represents the complex conjugate and $\overline{\cdot}$ expresses an ensemble mean. Hence, the Fourier transformation of Eq.(9) yields

$$S(\kappa,t)=(A_1 d)^2 \overline{(F[f_X]-1) \cdot (F^*[f_X]-1)} \cdot S_N(\kappa,t) \qquad (11)$$

where $S_N(\kappa,t)$ is a wave number spectrum corresponding to the spatial correlation of $\{N_S(x,t)\}$. As the spatial fluctuation of $\{N_S(x,t)\}$ has not been clarified, an exponential correlation model is adopted here for convenience' sake. Namely,

$$R_{NN}(\xi,t)=\mathrm{Var}[N_S(x,t)] \cdot \exp(-\alpha_0 \xi) \qquad (12)$$

where $R_{NN}(\xi,t)$ is an autocorrelation of the spatial fluctuation of $\{N_S(x,t)\}$, α_0 is a reciprocal of relaxation distance, and $\mathrm{Var}[N_S(x,t)]$ represents the variance of $\{N_S(x,t)\}$, which is given by

$$\mathrm{Var}[N_S(x,t)]=E[\{N_S(x,t)\}^2]-\{E[N_S(x,t)]\}^2 \qquad (13)$$

$$E[\{N_S(x,t)\}^k]=\int_0^t F^{-1}[1-\phi_T) \sum_{n=0}^{\infty} n^k \phi_T^n] d\tau \qquad (14)$$

$$\phi_T \equiv F[f_T(\tau)] \qquad (15)$$

where $F[\cdot]$ represents here the Fourier transformation with respect to τ, and F^{-1} represents its inverse transformation. Under flat bed condition, $f_T(\tau)$ is given by Eq.(6) and then,

$$\mathrm{Var}[N_S(x,t)] = p_S t \qquad (16)$$

On the other hand, $f_X(\xi)$ is given by Eq.(7). Consequently, we can obtain

$$S(\kappa,t)=4\alpha_0 p_S (A_1 d)^2 \{\kappa^2/[(\kappa^2+\alpha_0^2)(\kappa^2+1/\Lambda^2)]\} \cdot t \qquad (17)$$

This indicates that $S(\kappa,t)$ can increase from zero in proportion to time and that $S(\kappa) \sim \kappa^{-2}$ in the range of higher wave number. It is noticed that it is different from $S(\kappa) \sim \kappa^{-3}$ as clarified for fully-developed sand waves (Hino 1968). Moreover, it is clarified that the variance of bed elevation, σ_y^2, increases in proportion to the elapsed time from initially flattened bed.

As a result of this analysis, the following can be expected: Random repetitions of sand dislodgements and depositions bring about an initial bed disturbance as a germ or a trigger of sand waves, and it is considerably irregular.

5 INSTABILITY ANALYSIS OF FOURIER TRANSFORMED SAND BED

The preceding analysis has indicated that initial bed disturbance appears due to random motion of bed materials, and it must depend upon bed instability mechanism whether this disturbance grows up to sand waves or not. Since an initial bed disturbance is considerably random, it is significant to conduct a bed instability analysis for Fourier transformed version. In this stage of sand wave development, the flow is perturbed by bed disturbance and sediment pick-up rate varies spatially. Thus the interaction between the flow over disturbed bed and the bed load motion under non-equilibrium state plays an important role for this stage.

As a flow model over a disturbed bed, the following resistance formula is introduced, at first.

$$\tau(x)=\beta\rho U^2 [1-\alpha(\partial h/\partial x)] \qquad (18)$$

where β: a kind of resistance coefficient, α: a correction factor, U: local flow velocity and h: local flow depth. When the spatial fluctuation of bed shear stress is represented by

$$\tau(x)/\tau_0 = 1+r_\tau \psi_\tau(x) \qquad (19)$$

the following equation can be obtained by Fourier transformation of Eqs.(2) and (3).

$$\partial F[y(x)]/\partial t=A_1 d r_p p_{s0} F[r_\tau \psi_\tau(x)]\{F[f_X]-1\} \qquad (20)$$

where the subscript 0 represents the value corresponding to undisturbed state, and

$$r_p \equiv \partial p_{s*}/\partial \tau_* \big|_{\tau_{*0}} \qquad (21)$$

This is given by Eq.(4).

Applying Eq.(10) to this case, the spectral evolution due to instability mechanism is given as

$$S(\kappa,t)=S(\kappa,0) \cdot \exp\{[(2r_p A_1/\lambda)] p_{s0} t \cdot \Gamma^*(\kappa h_0)\} \qquad (22)$$

where

$$\Gamma^*(\kappa h_0)=-\{(\kappa\Lambda)^2/[1+(\kappa\Lambda)^2]\}[\alpha(R_*-1)+2\kappa\Lambda F_*] \qquad (23)$$

$$u(x) \equiv \kappa F_* \cdot y(x), \quad \xi(x) \equiv R_* \cdot y(x) \qquad (24)$$

and, $u(x)$ is the perturbation of flow velocity and $\xi(x)$ represents the water surface fluctuation. $\Gamma^*(\kappa h_0)$ can be regarded as a dimensionless growth rate of sand bed spectrum for each wave number. When a potential flow theory is applied to estimate the local flow velocity and the depth,

$$F_*=(1-Fr^2\kappa h_0\tanh\kappa h_0)/(\tanh\kappa h_0-Fr^2\kappa h_0) \quad (25)$$

$$R_*=Fr^2\kappa h_0\,\mathrm{sech}\,\kappa h_0/(Fr^2\kappa h_0-\tanh\kappa h_0) \quad (26)$$

where Fr is the Froude number for undisturbed flow ($Fr\equiv U_0/\sqrt{gh_0}$). By substituting Eqs. (25) and (26) into Eq.(23), the spectral evolution for each wave number can be predicted. This type of spectral evolution has a feature that the growth rate becomes infinitive at the wave number which satisfies the following condition.

$$Fr^2=\tanh\kappa h_0/\kappa h_0 \quad (27)$$

The above condition is well known as Airy's relation and it gives the critical Froude number of flow over wavy bed. This fact leads us to the intuition that the spectral density for this wave number may increase extraordinarily, and this is consistent to the experimental observation that regular sand waves appear at early stage of sand wave development from initially flattened bed.

The calculated relation between Γ^* and κh_0 is shown in Fig.6, where the followings are assumed:

$$\Lambda/h_0=1.0; \quad \alpha Fr^2=1.2 \quad (28)$$

These assumptions lead to an identical result with previous ones obtained by instability analysis for regular bed waves (Hayashi 1970).

By the analysis explained in this chapter, it is concluded that the instability mechanism may play a role for wave selection process, and that quite random initial bed disturbance caused by the mechanism explained in Chapter 4 grows up to one like sand waves by the mechanism mentioned in this chapter.

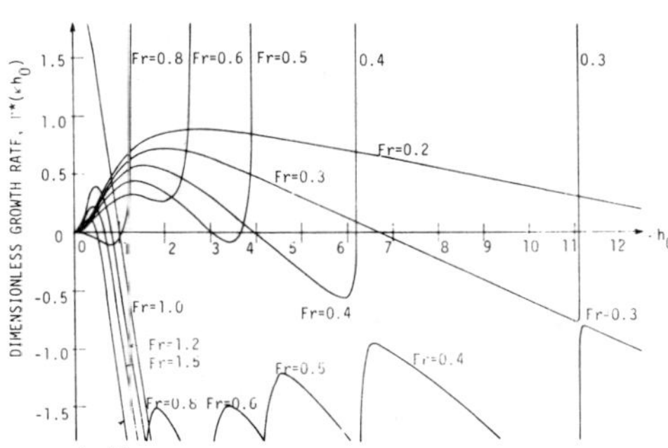

Fig.6. Spectral evoluting rate obtained by instability analysis.

6 EXPERIMENTAL INVESTIGATION ON SAND WAVE DEVELOPMENT FROM AN INITIALLY FLATTENED BED

Some experiments were conducted to inspect the mechanism of sand wave formation from an initially flattened bed. The experimental flume was 9m long and 33cm wide, and three kinds of sands were used as bed materials. The experimental conditions are showin in Table 1, where the hydraulic conditions were measured at 5min. after the flow was introduced to the alluvial bed. The time variation of longitudinal bed profile was measured by a super-sonic sounder which moved at the constant speed (30cm/sec) along the experimental flume, and the time variation of wave number spectrum of bed surface was calculated by M.E.M. (maximum entropy method).

Table 1. Experimental conditions.

RUN	i	d(cm)	h(cm)	U(cm/s)	Fr	τ_*	Re_*	h/d
A-2	0.00200	0.042	5.55	26.81	0.364	0.165	13.9	132.1
A-3	0.00200	0.042	8.76	32.75	0.353	0.261	17.4	208.6
A-5	0.00333	0.042	4.74	36.01	0.528	0.233	16.5	112.9
A-6	0.00333	0.042	7.32	39.33	0.464	0.359	20.5	174.3
A-7	0.00333	0.042	9.50	43.40	0.480	0.467	23.3	226.2
B-1	0.00200	0.050	4.97	35.81	0.513	0.128	14.2	99.4
B-2	0.00200	0.050	5.88	42.05	0.554	0.152	15.4	117.6
B-3	0.00200	0.050	10.50	50.29	0.496	0.271	20.6	210.0
C-1	0.00200	0.070	7.0	39.6	0.478	0.124	25.9	100.0
C-2	0.00200	0.070	10.0	46.4	0.469	0.177	31.0	142.9
C-6	0.00125	0.070	11.5	51.5	0.485	0.128	26.3	164.3
C-7	0.00125	0.070	7.0	50.1	0.605	0.207	33.5	100.0
C-9	0.00333	0.070	12.5	51.7	0.467	0.370	44.7	178.6

N.B. i: bed slope, Re_*: grain-size Reynolds number.

The time variation of standard deviation of bed elevation is shown in Fig.7, and the mean wave height and length, which have been calculated according to the definition by crest-to-crest method, are shown in Fig.8. The standard deviation of bed elevation and the mean wave height increase in proportion to the square root of the elapsed time, and the mean wave length also increase in the same manner after a definite period during it keeps constant value.

The tendency of increasing of σ_y is consistent to the model described in Chapter 4. Moreover, the arguments in Chapters 4 and 5 that the wave length does not change are justified by the experimental results at initial stage.

An example of spectral evolution of sand waves is shown in Fig.9. According to this figure, at the earliest stage the spectral density function is a gradually decreasing function in higher wave number range (see the result of 30sec.), and the next stage an appreciable spectral peak is observed (see the result of 2min.). This spectral peak becomes weakened with time and the wave number corresponding to the peak is gradually shifted to the lower wave number range. The initially appeared spectral peak must correspond to the infinitive growth rate at the wave number which satisfies Eq.(27) as expected by the analysis in Chapter 5.

The time variation of the wave number corresponding to the spectral peak is shown in Fig.10, and the initial and the final values are inspected. As to the former, Fig.11 is obtained, where the curve corresponding to Eq.(27) is also indicated. Meanwhile, as to the latter, Fig.12 is obtained, and this may correspond to the so-called critical wave number of the "-3 power law" (Hino 1968, Nakagawa & Tsujimoto 1979a).

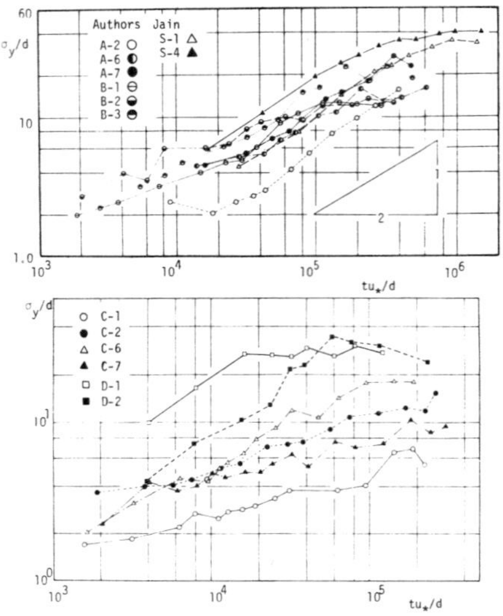

Fig.7. Time variation of standard deviation of bed elevation.

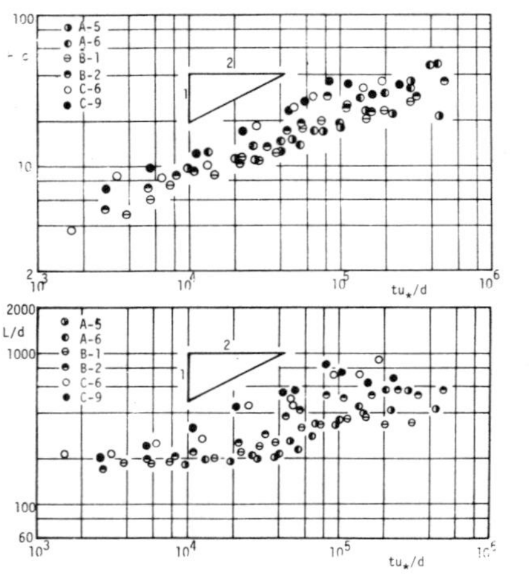

Fig.8. Time variation of mean wave height and wave length.

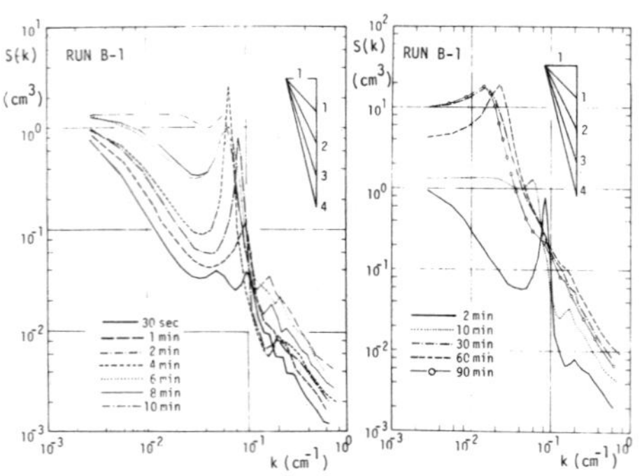

Fig.9. An example of spectral evolution of sand waves.

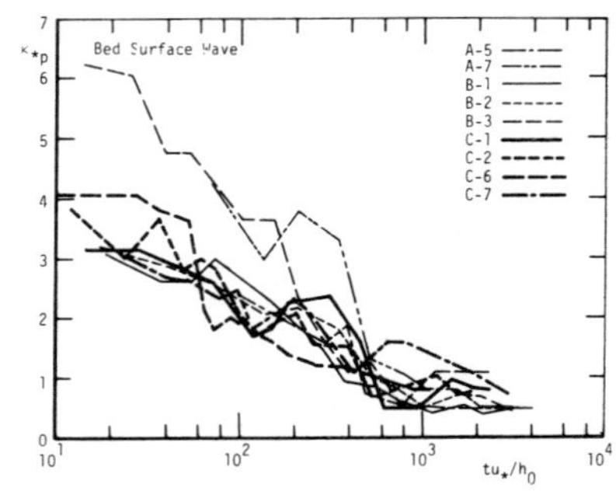

Fig.10. Time variation of wave number corresponding to spectral peak.

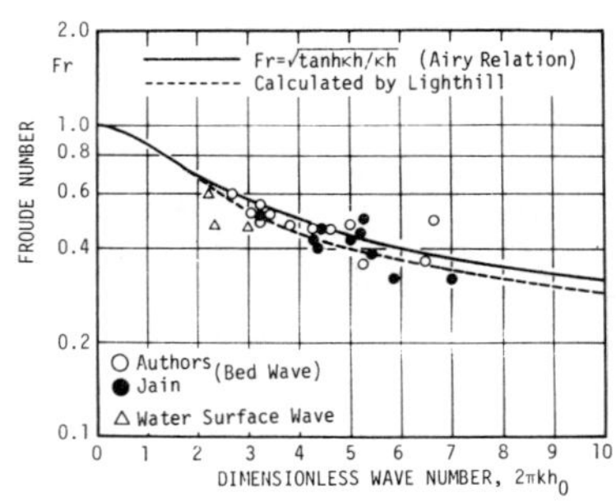

Fig.11. Wave number corresponding to initial spectral peak of sand waves.

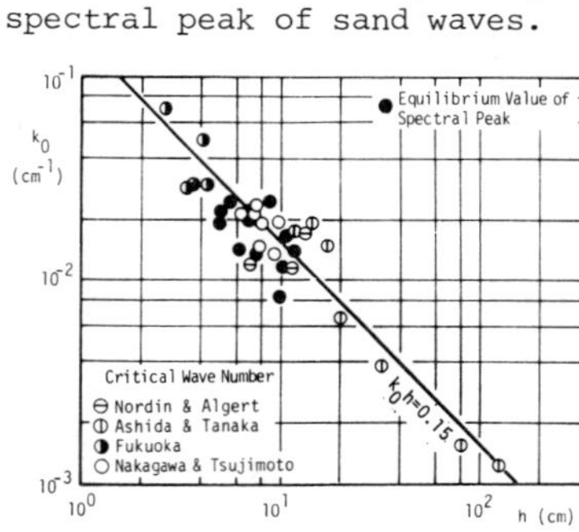

Fig.12. Equilibrium wave number corresponding to spectral peak.

The shift of the spectral peak may be caused by variance cascade process from higher to lower range of wave number, and an elementary event of this phenomenon must depend upon the behavior of individual sand waves such as unification among them. In such a stage, the effect of bed instability mechanism is hardly recognized, and sediment transport is strongly influenced by bed geometry and the growth of sand waves is followed by similar increasings of wave height and length. The above mentioned characteristics are recognized for all the experiments, and other some examples of spectral evolution of sand waves are shown in Fig.13. In these experimental runs, the spectral evolutions of water surface accompanied with sand wave development have also inspected as shown in Fig.14, where an appreciable spectral peak is observed for each run. The wave number of water surface corresponding to the spectral peak does not change with time, and this also satisfies Eq.(27) (see Fig.11), which is an essential characteristic relation of flow. Hence, the initial peak of sand wave spectrum is caused by the characteristics of water surface through the mechanism of sand bed instability. However, sand bed is deformed thereafter through variance cascade mechanism due to behaviors of individual sand waves,

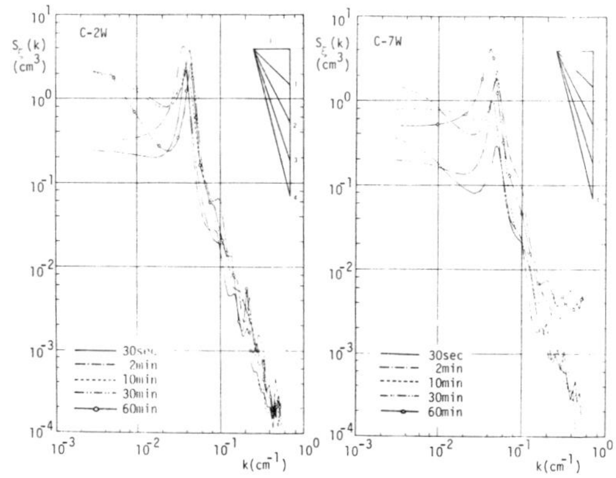

Fig.14. Examples of time variation of wave number spectra of water surface.

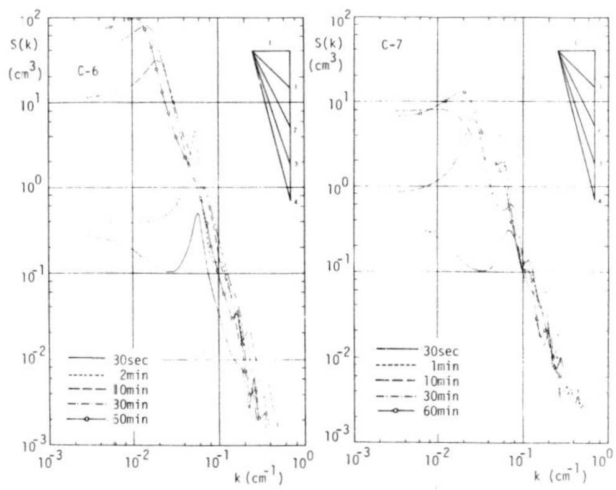

Fig.13. Examples of spectral evolution of sand waves.

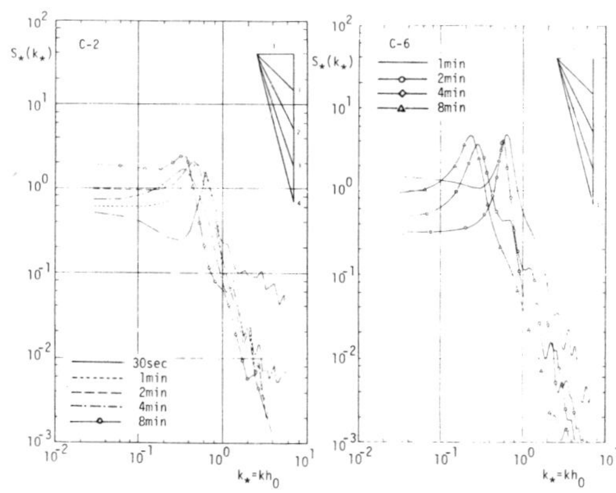

Fig.15. Normalized spectral evolution of sand waves.

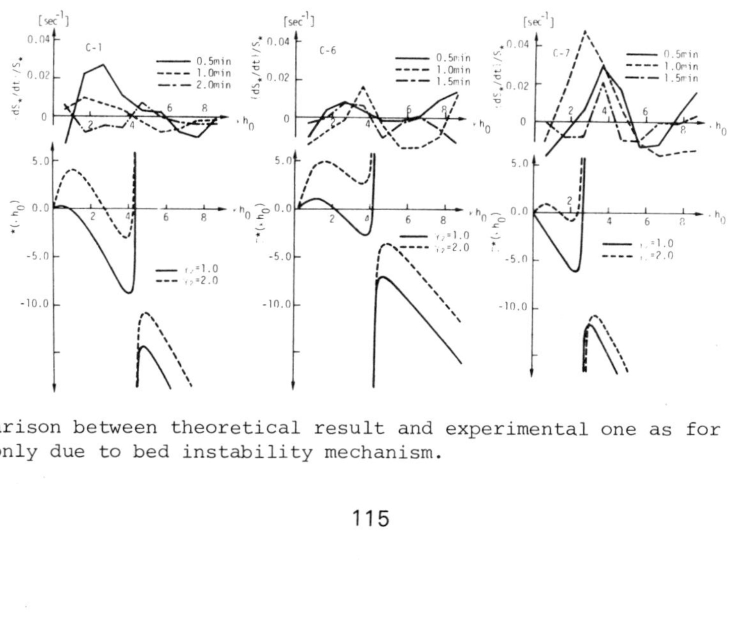

Fig.16. Comparison between theoretical result and experimental one as for spectral evoluting rate only due to bed instability mechanism.

and then the spectral peak shifts to the lower wave number range.

By the way, the wave selection process by instability mechanism will be inspected. The spectral density for each wave number increases with time in almost whole range of wave number against the theoretical result of instability analysis for any case. The reason of this fact is that the variance production due to random repetitions of sand dislodgements and depositions is continuously supplied even in this stage of sand wave development, which is explained as an incipient process of bed irregularity. The normalized spectral evolution must be significant to inspect the time variation of spectrum only due to instability mechanism, because the time variation of variance of bed elevation increases in proportion to time and this tendency can be explained only by the variance production model mentioned in Chapter 4. In other words, the instability is concerned only with wave selection process.

Wave number spectrum can be normalized as follows:

$$S(\kappa_*) \equiv S(\kappa)/\sigma_y^2 h_0, \quad \kappa_* \equiv \kappa h_0 \tag{29}$$

Some examples of normalized spectral evolutions of sand waves are shown in Fig.15 ($k \equiv \kappa/2\pi$: wave number). The experimental results on spectral evoluting rate for each wave number only due to instability mechanism, which are obtained from normalized spectral evolution data at a few minutes after introducing the flow to the initially flattened alluvial bed, are shown in Fig.16 to compare them with the theoretical results given by Eq.(23). On calculation of the theoretical curves, the mean step length is estimated by Eq.(8) and bed load transport formula by Meyer-Peter & Müller (see the curve (A) in Fig.5), and the value of the parameter $\gamma_2 \equiv \alpha Fr^2$ is chosen as 1.0 and 2.0. According to Fig.16, it can be recognized that the spectral evolution due to bed instability mechanism is regarded as a wave selection process and that this process can be considerably well explained by linear instability analysis for Fourier transformed sand bed derived in Chapter 5.

As an additional remark, this method can be applied to sand wave formation in a rectangular closed conduit. Examples of experimentally obtained spectral evolution of sand waves in a rectangular closed conduit are shown in Fig.17. The cross section of the experimental conduit is 31cm×7.5cm (a distance from alluvial bed to the upper wall). According to Fig.17, any spectral peaks are hardly observed. The growth rate of normalized spectral density for each wave number, which may depend only upon instability mechanism as expected in case of open channels, is shown in Fig.18. The variance production process in case of a closed conduit may be similar to that in case of an open channel, while instability mechanism must be very different each other. The theoretical result of instability analysis for Fourier transformed sand bed is

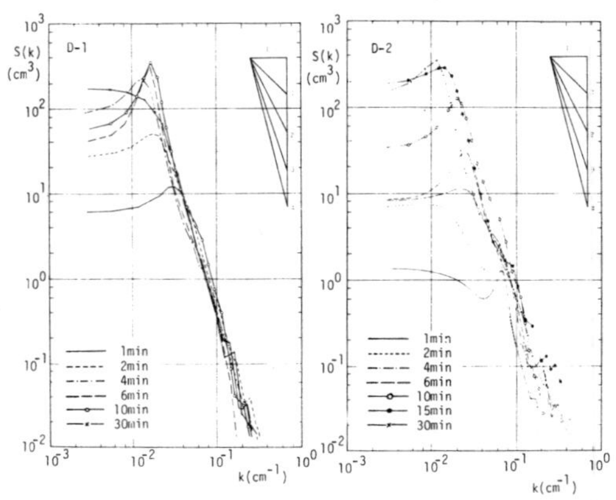

Fig.17. Examples of spectral evolution of sand waves in a rectangular closed conduit.

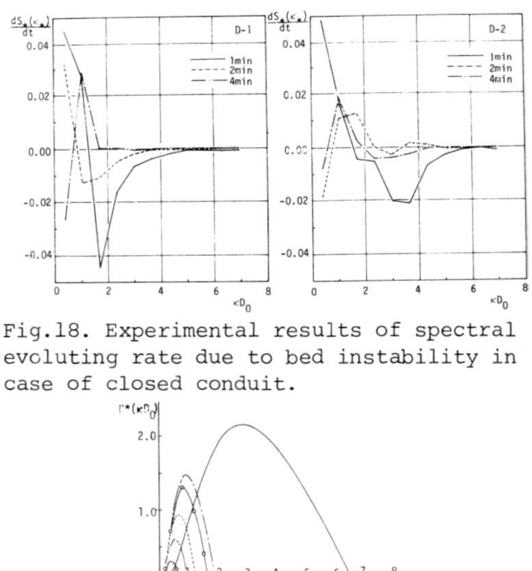

Fig.18. Experimental results of spectral evoluting rate due to bed instability in case of closed conduit.

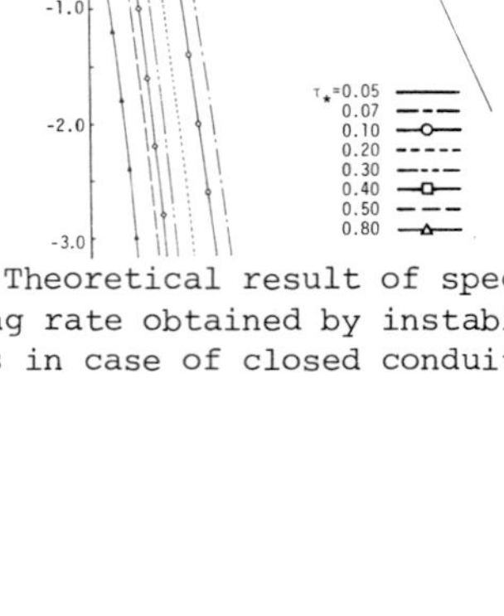

Fig.19. Theoretical result of spectral evoluting rate obtained by instability analysis in case of closed conduit.

shown in Fig.19, which is given by Eq.(23). In case of a rectangular closed conduit, F_* and R_* are given as follows also applying a potential flow theory.

$$F_*=\coth \kappa D_0, \quad R_*=0 \tag{30}$$

where D_0 is the distance from undisturbed sand bed to the upper wall of the flume and is introduced instead of h_0. When it is assumed that $\alpha=6.0$, the theoretical and the experimental results are qualitatively well consistent as shown in Figs.18 and 19.

7 CONCLUSION

The main results of this study are summarized as follows:

Random repetitions of sand dislodgements and depositions cause an initial bed surface disturbance as a germ of sand waves, and it continuously supplies the variance production of bed elevation. Meanwhile, it is due to the interaction between flow properties over a disturbed bed and non-equilibrium bed load transport whether such an initial bed surface irregularity grows up to sand waves or not, and it must depend upon bed instability mechanism. An instability analysis of Fourier transformed version is significant and the spectral evolution of sand bed for each wave number can be estimated by this analysis. In this analysis, bed load motion is also represented by a model constituted of sediment pick-up rate and step length, where essential characteristics of bed load transport process are preserved. One of the characteristics of the results of instability analysis is that there exists a wave number at which the spectral density increases extraordinarily with time, and it is consistent to the fact that comparatively regular waves are observed in the early stage of sand wave development in the laboratory. Moreover, it is expected that bed instability mechanism plays as wave selection process. The above mentioned theoretical results have been compared with some flume experiments as for sand wave development from an initially flattened bed, and most of theoretically expected properties have been verified at least qualitatively. The problems to be investigated in future are to make more accurate quantitative prediction of the phenomena possible and to clarify the all stage of sand wave development. The latter will be available to predict the behaviors of sand waves and subsequent flow resistance under unsteady conditions.

REFERENCES

Ashida, K. & M.Michiue 1972. Study on hydraulic resistance and bed-load transport rate in alluvial streams. Proc. JSCE, No.206:59-69, (in Japanese).

Bagnold, R.A. 1956. The flow of cohesionless grains in fluids. Phil.Trans. Royal Society of London, A-249, No.964.

Einstein, H.A. 1937. Der Geschiebetrieb als Wahrscheinlichkeitsproblem. Mitteilung der Versuchsanstalt für Wasserbau an der Eidgenössische Technische Hochschule in Zurich, Verlag Rascher & Co.

Hayashi, T. 1970. Formation of dunes and antidunes in open channels. Proc. ASCE, Vol.96, HY 2:357-366.

Hino, M. 1968. Equilibrium-range spectra of sand waves formed by flowing water. Journal of Fluid Mechanics, Vol.34, Pt. 3:565-573.

Jain, S.C. & J.F.Kennedy 1974. The spectral evolution of sedimentary bed forms. Journal of Fluid Mechanics, Vol.63, Pt. 2:301-314.

Kennedy, J.F. 1963. The mechanics of dunes and antidunes in erodible-bed channels. Journal of Fluid Mechanics, Vol.16, Pt. 4:521-544.

Meyer-Peter, E. & R.Müller 1948. Formulas for bed-load transport. Proc. 2nd Congr. IAHR, Stockholm, Sweden.

Nakagawa, H. & T.Tsujimoto 1976a. On probabilistic characteristics of motion of individual sediment particles on stream beds. Proc. 2nd International Symposium on Stochastic Hydraulics, Lund, Sweden : 293-316.

Nakagawa, H. & T.Tsujimoto 1976b. Experimental study on evolution of sand waves from an initially flattened bed. Annuals, DPRI, Kyoto Univ., No.19B-2:289-309, (in Japanese).

Nakagawa, H. & T.Tsujimoto 1979a. Study on statistical modeling of bed configurations in lower regimes. Proc. JSCE, No.290:79-87, (in Japanese).

Nakagawa, H. & T.Tsujimoto 1979b. On generalized stochastic model for bed load transport and sediment dispersion. Proc. JSCE, No.291:73-83, (in Japanese).

Nakagawa, H. & T.Tsujimoto 1980a. Stochastic study on origin of small scale bed forms related to probabilistic characteristics of bed load movements. Proc. 3rd International Symposium on Stochastic Hydraulics, Tokyo, Japan:359-370.

Nakagawa, H. & T.Tsujimoto 1980b. Sand bed instability due to bed load motion. Proc. ASCE, Vol.106, HY 12:2029-2051.

Nakagawa, H. & T.Tsujimoto 1981. Lag distance or phase shift as a cause of sand bed instability. Proc. JSCE, No.305:139-142, (in Japanese).

Euromech 156: Mechanics of Sediment Transport / Istanbul / 12-14 July 1982

The formation of dunes in open channel flow on an initially flattened erodible bed

B.DE JONG
Twente University of Technology, Enschede, Netherlands

ABSTRACT: A triggering mechanism is proposed for the formation of dunes on an initially flattened erodible bed based on the generation of internal waves in a flow which is horizontally stratified with respect to its velocity and sediment concentration. It is proposed that the free internal waves generate the dunes in the initial phase of the evolution of the sand wave bed. A numerical example has been given to demonstrate the effectiveness of the concept.

1 INTRODUCTION

The last two decades a considerable amount of research has been performed on the formation of sand wave patterns in erodible channels. In most theoretical studies the stability of the sand bed is considered by using a potential flow model or a rotational flow model, see e.g. Kennedy (1963), Engelund (1970) and a more recent study of Richards (1980). A review of these approaches is found in Reynolds (1976). After studying the papers in this field it becomes readily clear that many questions are still open and extensive further research has to be done to answer these in a satisfactory way. One of the interesting questions concerns the triggering mechanism responsible for the formation of ripples and dunes when the flow is started with a flat bed. As discussed by Richards (1980) it is unlikely that the "bursts of high shear stress" are the leading mechanism in this process as has been suggested by Williams & Kemp (1971).

In our paper a mechanism for the generation of sand waves is proposed based on a continuous horizontal stratification of the velocity and the density which may be assumed for the sediment-fluid mixture flowing with low Froude number. From other fields of physics we know that these flow conditions are liable to the development of internal waves. These waves are a very common phenomenon in meteorology and studied extensively in work on this subject, see e.g. the books of Lighthill (1978) and Turner (1979) and the papers they are referring to, especially those of Sawyer (1960) and Scorer (1949).

In section 2 the mechanism of the generation of dunes is formulated in relation to the general properties of internal waves. We further explain in this section how internal waves generate dunes and ripples with specific wavenumbers which are responsible for the familiar peaks in the bed-wave spectrum as observed in experiments. In section 3 we derive the boundary value problem describing the internal wave motions in a steady stratified two-dimensional open channel flow. In section 4 we show that for some well-chosen profiles for the density and the velocity, theory and experiments fit well together.

2 THE PHYSICAL BACKGROUND OF THE GENERATION OF DUNES AND RIPPLES

We consider a two-dimensional open channel with a loose boundary of erodible material in which water flows under steady conditions. We consider the flow to have a low Froude number and to be non-turbulent. Due to interaction of the flow with the bed material sediment particles will be suspended and deposited. For that reason the flow can be conceived as horizontally stratified both with respect to its velocity and its sediment concentration. It is clear that due to the presence of the sediment the fluid-sediment mixture behaves dynamically as a fluid with a higher density than the pure fluid. When the accelerations in the fluid are not too large and, consequently, the velocities of the sediment particles are approximately equal to the velocity of the surrounding fluid, the fluid-sediment mixture will behave gravitationally approximately as a fluid which has locally the same density as the local density of the fluid-sediment mixture as calculated in a proportional way from the sediment concentration and the densities of the fluid and the sediment particles. As a result we have a horizontally stratified flow in which both the density and the velocity are a function of the depth. Under these circumstances internal waves may be generated which have been extensively studied in relation to meteorological phenomena, see e.g. Turner (1979), Sawyer (1960) and Scorer (1949). An important item discussed in the literature on this subject is the generation of so-called lee-waves by an obstacle in a stratified wind, e.g. a mountain ridge at the bottom of the atmosphere. These waves are in the lee of the mountain ridge and are steady with respect to a fixed coordinate axis. In studying the leewaves it is important to determine the free waves of the flow. These are the waves corresponding with the natural frequencies of the system. These waves can exist with the obstacle removed and are also steady relative to fixed coordinate axes which implies that the upstream component of the phase velocity equals the fluid velocity. The group velocity (which for internal waves has a direction perpendicular to the phase velocity)has an upstream component smaller than the fluid velocity so that energy is being swept downstream, in the lee direction. Whether or not free waves can exist in a channel flow depends on the nature of the stratification. It has been shown for example that in a channel with a rigid boundary above and below only a finite number of wave modes can be generated, Turner (1979). In open channel flow with a loose erodible bed which is initially flat we may expect that for certain stratifications free wave modes may be generated, for example triggered by small irregularities at the bottom or at the free surface. These stationary free waves have locally upward and downward velocity components. On places where the waves generate a downward velocity component the net downward velocity of the sediment is higher than on the places where this velocity component has an upward direction. As a result the net deposition of sediment on the first mentioned places is larger than on the last mentioned. In this way dunes will be formed with the same wave numbers as those of the free waves. In experimental work, where the evolution of the bed wave spectrum is analysed under steady flow conditions starting from a flat bed, these effects of the free waves are observed as sharp peaks in the bed wave spectrum in the initial phase of the evolution, see Jain & Kennedy (1971). However, after some time, the density and velocity stratification will change due to an increase of the bottom roughness. Consequently, the wave numbers of the free modes will also change. In experiments this phenomenon is observed as a shift in the peaks of the sand wave spectrum.

As further points of interest we mention that the wave number components of the bed wave spectrum unequal to the free modes of the existing stratification will generate forced internal waves, while there may also be an exchange of energy between various internal wave modes due to non-linearities. Turner has noted that even weakly non-linear processes may produce these phenomena. Another point of interest is that imposed disturbances at the free surface, e.g. wind waves, can be very effective generators of internal waves, see Turner (1979). These last-mentioned items may be responsible for the fact that in the long run the bed wave spectrum becomes more smooth.

We, finally, observe that internal waves are essentially rotational phenomena. Consequently, three-dimensional velocity components are produced in the fluid flow which generate the familiar two-dimensional nature of the dune profiles. It will be clear that these rotational phenomena become more important in the long run when the vertical velocity gradient of the undisturbed flow increases due to an increase of the bottom roughness.

In this paper we consider only flows with low Froude numbers. For higher Froude numbers non-linearities may produce instabilities of the internal waves which generate turbulence, see Turner (1979).

3. INTERNAL WAVES IN A TWO-DIMENSIONAL STRATIFIED CHANNEL FLOW

In this section the boundary value problem governing the internal wave motions will be formulated. The method of solution will be indicated together with the properties of some solutions.

3.1 Equations of motions

We consider a two-dimensional horizontal channel which extends from $x = -\infty$ to $x = +\infty$. The x-axis coincides with the mean level of the bottom. The z-axis is vertical and positive in upward direction. The undisturbed free surface of the fluid is at $z = d$. The undisturbed velocity which is in the positive x-direction has a vertical distribution given by $U = U(z)$. The undisturbed density and pressure distribution are given by $\rho_o = \rho_o(z)$ and $p_o = p_o(z)$. Neglecting viscosity we obtain for the momentum equations the familiar Euler equations of motion

$$\rho \frac{D\underline{u}}{Dt} = - \nabla p + \rho \underline{g} \quad (1)$$

in which $\rho = \rho_o + \rho_e$, $p = p_o + p_e$, $\underline{u} = (U+u, w)$ and $\underline{g} = (o, -g)$, where ρ_e and p_e denote the disturbed density and pressure, u and w the disturbed velocities in the x- and z-direction respectively and g the acceleration of gravity. For the stationary and linearized case these equations may be reduced to

$$\rho_o (U \frac{\partial u}{\partial x} + w \frac{dU}{dz}) = - \frac{\partial p_e}{\partial x} \quad (2a)$$

$$\rho_o U \frac{\partial w}{\partial x} + \frac{\partial p_e}{\partial z} + \rho_e g = o \quad (2b)$$

In the last equation the hydrostatic balance equation $\nabla p_o = \rho_o g$ has been subtracted out. For an incompressible fluid we find for the linearized form of the continuity equation

$$\frac{\partial}{\partial x} (\rho_o u) + \frac{\partial}{\partial z} (\rho_o w) = - U \frac{\partial \rho_e}{\partial x} \quad (3)$$

We further assume that the internal wave motions constitute a reversible process which means in its full form that the rate of change of pressure and the rate of change of density following a particle satisfy the equation

$$\frac{D\rho}{Dt} = \frac{1}{c^2} \frac{Dp}{Dt}$$

in which c is the local sound speed. For an incompressible fluid this equation reduces to

$$\frac{D\rho}{Dt} = o$$

which means that there is no diffusion of sediment particles when a fluid element of the sediment-fluid mixture is moved from its undisturbed position to a higher or lower level. For the stationary and linearized case this equation reduces to

$$U \frac{\partial \rho_e}{\partial x} + w \frac{d\rho_o}{dz} = o$$

which we rewrite in the form

$$\frac{Ug}{\rho_o} \frac{\partial \rho_e}{\partial x} - N^2 w = o \quad (4)$$

where $N = (- \frac{g}{\rho_o} \frac{d\rho_o}{dz})^{\frac{1}{2}}$ represents the familiar Väisälä-Brunt frequency.

Eqs. (2), (3) and (4) are the fundamental equations for the internal wave motions in an incompressible non-diffusive stratified fluid.

We eliminate ρ_e from eqs. (3) and (4) and eqs. (2b) and (4). This yields respectively

$$\frac{\partial u}{\partial x} + \frac{\partial w}{\partial z} = o \quad (5)$$

$$U \frac{\partial^2 w}{\partial x^2} + \frac{1}{\rho_o} \frac{\partial^2 p_e}{\partial x \partial z} + \frac{N^2}{U} w = o \quad (6)$$

We further eliminate p_e from eqs. (2a) and (6)

$$\frac{d}{dz} (\rho_o U) \frac{\partial u}{\partial x} + \rho_o U \frac{\partial^2 u}{\partial x \partial z} + w \frac{d}{dz} (\rho_o \frac{dU}{dz}) + + \rho_o \frac{dU}{dz} \frac{\partial w}{\partial z} = \rho_o U \frac{\partial^2 w}{\partial x^2} + \frac{\rho_o N^2}{U} w \quad (7)$$

Eqs. (5) and (7) are two equations for the disturbed velocity components u and w. We satisfy eq. (5) by introducing a streamfunction ψ defined by

$$u = \frac{\partial \psi}{\partial z} \quad \text{and} \quad w = - \frac{\partial \psi}{\partial x} \quad (8)$$

By substituting these expressions in (7) we obtain the following equation for ψ

$$\frac{\partial}{\partial x} \left[\frac{\partial^2 \psi}{\partial z^2} + \frac{\partial^2 \psi}{\partial x^2} + \frac{1}{\rho_o} \frac{d\rho_o}{dz} \frac{\partial \psi}{\partial z} + \{ \frac{N^2}{U^2} - \frac{1}{\rho_o U} \frac{d\rho_o}{dz} \frac{dU}{dz} - \frac{1}{U} \frac{d^2 U}{dz^2} \} \psi \right] = o \quad (9)$$

We restrict our discussions to disturbances which are harmonic in the x-direction with wave number k. Then the x-variation can be represented by a factor e^{ikx}. In view of this we introduce a function $\psi_k = \psi_k(z)$ related to the streamfunction $\psi(x,z)$ by

$$\psi(x,z) = (\frac{\rho_{oo}}{\rho_o(z)})^{\frac{1}{2}} \psi_k(z) e^{ikx} \quad (10)$$

in which $\rho_{oo} = \rho_o(o)$ is the density at the bottom. Then eq. (9) is reduced to

$$\frac{d^2\psi_k}{dz^2} + \{ \frac{N^2}{U^2} - \frac{1}{U}\frac{d^2U}{dz^2} - \frac{1}{\rho_o U}\frac{d\rho_o}{dz}\frac{dU}{dz} + \frac{1}{4\rho_o^2}(\frac{d\rho_o}{dz})^2 - \frac{1}{2\rho_o}\frac{d^2\rho_o}{dz^2} - k^2\}\psi_k = o \quad (11)$$

Setting

$$p(z) = \frac{N^2}{U^2} - \frac{1}{U}\frac{d^2U}{dz^2} - \frac{1}{\rho_o U}\frac{d\rho_o}{dz}\frac{dU}{dz} + \frac{1}{4\rho_o^2}(\frac{d\rho_o}{dz})^2 - \frac{1}{2\rho_o}\frac{d^2\rho_o}{dz^2} \quad (12)$$

eq. (11) may be written as

$$\frac{d^2\psi_k}{dz^2} + (p(z)-k^2)\psi_k = o \quad (13)$$

3.2 Boundary condition at the bottom

For bottom elevations h = h(x) which are small as compared to the total water depth d, the linearized bottom condition $w = U(dh/dx)$ can be applied at the level z = o. Harmonic bottom elevations $h(x) = h_k e^{ikx}$ generate in the linearized case flow disturbances which are harmonic in the x-direction with wave number k. In view of eqs. (8) and (10) we find

$$U_o h_k + \psi_k = o, \quad z = o \quad (14)$$

where $U_o = U(o)$.

3.3 Boundary condition at the free surface

Since internal waves are rotational motions it is not self-evident that the familiar "free surface condition" of the theory of irrotational surface waves may be applied here. However, since the free surface is a reflecting boundary for the sediment particles in the fluid we may assume as a first approximation that the fluid-sediment mixture has a uniform density in a layer at the surface of the fluid. This means for an inviscid fluid that there is no generation of vorticity in that layer, see e.g. Turner (1979). So as a first approximation the fluid motions near the free surface may be considered as a potential flow. For this reason we derive the condition at the free surface in a similar way as for a potential flow.

Representing the surface elevations by $\eta = \eta(x,t)$ we find for the stationary kinematic condition

$$\frac{\partial\eta}{\partial x} = \frac{w}{U+u} = -\frac{\partial\psi/\partial x}{U+\partial\psi/\partial z}, \quad z = d \quad (15)$$

The stationary Bernoulli equation for constant atmospheric pressure has the form

$$g\eta + Uu + \frac{1}{2}u^2 + \frac{1}{2}w^2 = o, \quad z = d \quad (16)$$

or

$$g\eta + U\partial\psi/\partial z + \frac{1}{2}(\partial\psi/\partial z)^2 + \frac{1}{2}(\partial\psi/\partial x)^2 = o, \quad z = d \quad (17)$$

Eliminating η from (15) and (17) and linearizing yields

$$-g(\partial\psi/\partial x) + U^2(\partial^2\psi/\partial x\partial z) = o, \quad z = d \quad (18)$$

We substitute eq. (10) for ψ in which $\rho_o(z)$ is now a constant equal to the density at the surface. We obtain finally

$$-g\psi_k + U^2(d\psi_k/dz) = o, \quad z = d \quad (19)$$

3.4 Solution of the boundary value problem. Free waves and trapped waves

We rewrite eq. (13) with its boundary conditions (14) and (19) in terms of a dimensionless vertical coordinate z' = z/d where d is the depth of the fluid. To simplify the notations we delete the prime in our further discussions. The boundary value problem then becomes

$$\frac{d^2\psi_k}{dz^2} + (p(z)-k^2)\psi_k = o, \quad 0 \le z \le 1 \quad (20)$$

$$U_o h_k + \psi_k = o, \quad z = o \quad (21)$$

$$-\frac{g}{U^2}\psi_k + \frac{d\psi_k}{dz} = o, \quad z = 1 \quad (22)$$

in which k denotes the dimensionless wave number, (made dimensionless with the depth d). In terms of the dimensionless depth coordinate z the function p(z) preserves the same shape. We write

$$p(z) = \frac{N^2}{U^2} - \frac{1}{U}\frac{d^2U}{dz^2} - (\frac{1}{\rho_o}\frac{d\rho_o}{dz})(\frac{1}{U}\frac{dU}{dz}) + \frac{1}{4}(\frac{1}{\rho_o}\frac{d\rho_o}{dz})^2 - \frac{1}{2}(\frac{1}{\rho_o}\frac{d^2\rho_o}{dz^2}) \quad (23)$$

where $N^2 = -\frac{g}{\rho_o}\frac{d\rho_o}{dz}$

Consequently, the solution ψ_k as a function of the dimensionless z is determined by the dimensionless numbers

$$\frac{U}{\sqrt{g}}, \ \frac{1}{\rho_o}\frac{d\rho_o}{dz}, \ \frac{1}{U}\frac{dU}{dz}, \ \frac{1}{U}\frac{d^2U}{dz^2} \text{ and } \frac{1}{\rho_o}\frac{d^2\rho_o}{dz^2} \qquad (24)$$

In dimensional z these numbers are

$$\frac{U}{\sqrt{gd}}, \ \frac{d}{\rho_o}\frac{d\rho_o}{dz}, \ \frac{d}{U}\frac{dU}{dz}, \ \frac{d^2}{U^2}\frac{d^2U}{dz^2} \text{ and } \frac{d^2}{\rho_o}\frac{d^2\rho_o}{dz^2} \qquad (25)$$

We recognize $U/\sqrt{gd}$ as the (internal) Froude number in which U is the local undisturbed velocity. The reciprocal of the second number $(-\rho_o/d)/(d\rho_o/dz)$ is in the literature sometimes referred to as the sediment transfer coefficient. In many practical applications the major contribution to the value of p(z) comes from the first term N^2/U^2 in the right-hand side of (23). In this case the stratification of the flow is characterized by the dimensionless numbers $U/\sqrt{g}$ and $(1/\rho_o)(d\rho_o/dz)$, (or in dimensional z by $U/\sqrt{gd}$ and $(d/\rho_o)(d\rho_o/dz)$).

The solution of the boundary value problem (20), (21) and (22) in which we assumed a sinusoidally corrugated bottom is obtained by solving first the following initial value problem for some function f

$$\frac{d^2f}{dz^2} + (p(z)-k^2)f = o \qquad (26)$$

$$-\frac{g}{U^2}f + \frac{df}{dz} = o \text{ and } f = 1 \ , \ z = 1 \qquad (27)$$

The solution of the boundary value problem is

$$\psi_k(z) = -h_k U_o \frac{f(z,k)}{f(o,k)} \qquad (28)$$

in which the dependence of f on the dimensionless wave number k has been explicitly denoted. Combining with eq. (10) we find for the streamfunction

$$\psi(x,z) = -\left(\frac{\rho_{oo}}{\rho_o(z)}\right)^{\frac{1}{2}} h_k U_o \frac{f(z,k)}{f(o,k)} e^{ikx} \qquad (29)$$

In view of the linearity of the problem it is clear that a bottom profile which may be expressed as a Fourier decomposition generates a solution which is a simple superposition of solutions with the shape (29) corresponding to the various components in the bottom profile. As has been discussed by Scorer (1949) the values of k which yield f(o,k) = 0 correspond with the free waves of the fluid flow. These are the infinite solutions for the streamfunction in a flow without damping.

Another feature that will become important in our further discussions is the phenomenon of the so-called "trapped waves". As has been discussed in various texts on differential equations eq. (20) has a harmonic solution in the depth range where $p(z) > k^2$ while for the z-values where $p(z) < k^2$ only exponentially decreasing solutions can exist, (the wavy solutions are trapped in the region where $p(z) > k^2$).

4 NUMERICAL EXAMPLES AND DISCUSSION

In this section we consider two of the flow configurations which have been investigated by Jain & Kennedy (1971). They have studied in a flume, starting from a flat bed, the evolution of the bed profile for several values of the mean velocity and the fluid depth. In the initial phase of the bed wave evolution they observe two distinct peaks in the bed wave spectrum. Our procedure will be to choose the velocity and density distribution in such a way that the flow stratification has free waves for wave numbers corresponding with the observed peaks in the bed wave spectrum.

For the undisturbed velocity distribution we choose

$$U(z) = u*\left[\sqrt{1-z'} - \sqrt{1-z_o} + \ln\{(1-\sqrt{1-z'})/(1-\sqrt{1-z_o})\}\right] \qquad (30)$$

in which z' is a linear function of the dimensionless depth coordinate z

$$z' = z_o + z_{vb} + (1-z_o-z_{vb}-z_{vs})z, \ o \le z \le 1 \qquad (31)$$

This coordinate transformation has been introduced to avoid singularities in the first and second derivatives at the free surface and to allow values unequal zero at the bottom. For $z_{vb} = z_{vs} = o$ we have the Von Karman velocity profile. The profile given by eqs. (30) and (31) has much similarity with the profile suggested by Hunt (1954) which has been shown by him to give good agreement with observational data. The quantities z_o, z_{vb} and z_{vs} are parameters for which we have to choose appropriate values. For a given set of values for z_o, z_{vb} and z_{vs} the value of u^* is determined from the prescribed mean velocity in the channel.

The density distribution of the fluid-sediment mixture is derived from a formula of Rouse for the concentration C(z) of the sediment in the fluid, see e.g. Raudkivi (1967)

$$C(z) = C_a \left(\frac{1-z}{z}\frac{a}{1-a}\right)^Z \qquad (32)$$

where C_a is the concentration at the level $z = a$. The exponent Z is a positive constant. In deriving the undisturbed density distribution we first note that the relation between the concentration and the density is a linear one. Further, the density distribution appears in eq. (23) only in the combinations $(d\rho_o/dz)/\rho_o$ and $(d^2\rho_o/dz^2)/\rho_o$ which means that only the shape of the density distribution plays a role in the determination of the free wave modes. For these reasons we assume for the density distribution a shape which is similar to eq. (32)

$$\rho_o(z) = c \left\{ \frac{1-z''}{z''} \right\}^Z \qquad (33)$$

in which z" relates to the dimensionless depth coordinate z by

$$z'' = z_{db} + (1-z_{ds}-z_{db})z, \quad o \leq z \leq 1 \qquad (34)$$

This coordinate transformation has again been introduced to avoid singularities at the bottom and the free surface. The constant c has no importance for our further discussions. Similar to the parameters z_o, z_{vb} and z_{vs} in the velocity expression (30), the parameters Z, z_{db} and z_{ds} in the density expression (33) should be chosen in such a way that the free wave modes for this stratification coincide with the peaks observed in the bed wave spectrum in the initial phase of the evolution. This procedure has been carried out for the runs S-8 and S-9 described in the paper of Jain & Kennedy. In Table 1 the relevant hydraulic conditions are summarized together with the dimensionless wave numbers where the peaks in the spectrum are observed.

In table 2 we summarized the values of the velocity and density parameters and the horizontal components of the free wave modes calculated for these stratifications on the basis of the analysis of section 3. It is seen that except for the free wave modes k_1 and k_2 corresponding with the two observed peaks in the bed wave spectrum a third free wave mode may exist for the given stratification. The velocity and density distributions are given in Figures 1 and 2. Figure 3 shows the behaviour of the function p(z) appearing in eqs. (20) and (26) together with the first term N^2/U^2 in the expression (23) for p(z). It is seen that only this first term gives a substantial contribution. The other terms are negligible small. We further depicted in Figure 3 as vertical broken lines the values of k_1^2, k_2^2 and k_3^2, representing the value of the term k^2 in eq. (20) for the various values of the horizontal component of the free wave modes. It is seen that for both runs the factor $p(z)-k^2$ of eq. (20) has a positive value over the whole dept range when $k = k_1$ and $k = k_2$. This means that eq. (20) has everywhere a harmonic solution. However, for $k = k_3$ the factor $p(z)-k^2$ is positive only in narrow regions near the bottom and the free surface while in the middle region in between $p(z)-k^2$ is negative. The wave number $k = k_3$ corresponds with a trapped wave with harmonic solutions for eq. (20) in the regions near the two boundaries and a negative exponential solution in the region in between. Consequently, the free wave mode with horizontal wave number component k_3 has only significance in the above mentioned regions near the boundaries and have therefore less importance for the deposition of material than the first two waves. It is possible that Jain & Kennedy failed

Table 1. Summary of hydraulic conditions. Spectral peaks at small times,(Jain & Kennedy ,1971).

Run No	Depth d m	Mean velocity mps	Froude No F	Water Temp °C	First peak* k	Second peak* k
S-8	0.0771	0.3847	0.442	21.0	4.19	7.54
S-9	0.0756	0.3490	0.405	24.0	4.27	7.41

*$k = 2\pi d/\lambda$, where λ is the wave length of the bed wave component.

Table 2. Velocity and density parameters. Free wave modes.

Run No	Velocity parameters			Density parameters			Horizontal component of free wave modes *		
	z_o	z_{vb}	z_{vs}	Z	z_{db}	z_{ds}	k_1	k_2	k_3
S-8	0.001	0.05	0.01	0.27	0.05	0.01	4.19	7.54	9.62
S-9	0.0015	0.05	0.01	0.205	0.025	0.01	4.27	7.41	10.60

*$k_i = 2\pi d/\lambda_i$, where λ_i is the horizontal wave length component of the free wave mode.

to observe a third peak in the bed wave spectrum because the third free wave mode is less effective than the other two. However, another reason might be that the shape of $\rho_o(z)$ is less accurate near the bottom and the free surface. Expression (33) has been stated on the basis of eq. (32) which gives a zero value for the concentration at the free surface and an infinite value at the bottom. Obviously, formula (32) gives less realistic values near the bottom and the free surface. This fact has been discussed by several authors, see e.g. Hunt (1954). It appears that another more realistic shape for the density distribution should be used to answer this question in an appropriate way.

Richards defines dunes as the bed forms which depend on the whole flow depth while ripples are the bed forms depending solely on the local bed properties, (and therefore on the local sediment concentration near the bottom). According to this classification we should define the bed forms generated by the waves with horizontal wave number component k_1 and k_2 as dunes and the bed form generated by the wave with the wave number component k_3 as a ripple.

As a final remark we mention that Nordin (1971) has observed also peaks in the bed wave spectrum for a fully developed bed. For some flume experiments he observed two peaks in the lower wave number range and some additional peaks for higher wave numbers. It is tempting to identify these peaks also with free wave modes which may exist for the equilibrium stratification. It is clear that extensive research is necessary to obtain any definite answers to this speculation.

REFERENCES

Engelund, Frank 1970. Instability of erodible beds. J. Fluid Mech., vol. 42, part 2, pp 225-244.

Hunt, J.N. 1954. The turbulent transport of suspended sediment in open channels. Proc. R. Soc. London, Series A, 224, pp 322-335.

Jain, Subhash C. & Kennedy John F. 1971. The growth of sand waves. In Chao-Lin Chiu (ed.), Stochastic Hydraulics, pp 449-471. Pittsburgh: University of Pittsburgh.

Kennedy, John F. 1963. The mechanics of dunes and antidunes in erodible-bed channels. J. Fluid Mech. vol 16, part 4, pp 521-546.

Lighthill, James 1980. Waves in fluids. Cambridge: Cambridge University press.

Nordin, Carl F. 1971. Sediment transport in alluvial channels. Geological survey professional paper 562-F. Washington: United States Government printing office.

Raudkivi, A.J. 1976. Loose boundary hydraulics, (2nd edn). Oxford: Pergamon Press.

Reynolds, A.J. 1976. A decade's investigation of the stability erodible stream beds. Nordic Hydrology, 7, pp 161-180.

Richards, Kelvin J. 1980. The formation of ripples and dunes on an erodible bed. J. Fluid Mech., vol 99, part 3, pp 597-618.

Sawyer, J.S. 1960. Numerical calculation of the displacements of a stratified airstream crossing a ridge of small height. Quart. J. Roy. Met.Soc. 86, pp 326-345.

Scorer, R.S. 1949. Theory of waves in the lee of mountains. Quart. J. Roy. Met.Soc. 75, pp 417-428.

Williams, P.B. and Kemp, P.H. 1972. Initiation of ripples by artificial disturbances. J. Hyd.Div., Proc.A.S.C.E. HY6, 98, pp 1057-1070.

Turner, J.S. 1979. Boyancy effects in fluids. Cambridge: Cambridge University Press.

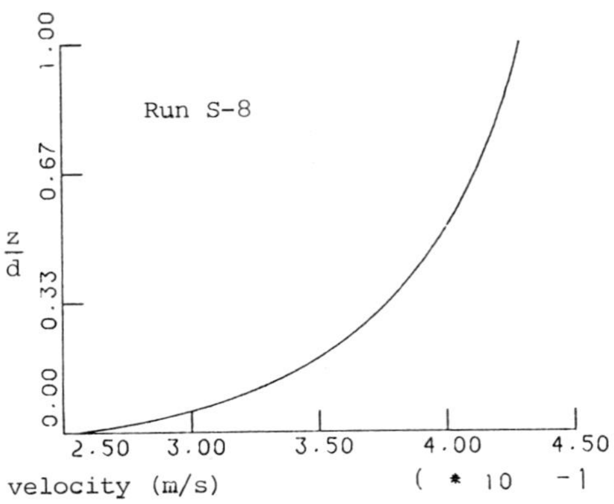

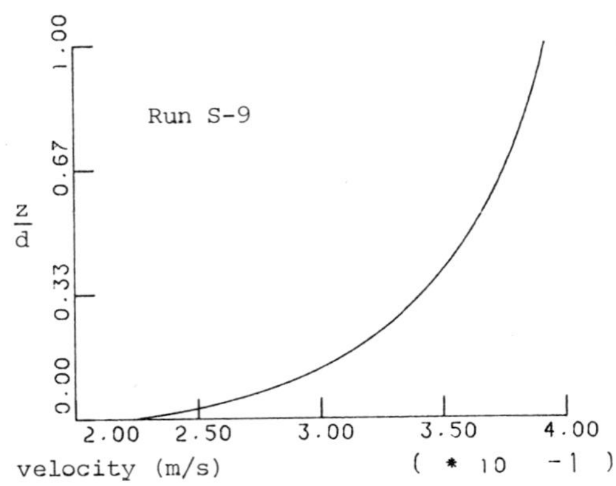

Fig. 1a. Fig. 1b.
Velocity distribution as function of dimensionless depth z/d.

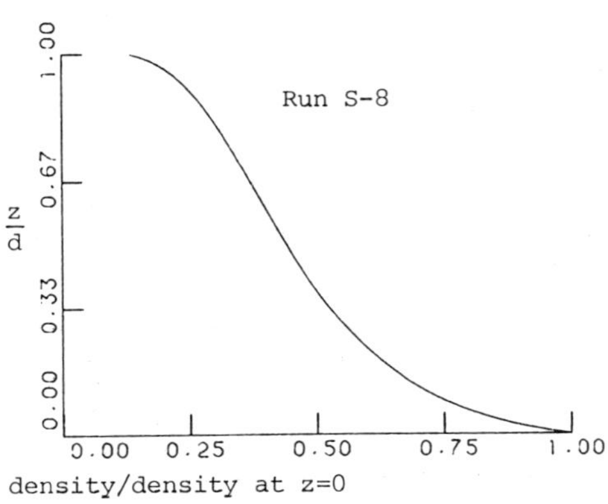

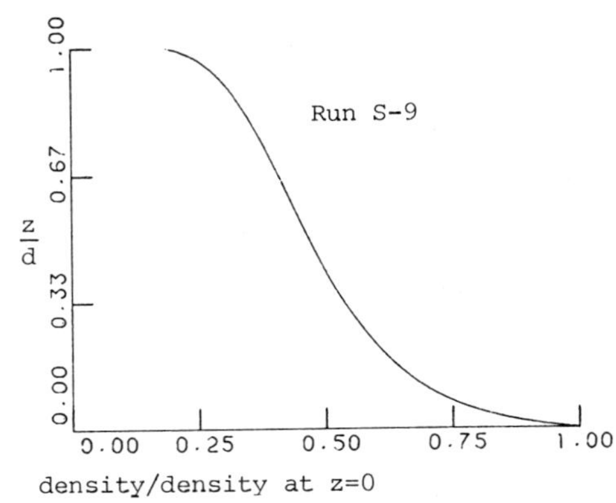

Fig. 2a. Fig. 2b.
Normalized density distribution as function of dimensionless depth z/d.

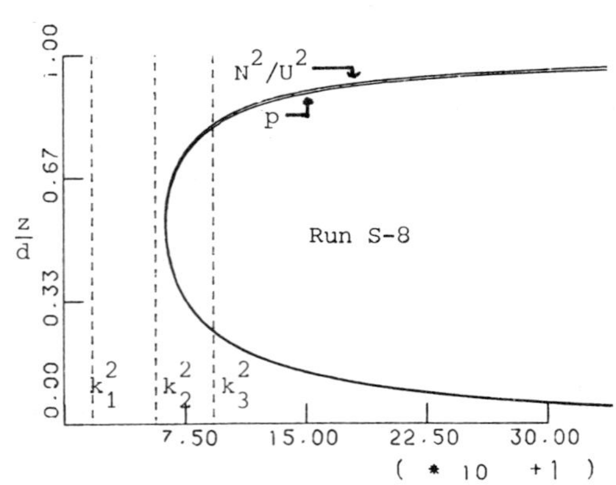

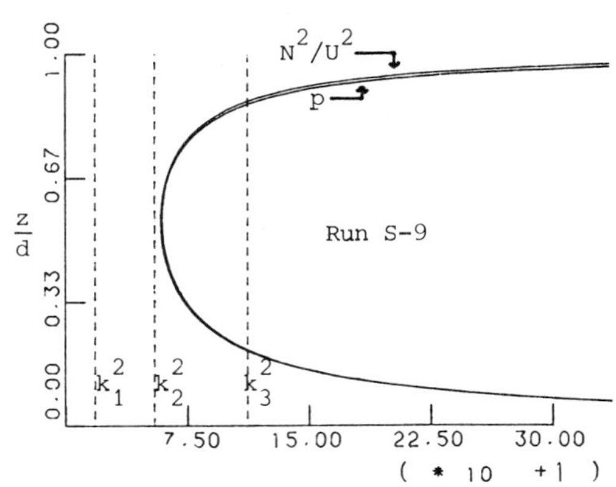

Fig. 3a. Fig. 3b.
p and N^2/U^2 of. eq.(12) as function of dimensionless depth z/d.
k_1^2, k_2^2 and k_3^2 are the squares of the free wave numbers.

Euromech 156: Mechanics of Sediment Transport / Istanbul / 12-14 July 1982

Turbulent flow over ripples and their effective roughness

K.J.RICHARDS
Institute of Oceanographic Sciences, Wormley, Godalming, UK

ABSTRACT: A numerical model of turbulent flow over ripples and the related sediment transport is described. The results from the model are used to investigate the dependency of the effective roughness length of a rippled bed on both ripple size and shape.

1 INTRODUCTION

Most measurements taken in the natural environment of flow close to a river or sea bed and the related sediment transport are restricted to a single current meter placed, say, one metre above the surface. The stress on the surface is then calculated by either assuming a logarithmic velocity profile and prescribing a roughness length or using a drag co-efficient. The presence of ripples on the bed will influence the flow close to the bed and, in particular, the surface stress distribution. The ripples will increase the roughness of the surface. The integrated shear stress over the surface will be reduced due to the form drag of the ripples. These factors will affect the transport of sediment over the bed in both the initisation of sediment movement and the transport rate.

The need to be able to parameterise these effects is very apparent. Studies have been made to relate the properties of a rippled bed to the flow measured some way above it. In particular, Kachel and Sternberg (1971) studied the movement of ripples and used their phase velocities to give the transport of sediment. It is not obvious, however, how much of the sediment is transported with the ripple and what percentage is carried over them. More recently Dyer (1980) has taken detailed flow measurements over a rippled bed in a tidal regime and finds that the roughness length varies by two orders of magnitude over a single tidal cycle.

This study uses a numerical model to investigate the steady flow over a rippled surface and the related sediment transport. Initial investigations using a similar model were carried out by Taylor et al (1976). The model has been used previously by Richards and Taylor (1981) to study the dynamics of larger scale dunes where the depth of water is important. The importance of modelling the turbulence in the flow to give the development of an erodible bed has been demonstrated by the stability analysis of Richards (1980). The results presented here are designed to indicate some of the integral properties of a rippled bed, in particular the effective roughness of the bed and the total sediment transport over the ripples.

2 MODEL AND TURBULENCE CLOSURE

The flow considered is that of a uniform flow over a periodic, two-dimensional wavy bed of wavelength L driven by a constant stress, ρu^{*2}, at its upper boundary (Figure 1). The upper boundary is taken sufficiently high so as not to influence the flow close to the bed. The flow is assumed to be hydraulically rough so that the drag on the surface can be characterised by a local roughness length, z_o, taken in this study to be constant over the ripple. In Cartesian coordinates the equations of motion are

$$U\frac{\partial U}{\partial x} + W\frac{\partial U}{\partial z} = -\frac{1}{\rho}\frac{\partial p}{\partial x} - \frac{\partial}{\partial x}\overline{u'^2} - \frac{\partial}{\partial z}\overline{u'w'} \quad (1)$$

$$U\frac{\partial W}{\partial x} + W\frac{\partial W}{\partial z} = -\frac{1}{\rho}\frac{\partial p}{\partial z} - \frac{\partial}{\partial x}\overline{u'w'} - \frac{\partial}{\partial z}\overline{w'^2} \quad (2)$$

$$\frac{\partial U}{\partial x} + \frac{\partial W}{\partial z} = 0 \quad (3)$$

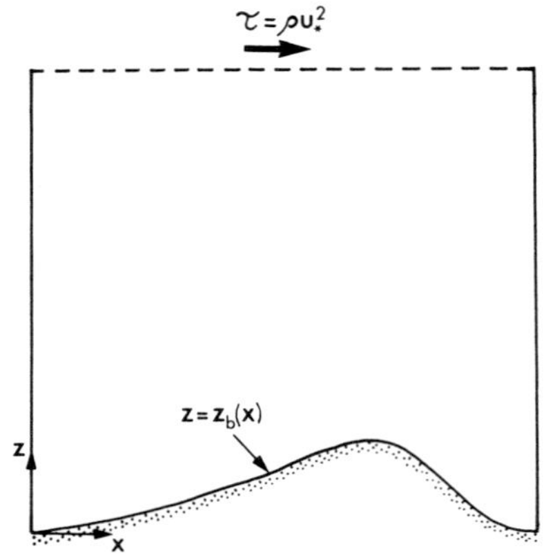

Fig. 1 The flow considered

where the coordinate system is defined in Figure 1, U, W are velocity components, p the pressure and g the acceleration due to gravity.

The Reynolds stress terms $\overline{u'w'}$ etc. are modelled using an isotropic eddy viscosity approach, i.e.

$$-\overline{u_i u_j} + \frac{2}{3} k\delta_{ij} = K\left(\frac{\partial u_i}{\partial x_j} + \frac{\partial u_j}{\partial x_i}\right) \qquad (4)$$

where k is the mean turbulent kinetic energy. The eddy viscosity K is defined by

$$K = (\lambda k)^{1/2} \ell \qquad (5)$$

where $1/\lambda$ is the equilibrium value of k/u^{*2} and u^* the friction velocity. The mixing length ℓ is a prescribed function of position, it being proportional to the normal distance from the surface. The turbulent kinetic energy is determined by a transport equation.

Britter, Hunt and Richards (1981) have investigated the applicability of turbulence models to flow over topography. Their findings can be summarised as the following. To calculate the changes in the mean flow or turbulent stresses close to the ground we can apply an eddy viscosity approach. The mean flow away from the surface can be calculated using inviscid theory, while rapid distortion theory is needed to estimate the changes in turbulence structure. We shall restrict ourselves to study the bed load only so that the use of an eddy viscosity is appropriate. It should be noted that if suspended sediment is modelled then care is needed in modelling the turbulence and Equation 4 is inadequate.

The equations are solved numerically using a surface following coordinate system. Full details are given in Richards and Taylor (1981).

The model predicts the surface distribution of shear stress over the wave. We shall assume that the sediment is transported as bed-load only. The sediment transport can then be related to the local bed shear stress. For a bed of local slope α Bagnold's (1966) formula for bed load transport, q_b is

$$q_b = \frac{8.5}{\gamma g} \frac{e_b \tau^{3/2}}{(\tan\phi + \tan\alpha)\cos\alpha} \qquad (6)$$

where τ is the local bed shear stress, e_b is an efficiency factor of order 0.1, ϕ the angle of frictional resistence and $\gamma = (\rho s - \rho)/\rho$.

The growth rate of the bed, $\partial zb/\partial t$, can be related to the local bed-load transport using the sediment continuity equation

$$\frac{\partial qb}{\partial x} = -(1-n)\frac{\partial zb}{\partial t} \qquad (7)$$

where n is the porosity of the bed. Results will be presented of the non-dimensional erosion rate

$$\sigma = \frac{-zo\,\alpha g(1-n)}{e_b\, u^{*3}} \frac{\partial z_b}{\partial t} \qquad (8)$$

3 AN EXAMPLE

The results of the model for the flow over a particular ripple are shown in Figure 2. The ripple is asymmetric with L/z_o = 10,000 and ak = 0.16, where k is the wavenumber of the ripple. With z_o = 0.005 cm this gives L = 50 cm and a crest to trough height of 2.5 cm. The maximum slope on the lee side of the ripple is 0.32. The bed shear stress is shown in Figure 2a. The stress reaches a maximum at the crest of $\tau/\rho u^{*2}$ = 1.65 and is close to separating on the lee side. The erosion rate curve, taking $\tan\phi$ = 0.6, (figure 2b) shows zero erosion at the crest with erosion of the upstream slope and deposition on the lee slope of the ripple.

For the ripple to propagate forward without change in shape with a velocity c we require

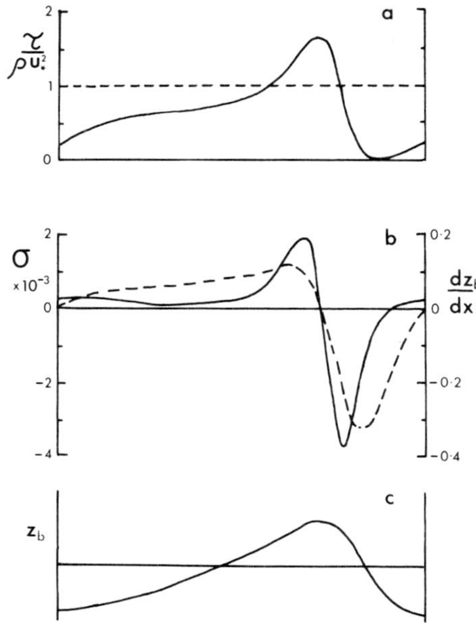

Figure 2. Bed shear stress and erosion rate for flow over an asymmetric wave; $L/z_o = 10{,}000$, $a/z_o = 250$. (a) Shear stress distribution; (b) Erosion rate (dashed curve local bed slope); (c) Wave shape (vertical scale x 5).

$$\frac{\partial cb}{\partial x} = c\ (1-n)\ \frac{\partial zb}{\partial x} \qquad (9)$$

The bed slope $\partial z_b/\partial x$ is shown in Figure 2b. Comparing this with the erosion rate we find that the ripple will become more asymmetric with an increase in the lee slope and consequent separation of the flow. Taking $u^* = 3\ cm\ s^{-1}$ the propagation speed of the ripple can be estimated to be approximately $2.5 \times 10^{-2}\ cm\ s^{-1}$. This compares with observed ripple velocities of between $0.25 \times 10^{-2}\ cm\ s^{-1}$ and $2.8 \times 10^{-2}\ cm\ s^{-1}$ (Kachel and Sternberg, 1971) for friction velocities between 2.5 and $4.5\ cm\ s^{-1}$.

It should be noted that Equation 6 is applicable only when $\tau >> \tau_{cr}$. Clearly this is not so when the flow is close to separating and a more sophisticated transport model will need to be used in future studies.

The maximum and minimum shear stress are plotted as a function of ak in Figure 3 for the ripple shape used above. This shows a maximum in τ max at ak = 0.16 and the flow separating a little above this value. As the threshold for sediment movement will be dependent on τ max we find that the critical shear stress will be reduced when ripples are present from the value for a flat bed. From Figure 3, τ_{cr} will be 0.6 of the flat bed value when ak = 0.16.

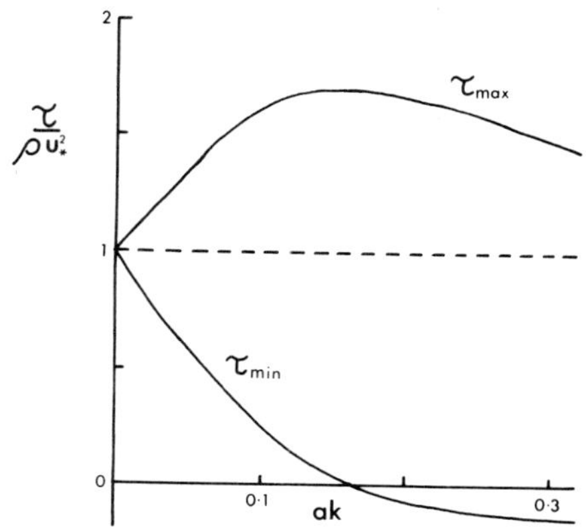

Figure 3. Variation of maximum and minimum values of the bed shear stress with wave slope for an asymmetric wave; $L/z_o = 10{,}000$.

4 DRAG AND EFFECTIVE ROUGHNESS

The total stress on the bed is comprised of the form drag of the ripple and the skin friction. As the amplitude of the ripple increases the contribution of the form drag increases from zero for a flat bed to become a significant proportion of the total stress. The form drag of the ripple considered in the previous section was 34% of the total, thus reducing the skin friction which transports the sediment.

The results from a number of runs of the model for flow over a ripple of varying amplitude and shape are summarised in Tables 1-3. In all cases the ripple wavelength was taken to be $L/z_o = 10{,}000$. The three ripple shapes are a sine wave, an asymmetric wave and an asymmetric wave with a reverse flow over it. The maximum slope of the asymmetric wave is approximately 2 ak. The flow separates with ak = 0.32 for both the sine wave and the asymmetric wave, but not for the reverse flow case. The integrated bed shear $\langle\tau\rangle$

TABLE 1. Integral properties of flow over a sine ripple; L/z_o = 10,000, z_o = 0.005 cm

ak	$\langle\tau\rangle$	$\langle q_b\rangle/q_{bo}$	z_1/z_o	C_{100} ($\times 10^{-3}$)	G ($\times 10^{-2}$)
0.03	0.98	0.98	1.05	1.65	1.80
0.06	0.95	0.94	1.21	1.70	1.80
0.16	0.74	0.71	2.74	2.02	1.53
0.32	0.35	0.38	16.7	3.19	1.52

TABLE 2. Integral properties of flow over an asymmetric ripple; L/z_o = 10,000, z_o = 0.05 cm

ak	$\langle\tau\rangle$	$\langle q_b\rangle/q_{bo}$	z_1/z_o	C_{100} ($\times 10^{-3}$)	G ($\times 10^{-2}$)
0.03	0.98	0.97	1.05	1.65	1.80
0.06	0.93	0.92	1.25	1.70	1.80
0.16	0.66	0.62	3.61	2.15	2.04
0.32	0.31	0.27	22.8	3.48	1.81

TABLE 3. Integral properties of reverse flow over an asymmetric ripple; L/z_o = 10,000, z_o = 0.05 cm

ak	$\langle\tau\rangle$	$\langle q_b\rangle/q_{bo}$	z_1/z_o	C_{100} ($\times 10^{-3}$)	G ($\times 10^{-2}$)
0.03	0.98	0.97	1.05	1.65	1.80
0.06	0.93	0.92	1.23	1.70	1.80
0.16	0.67	0.64	3.32	2.01	1.83
0.32	0.24	0.23	23.5	3.52	1.84

TABLE 4. Integral properties of flow over a sine wave; L/z_o = 10,000, z_o = 0.05 cm

ak	$\langle\tau\rangle$	$\langle q\rangle/q_{bo}$	z_1/z_o	C_{100} ($\times 10^{-3}$)	G ($\times 10^{-2}$)
0.16	0.78	0.77	1.72	3.21	1.73
0.32	0.43	0.42	4.91	4.43	1.62

is expressed as a fraction of the total stress, the remainder being form drag. This shows a strong dependence on wave slope for a given wave shape, and is slightly lower for the asymmetric wave. Also shown in the tables is the integrated bed load over the ripple $\langle q_b\rangle$ divided by the bed load over a flat bed q_{bo}. This is very similar in value to $\langle\tau\rangle$. The bed load decreases with ripple height and is only 27% of the flat bed value for the asymmetric ripple with ak = 0.32.

The effective roughness of the ripple, z_1 was calculated by taking the velocity at the top of the integration region and assuming a logarithmic velocity profile, i.e.

$$U_i = \frac{u^*}{\kappa} \ln \frac{z}{z_1} \qquad (10)$$

where κ is von Karman's constant taken to be 0.4. The ratio z_1/z_o is shown in Tables 1-3 for the three ripple shapes. Again these are highly dependent on the slope of the wave and reach a value of z_1/z_o = 23 for the flow over the asymmetric wave with ak = 0.32.

The drag of the surface can also be expressed as a drag coefficient C_{100} which relates the total stress on the bed to the velocity at a height 100 cm above the bed.

Thus

$$C_{100} = \frac{u^{*2}}{U^2 100} \qquad (11)$$

In terms of the roughness length

$$C_{100} = \frac{\kappa}{(\ln z/z_1)^2} \quad \text{at } z = 100 \text{ cm}$$

Taking a value of $z_o = 0.005$ cm for the flat bed the drag coefficient for the flat bed $C^o_{100} = 1.63 \times 10^{-3}$.

The drag coefficient can be related to the slope of the ripple by

$$C_{100} = C^o_{100} + (ak)^2 G \qquad (12)$$

where in general G is a function of L, z_o, ak and wave shape. The usefulness of this expression will depend on how little G varies with these. As shown in Tables 1-3 the variation of G is not unduly large, being between 1.5×10^{-2} and 2.0×10^{-2} for the wave shapes and slopes considered.

For a fixed bed with no sediment transport and for hydraulically rough flow the roughness length z_o can be related to the grain diameter of the sediment, d, by $z_o = d/30$ (Nikuradse, 1933). Moving sediment will affect this value. Smith (1977) has proposed that z_o is related to the excess shear stress above the value for sediment movement by

$$z_o = \frac{\alpha(\tau - \tau_{cr})}{(\rho_s - \rho) g} + z_n \qquad (13)$$

where z_n is the Nikuradse roughness length. He finds a value of $\alpha = 26.4$ gives good agreement with observation. With values of u* varying between 1.7 and 4.6 cms this gives values of z_o between 0.01 and 0.32 cm, for sediment of grain size 0.3 mm, an order of magnitude higher than the Nikuradse value.

Two runs of the model have been performed for flow over a sine wave with $L/z_o = 1{,}000$. This corresponds to the same ripple considered in Table 1 but with z_o increased to 0.05 cm. The flat bed drag coefficient is now $C^o_{100} = 2.77 \times 10^{-3}$. The results for two values of ak for this ripple are shown in Table 4. With ak = 0.32 the roughness length z_1 is increased to $4.9 z_o$. Note that with $z_o = 0.05$ cm this gives $z_1 = 0.25$ cm compared with $z_1 = 0.08$ cm for the smoother wave. The effective roughness of the ripple is therefore very dependent on the local roughness of the bed. The value of G, however, has only been increased by 10% of the smooth wave value. G is found to increase with the local bed roughness and with the asymmetry of the wave.

5 DISCUSSION

The results of the numerical model presented in this paper are preliminary but show that the presence of ripples on an erodible bed has a large influence on the dynamics of the flow close to the bed and the sediment transport. Direct comparison with observation is difficult due to the lack of detailed measurements taken of both the flow and bed conditions. Form drag estimates have been made for dunes. Flume measurements made at Fort Collins and reported by Engelund and Hansen (1972) show the form drag to be up to 66% of the total stress. The measurements of Smith (1977) show it to be as much as 75% for flow over sand waves in the Columbia River. These are comparable to the estimates given by the numerical model.

Dyer (1980) reports that z_o varied from 0.01 cm to 4.0 cm for flow over a rippled bed during a single tidal cycle. He suggests that half this variation is due to changes in the form drag of the ripples and half due to changes in the local roughness length. Again the model gives such large changes in the effective roughness length. Unfortunately Dyer's measurements were not related to the ripple size and shape at the time of measurement, although this is planned for the future.

REFERENCES

Bagnold, R.A. 1966. An approach to the sediment transport problem from general physics, U.S. Geol. Surv. Prof. Paper 422-21.

Britter, R.E. Hunt, J.C.R. and Richards, K.J., 1981. Air flow over a two-dimensional hill: studies of velocity speed-up, roughness effects and turbulence. Q.J.R. Met. Soc., 107, 91-110.

Dyer, K.R., 1980. Velocity profiles over a rippled bed and the threshold of movement of sand. Estuarine and Coastal Marine Science 10, 181-199.

Engelund, F. and Hansen, E. (1972). A monograph on sediment transport in alluvial streams. Danish Technical Press, Copenhagen.

Kachel, N.B. and Sternberg, R.W. 1971. Transport of bedload as ripples during an ebb current. Marine Geology 10, 229-244.

Nikuradse, J. 1933. Strömungs gesetz in rauhen Röhren. V.D.I. - Forshungsheft No. 361.
Richards, K.J., 1980. The formation of ripples and dunes on an erodible bed. J. Fluid Mechanics 99, 597-618.
Richards, K.J. and Taylor, P.A. 1981. A numerical model of flow over sand waves in water of finite depth. Geophys. J.R. Astr. Soc. 65, 103-128.
Smith, J.D. 1977. Modelling of sediment transport on continental shelves. In The Sea, 6, 539-577, ed. Goldberg, E.D., Wiley, London.
Taylor, P.A., Gent, P.R. and Keen, J.M. 1976. Some numerical solutions for turbulent boundary-layer flow above fixed, rough, wavy surfaces. Geophys. J.R., Astr. Soc. 44, 177-201.

Euromech 156: Mechanics of Sediment Transport / Istanbul / 12-14 July 1982

The prediction of bedforms and alluvial roughness

L.C.VAN RIJN
Delft Hydraulics Laboratory, Emmeloord, Netherlands

ABSTRACT: A method is presented which enables the computation of the bed-form dimensions and the equivalent roughness of the bed forms in the lower, transitional and upper flow regime.

1 INTRODUCTION

To predict the flow depth and sediment transport rate of an alluvial channel, the hydraulic roughness of the movable bed surface should be known as a function of the flow conditions and sediment properties. Also the prediction of the bed-form dimensions is of importance to determine the allowable navigation depth in a channel with a movable bed surface.
Although various (empirical) methods are available to estimate the hydraulic roughness (Engelund-Hansen, 1967 and White et al, 1979), no reliable method is available to predict the bed-form dimensions. Therefore, the attention is focussed on the prediction of the bed-form dimensions while also the equivalent roughness of the bed forms is considered.

2 CHARACTERISTIC PARAMETERS

It is assumed that the dimensions of the bed forms are controlled by the bed-load transport rate and can be described by the same dimensionless parameters as used to describe the bed-load transport rate, being (van Rijn, 1981):

particle parameter

$$D_* = D_{50} \left[\frac{(s-1)\,g}{\nu^2} \right]^{\frac{1}{3}} \tag{1}$$

in which: D_{50} = particle diameter of bed material, $s = \rho_s/\rho$ = specific density, ρ_s = sediment density, ρ = fluid density, g = acceleration of gravity and ν = kinematic viscosity coefficient.

transport stage parameter

$$T = \frac{(u'_*)^2 - (u_{*,cr})^2}{(u_{*,cr})^2} \tag{2}$$

in which: $u'_* = (g^{0.5}/C')\,\bar{u}$ = bed-shear velocity related to grains, $\bar{u}$ = mean flow velocity, $C' = 18 \log(12\,R_b/3\,D_{90})$ = Chézy-coefficient related to grains, R_b = hydraulic radius of the bed, D_{90} = particle diameter of bed material, $u_{*,cr}$ = critical bed-shear velocity for initiation of motion according to Shields.

3 BED-FORM DIMENSIONS

Using the equation of continuity for the bed-load particles and a simple relationship for the bed-load transport rate [van Rijn, 1982], the relative bed-form height and steepness can be expressed as:

$$\frac{\Delta}{h} = F\left(\frac{D_{50}}{h}, D_*, T\right) \tag{3}$$

$$\frac{\Delta}{\lambda} = F\left(\frac{D_{50}}{h}, D_*, T\right) \tag{4}$$

in which: Δ = bed-form height, λ = bed-form length, h = flow depth.

To determine these functional relationships, a large quantity of experimental bed-form data were analyzed. In all, 84 flume data with particle diameters ranging from 190-2300 μm and 22 field data with particle diameters ranging from 490-3600 μm, were used [van Rijn, 1982].
The influence of the D_*-parameter was found to be negligibly small. The best agreement

between measured and computed results was obtained for:

$$\frac{\Delta}{h} = 0.11 \left[\frac{D_{50}}{h}\right]^{0.3} \left[1 - e^{-0.5\,T}\right] \left[25 - T\right] \quad (4)$$

$$\frac{\Delta}{\lambda} = 0.015 \left[\frac{D_{50}}{h}\right]^{0.3} \left[1 - e^{-0.5\,T}\right] \left[25 - T\right] \quad (5)$$

It is assumed that for $T \lesssim 0$ and $T \gtrsim 25$ the bed is almost plane. Equation (4) and (5) as well as the error range of a factor 2, are shown in Figures 1 and 2. Nearly all data are within the plotted error range. From equations (4) and (5) it follows that:

$$\lambda = 7.3\ h \quad (6)$$

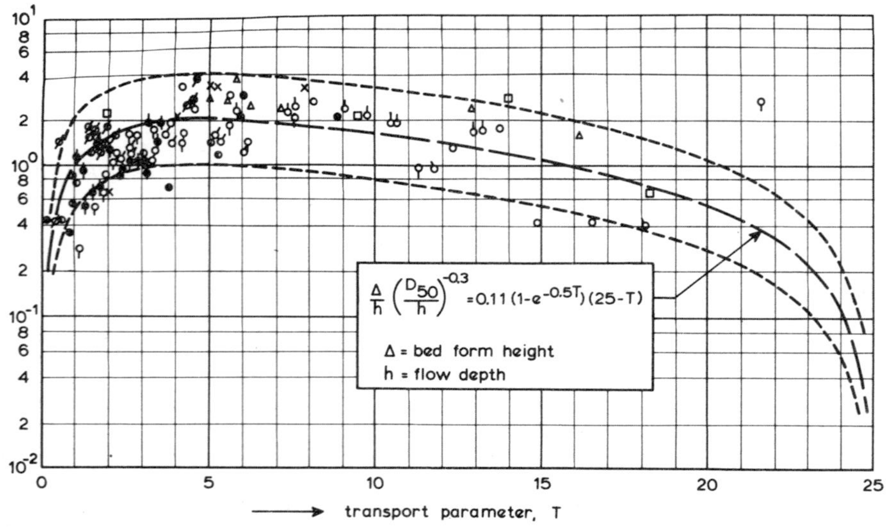

Fig. 1. Bed-form height as a function of particles size, flow depth and transport stage

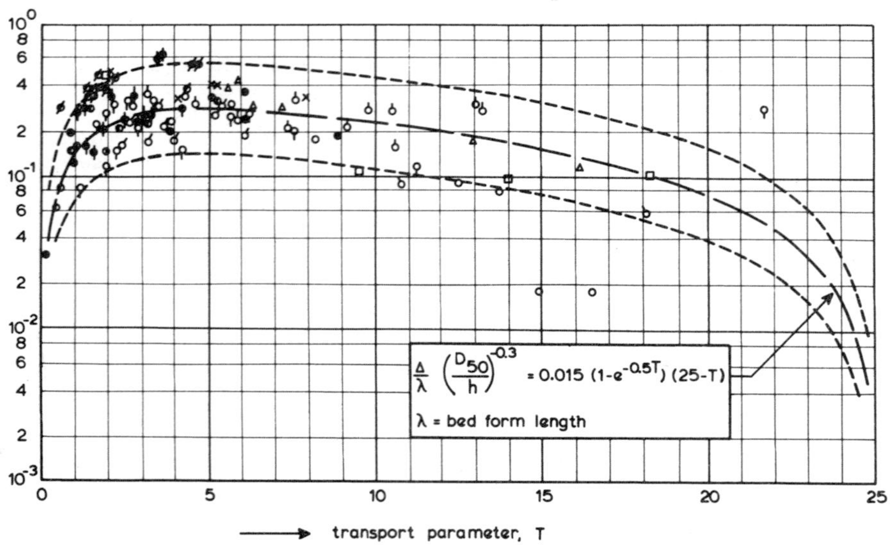

Fig. 2. Bed-form steepness as a function of particle size, flow depth and transport stage

4 EQUIVALENT ROUGHNESS OF BED FORMS

The hydraulic roughness of a movable bed surface is caused by grain roughness and form roughness, which can be expressed as:

$$k_s = F(D, \Delta, \frac{\Delta}{\lambda}) \quad (7)$$

Based on 40 flume and field data, the following expression is proposed:

$$k_s = 3\,D_{90} + 1.1\,\Delta\left[1 - e^{-25\psi}\right] \quad (8)$$

in which: $\psi = \Delta/\lambda$ = bed-form steepness

It may be noted that for $\psi = 0$ the plane bed value $k_s = 3\,D_{90}$ is obtained.
Figure 3 shows the equivalent form roughness as a function of bed-form dimensions.

Table 1. Comparison of prediction methods

758 flume data (score in error range ± 20%)	
1 Engelund-Hansen	65%
2 van Rijn	56%
3 White et al	54%

786 field data (score in error range ± 20%)	
1 van Rijn	74%
2 White et al	58%
3 Engelund-Hansen	47%

For field conditions the proposed method according to equations (4), (5) and (8) is superior to both other methods.

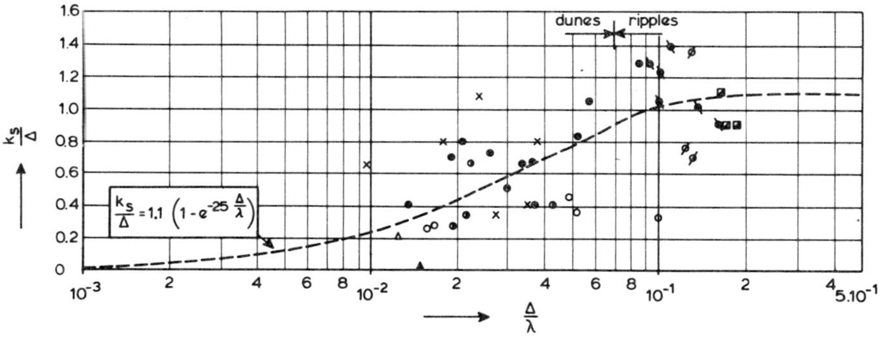

Fig. 3. Equivalent form roughness as a function of bed-form height and steepness

The Chézy-coefficient follows from:

$$C = 18\,\log\left(\frac{12\,R_b}{k_s}\right) \quad (9)$$

in which: R_b = hydraulic radius of the bed.

5 VERIFICATION

The proposed functions were verified by comparing predicted and measured Chézy-coefficients using 1544 flume and field data. For comparison also the methods of Engelund-Hansen (1967) and White et al (1979) were applied to the selected data. In Table 1 the accuracy of the three methods is given in terms of the score (in %) of the predicted values in the error range ± 20% of the measured values.

REFERENCES

Engelund, F. and Hansen, E., 1967. A monograph on sediment transport, Teknisk Forlag, Copenhagen, Denmark

Rijn, L.C. van, 1981. The computation of bed-load concentration and bed-load transport, Delft Hydraulics Laboratory, Report S 487-I, The Netherlands.

Rijn, L.C. van, 1982. The computation of bed load and suspended load transport, Delft Hydraulics Laboratory, Report S 487-II, The Netherlands.

Rijn, L.C. van, 1982. The prediction of bed forms, alluvial roughness and sediment transport, Delft Hydraulics Laboratory, Report S 487-III, The Netherlands.

White, W. et al, 1979. A new general method for predicting the frictional characteristics of alluvial streams, H.R.S. Wallingford, Report No. IT 187, England.

The mechanism of sediment transport on bed forms

J.C.C.DE RUITER
Delft Hydraulics Laboratory, Emmeloord, Netherlands

ABSTRACT: Forces exerted on bed grains by turbulent flow have been considered in some detail in order to quantify the enhanced entrainment of sediment particles in regions of strong turbulent agitation. This has resulted in a sediment pick-up function relating the entrainment rate to the time-average and the standard deviation of the instantaneous bed shear stresses. The expression is restricted to uniform cohesionless bed material, and has been calibrated with experimental pick-up rates for a plane bed. In combination with estimates for particle step lengths and a suitable flow model, local bed-load transport rates on river-bed dunes can be calculated. The relations allow for non-negligible erosion all over an upstream dune face, even though the time-averaged bed shear stress on the lower part may be well below the Shields critical value. The derived expressions will be applied in a numerical model for the prediction of bed-form behaviour under unsteady hydraulic conditions.

1 INTRODUCTION

Turbulent agitation is known to be an important mechanism for the entrainment of particles along the sloping surface of a river-bed dune, particularly in the region of flow reattachment downstream of the eddy formed in a dune trough (Sutherland 1967, Raudkivi 1976). Although the local time-averaged bed shear stress is small in such regions, relatively large instantaneous values due to strong turbulent velocity fluctuations near the bed occur frequently. Any attempt to explain or predict the detailed behaviour (growth, propagation or decay) of bed forms therefore should include a quantitative description of the effect of these fluctuations on local entrainment and transport rates.

A first approach to this problem is described in this paper. The study has been done as a part of a comprehensive research programme on bed forms and alluvial roughness (Klaassen 1978), commissioned by the Dutch Water Control and Public Works Department. In the next section, previous theoretical studies of the entrainment process are reexamined and an improved formula for the entrainment rate is derived. The relevant sediment parameters and flow variables are discussed in sections 3 and 4. Finally, it is shown in section 5 how the results can be applied in a mathematical model for the calculation of bed-form changes.

2 ENTRAINMENT RATE

Since the pioneering work of Einstein (1950), the usual starting-point in stochastic descriptions of sediment entrainment (Paintal 1971, Yalin 1977) is to write the average number of grains picked up from the bed per unit time and per unit area as

$$n_p = (4\eta/\pi D^2)\ P_c/T \qquad (1)$$

The factor in parentheses, in which η is a dimensionless coefficient and D = grain diameter, represents the density (i.e. number per unit plan area) of susceptible grains in the uppermost bed layer. Furthermore, P_c = probability of the instantaneous bed shear stress τ to exceed a certain critical value τ_c and T = characteristic time for the dislodgment process. T is mostly assumed to be proportional to either D/w (Einstein 1950, Christensen & Chiu 1973) or D/u_* (Paintal 1971, Yalin

1977), where w = fall velocity, $u_* = (\bar{\tau}/\rho)^{1/2}$, $\bar{\tau}$ = time-averaged bed shear stress and ρ = fluid density.

Although eq.(1) is dimensionally correct, it is not really convincing. In fact, any estimate of the time required for a grain to accelerate from rest and to leave its position on the bed will show that this time is much smaller than the typical "bursting" periods of the turbulence phenomena near the bed. The latter scale with the outer flow variables (Hinze 1975, p. 682) and are of the order 5h/U for open channel flow, where h = depth and U = mean velocity (Blinco & Simons 1975). In the wake of bed forms somewhat smaller periods may be introduced (Rifai & Smith 1971), but under most conditions the dislodgment time T is at least an order of magnitude smaller. For instance, estimating T from eq.(3) below one finds T < 0.03 sec for sand with D = 1 mm, if $\tau > 1.5\ \tau_c$ (say). Such "excursions" of the instantaneous shear stress τ are quite common under normal transport conditions, even if $\bar{\tau}$ is near of slightly below τ_c (see also section 4). Only for large grain sizes and small h/U ratios the inertia of a grain may prevent entrainment, in which case a more refined analysis would be required.

The above considerations lead to the conclusion that the characteristic time T in eq.(1) is not to be associated with any turbulence period and that, in fact, the variation of the instantaneous bed shear stress τ during the dislodgment process can be neglected. This means that T depends on τ rather than $\bar{\tau}$, an suggests to write the entrainment rate as

$$n_p = \frac{4\eta}{\pi D^2 \sigma\sqrt{2\pi}} \int_{\tau_c}^{\infty} \frac{1}{T(\tau)} \exp\left[-\tfrac{1}{2}\left(\frac{\tau-\bar{\tau}}{\sigma}\right)^2\right] d\tau \qquad (2)$$

This is a straightforward generalization of eq.(1), 1/T now being averaged over all instantaneous values τ larger than τ_c. For simplicity, the τ-distribution is assumed to be gaussian, with standard deviation σ. Other (more skewed) distributions have been reported in literature and can be used in future refinements of eq.(2).

In view of the irregular shape and packing of natural bed grains, it is clear that an exact description of the first stages of motion of an individual grain can not be given. However, the structure of eq.(2) is such that any representative approximation for the relationship $T(\tau)$ will reveal the typical dependence of n_p on the basic variables $\bar{\tau}$ and σ. Such an approximation can be obtained by describing the first stages of motion as a simple sliding motion. Assuming the instantaneous lift and drag forces on a grain to be proportional to τ, and defining T as the time during which a grain has moved (in a uniformly accelerated motion) over a distance ½D from its initial position, one can easily show (de Ruiter 1982) that, for $\tau > \tau_c$,

$$T(\tau) = (\rho_s D)^{\frac{1}{2}} [g\rho\Delta\ \mathrm{tg}\phi\ (\tau/\tau_c - 1)]^{-\frac{1}{2}} \qquad (3)$$

in which g = gravitational acceleration, $\Delta = (\rho_s - \rho)/\rho$, ρ_s = sediment density and $\mathrm{tg}\phi$ = friction coefficient. The typical distance ½D has been chosen because (loosely speaking) when a grain has moved over this distance, underlying grains become less protected and can be eroded in their turn. The exact value is of course unimportant: any deviation from the factor ½ can be absorbed in the coefficient η in eq. (2). A similar remark applies to shape factors (spherical particles have been assumed), added-fluid-mass corrections, etc. It is noted that a result analogous to eq. (3) can be obtained if one considers (cf. White 1940, Paintal 1971) a pivoting instead of a sliding particle.

Substitution of eq.(3) into eq.(2) and some rearranging leads to the result

$$n_p = \frac{4\eta}{\pi D^2} \left[\frac{\rho g \Delta\ \mathrm{tg}\phi\ \sigma}{\rho_s D \tau_c}\right]^{\frac{1}{2}} F_p(r) \qquad (4)$$

in which $r = (\bar{\tau} - \tau_c)/\sigma$ and F_p a function defined by

$$F_p(r) = (2\pi)^{-\frac{1}{2}} \int_0^{\infty} \sqrt{x}\ e^{-\frac{1}{2}(x-r)^2}\, dx \qquad (5)$$

The integral can be approximated by analytical or numerical means and the resulting function $F_p(r)$ is graphically shown in Figure 1, in which also the asymptotic behaviour for $r \to \pm\infty$ is indicated. Eq. (4) is valid for a plane bed. For a local bed inclination β with respect to the longitudinal (main flow) direction - as on the upstream face of a dune - the critical shear stress τ_c as occurring in the variable r has to be multiplied by the well-known Schoklitsch factor ($\cos\beta + \sin\beta/\mathrm{tg}\phi$). The quantity τ_c in the denominator in eq. (4) is not affected.

From a physical point of view, the basic idea in the above derivation is that for a given (instantaneous) bed shear stress the number of particles picked up per unit time is limited by the inertia of the grains, and by nothing else. (Note that eq. (3) is determined completely by the equation of motion of an accelerating particle.) In this respect, eq.(4) is considered as an improvement of eq.(1), in which T has been chosen rather arbitrarily in the past

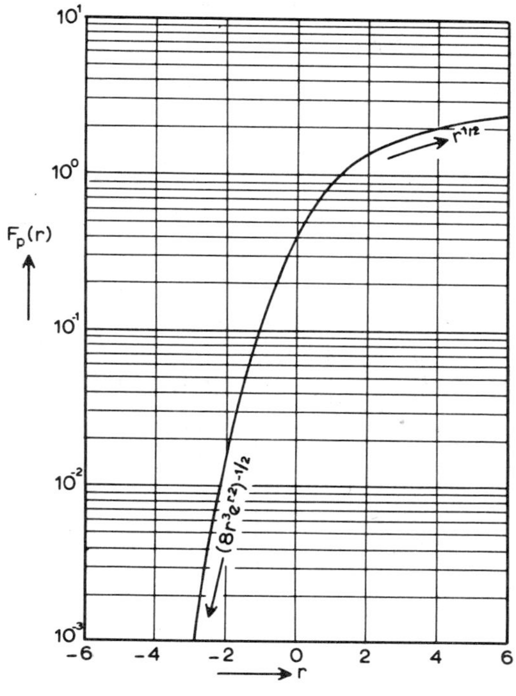

Fig. 1 Sediment pick-up function

and in which the quantity 1/T in fact has been excluded from the integration in eq. (2). Obviously, other approximations and assumptions are still involved. As in eq. (1), the flow and bed geometry are assumed to be two-dimensional in the sense that shear stress fluctuations and particle movement in the transverse direction are neglected. Effects of particle shape and position can only be accounted for in an average way, by experimental adjustment of the parameters η and τ_c (see section 3). No explicit assumptions have been made regarding the detailed turbulence structure near the bed or the precise nature of the forces on a grain. However, eqs. (2) and (3) imply the important assumption that the instantaneous lift and drag forces on a particle (though they may differ from grain to grain) are proportional to each other and vary in time according to a gaussian law. A more detailed discussion will be given elsewhere (de Ruiter 1982).

3 SEDIMENT PROPERTIES

The specification of the parameters τ_c and η requires some care. As used in eq. (3), τ_c is an absolute threshold value which, for a particle to be detached, must be exceeded by the instantaneous bed shear stress. This "instantaneous" or "particle" critical shear stress (Grass 1970) is larger than the usual Shields value (which is the critical value for the time-averaged shear stress). In fact, each individual grain has its own threshold value, which may depend on the protrusion of the grain, the configuration of neighbouring particles, etc. This would lead - in the spirit of the stochastic model of Paintal (1971) - to an averaging of eq. (4) over all possible τ_c values, but for the present it is assumed that for a given grain size a single value τ_c, representative for the uppermost grains, can be used (a similar remark applies to tgϕ). On dimensional grounds, one can expect a dependence of the dimensionless form θ_c on the particle Reynolds number R_{*c} or the dimensionless grain diameter D_*, using the definitions

$$\theta_c = \tau_c/(\rho_s-\rho)gD$$

$$R_{*c} = D(\tau_c/\rho)^{1/2}/\nu$$

$$D_* = D(\Delta g/\nu^2)^{1/3}$$

in which ν = kinematic viscosity. Typical θ_c values can be obtained from the experiments of White (1940), Grass (1970) and Yalin & Karahan (1979), which (being based on quite different methods) all give information about the instantaneous critical shear stress. The experimental results are indicated in Figure 2, together with a recent version of the well-known Shields curve (Mantz 1977, Yalin & Karahan 1979). A provisional estimate for the relationship $\theta_c(R_{*c})$ is indicated by the (bold) dashed line. Some theoretical support for this curve is given by de Ruiter (1982).

In view of the approximations involved a perhaps better procedure is to adjust the representative τ_c value in eq.(4) to experimentally observed plane-bed entrainment rates. Such entrainment experiments have recently been carried out by L.C. van Rijn at the Delft Hydraulics Laboratory. The experiments were done by measuring the erosion from a small (5 cm) sediment lift, built in a roughened flume bottom (for experimental details, see van Rijn 1982). With the choices θ_c = 0.086 and 0.082 for the two grain sizes used (D_{50} = 0.8 and 1.5 mm), the entrainment rates calculated from eq. (4) were within

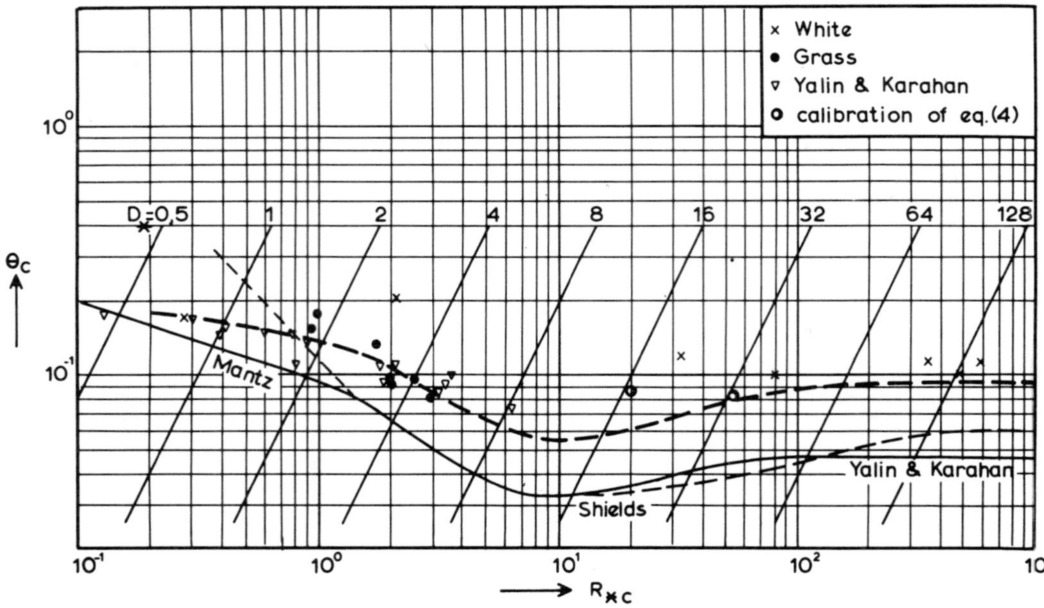

Fig. 2 Dimensionless critical shear stresses, ——— time-averaged (Shields), ----- instantaneous (estimate)

30% from the experimental results in a wide range of bed shear stresses ($\bar{\tau}/\tau_c = 0.7 - 2.1$). These θ_c values are indicated by open circles in Figure 2. The calculated entrainment rates for large $\bar{\tau}$ are not sensitive to θ_c and can be used to determine the constant coefficient η, which is estimated to be about 0.03 (if tgϕ is chosen as 0.5). This rather small value can be understood from the fact that only a small fraction of the grains in the top layer of a natural bed has a protrusion large enough to become unstable.

4 FLOW VARIABLES

In the presence of bed forms, the variables $\bar{\tau}$ and σ refer to the local bed shear stresses due to "skin friction", regardless of the form-drag contribution to the over-all bottom friction. Thus, eq.(4) can be used to calculate local entrainment rates on bed forms, provided that the local values of $\bar{\tau}$ and σ are known.

The distributions of $\bar{\tau}$ over two-dimensional bed forms has been studied by several authors in the past, but only for special cases. In general, one has to resort to numerical flow models. Advanced turbulence models (Puls 1981) can be used, although the required computer time seems to be prohibitive for long-term morphological computations. In connection with the present investigation, an improved version of an earlier developed simple boundary layer model (Klaassen 1978) is currently being tested and numerically adapted for future use in combination with the transport model described in the next section.

Much less is known with respect to the shear stress fluctuations on a sloping bed, and there is as yet no other alternative but making reasonable guesses for σ. For uniform flow over a plane bed, σ is proportional to the average value $\bar{\tau}$, the constant of proportionality depending on the grain Reynolds number $R_* = u_*D/\nu$. A survey of the relevant literature (de Ruiter 1982) indicates the ratio $\sigma/\bar{\tau}$ to vary from about 0.3 for small R_* (smooth limit) to about 0.6 for large R_* (fully rough limit). For intermediate values of R_*, a smooth interpolation has been used in the fitting procedure for θ_c described in the previous section. As for a sloping bed, ratios $\sigma/\bar{\tau}$ similar to those on a plane bed can be expected for the flat regions near the top of a dune. Moreover, there is some evidence that the absolute value of σ does not vary too much over a bed form and that the σ-value in a dune through is of the same order of magnitude as the value near a dune top. However, to get more insight, further experimental study is necessary.

5 LOCAL TRANSPORT MODEL

Locally on propagating dunes, a non-equilibrium situation exists in the sense that pick-up of sediment is not balanced by deposition. Restricting us to conditions under which bed-load transport prevails, the deposition rate n_d (defined analogously to n_p) is assumed to be given by $n_d = s/\lambda$, in which s is the local transport rate, expressed as number of particles passing per unit width and time, and λ the average particle steplength. Here, a steplength is defined as the distance travelled by a grain between two (short) periods of rest on the dune surface, i.e. the average length of the flat trajectories described by a bed-load particle in a rolling and saltating mode of motion. For a plane bed in equilibrium, the relation $n_d = s/\lambda$ is trivial. For a sloping bed, it is considered as a plausible local approximation provided that there are no strong variations in the bed level, $\bar{\tau}$, etc. over distances of the order λ (i.e. dune lengths must be much larger than λ). Experimental verification for a particular non-equilibrium situation is given by Nakagawa & Tsujimoto (1980).

Although there is no generally accepted relation for λ in terms of flow variables and sediment parameters, the available literature (Yalin 1977, van Rijn 1981) on steplengths and saltation lengths of particles in bed-load motion suggests that λ is more or less proportional to $\bar{\tau}$, at least for $\bar{\tau} > \tau_c$. Under critical conditions ($\bar{\tau} \leq \tau_c$), particles are moved by "bursts" with $\tau > \tau_c$, and λ is expected to depend on σ and τ_c as well and to approach a finite minimum value for $\bar{\tau} \rightarrow 0$). These considerations have been incorporated in a tentative expression for the relationship $\lambda = \lambda(\bar{\tau}, \sigma, \tau_c)$, in which λ is assumed to be proportional to the average of the instantaneous bed shear stresses larger than τ_c. This expression can be calibrated by comparing calculated plane-bed transport rates $s = n_p\lambda$ with well-known experimental values. Such a calibration, for various values of D_*, is in progress.

The continuity equation for sediment requires the local net erosion rate n_p-n_d to be equal to the derivative of s in the flow direction, i.e.

$$\frac{\partial s}{\partial x} = n_p - n_d = n_p - s/\lambda \qquad (6)$$

This simple differential equation for s can be solved numerically once n_p and λ (depending on x through the flow variables $\bar{\tau}$ and σ) are known. Thus, eq. (6) and the proposed expressions for n_p, n_d and λ constitute a mathematical bedload transport model, which can be used in combination with any suitable flow model to calculate local transport and erosion rates on a sloping bed and to perform morphological computations on bed forms. Preliminary calculations based on the before-mentioned boundary-layer model and provisional estimates for σ and λ show promising results with respect to transport rates, dune propagation and numerical stability.

It is finally noted that, in regions of flow reattachment, grains will be moved alternately in the downstream and the upstream direction. The pick-up by negative shear stress fluctuations ($\tau < -\tau_c$) can be calculated by a simple modification of eq. (4) and is easily accounted for in a generalized version of eq. (6). Although the time-averaged bed shear stress in and near dune throughs is small, σ is generally not small, so that large instantaneous values τ occur everywhere on a dune. Since this effect is accounted for in eq. (4), the transport model can predict sediment movement over the whole upstream face of a dune. This would be impossible with the use of conventional bed-load formulas based on $\bar{\tau}$ alone.

REFERENCES

Blinco, P.H. & Simons, D.B. 1975. Turbulent structure near smooth boundary. J.Engng. Mech. Div. ASCE, 101 (EM3): 241

Christensen, B.A. & Chiu, T.Y. 1973. Water and air transport of cohesionless materials. Proc. 15th Congr. IAHR, Istanbul, Paper A32

Einstein, H.A. 1950. The bed-load function for sediment transportation in open channel flows. Techn. Bull. No. 1026, U.S. Dept. of Agric., Washington D.C.

Grass, A.J. 1970. Initial instability of fine bed sands. J.Hydr.Div. ASCE, 96 (HY3): 619

Hinze, J.O. 1975. Turbulence (2nd ed.). New York: McGraw-Hill

Klaassen, G.J. 1978. Sediment transport and hydraulic roughness in relation to bed forms. Proc. Conf. "Advances in sediment transport", Jablonna (Poland) (also DHL-publication no. 213)

Mantz, P.A. 1977. Incipient transport of fine grains and flakes by fluids -extended Shields diagram. J. Hydr. Div. ASCE, 103 (HY6): 601

Nakagawa, H. & Tsujimoto, T. 1980. Sand bed instability due to bed load motion. J. Hydr.Div. ASCE, 106 (HY12): 2029
Paintal, A.S. 1971. A stochastic model of bed load transport. J.Hydr.Res. (IAHR) 9 (no. 4): 527
Puls, W. 1981. Numerical simulation of bed-form mechanics. Mitt.Inst. f. Meereskunde Univ. Hamburg, No. 24, Hamburg
Raudkivi, A.J. 1976. Loose boundary hydraulics (2nd ed.). Oxford: Pergamon Press
Rifai, M.F. & Smith, K.V.H. 1971. Flow over triangular elements simulating dunes J. Hydr.Div. ASCE, 97 (HY7): 963
Rijn, L.C. van, 1981. Computation of bed-load concentration and bed-load transport. Delft Hydraulics Laboratory,Report S478-I
Rijn, L.C. van, 1982. Entrainment of sediment particles. Delft Hydraulics Laboratory, Report M1521-IV (in preparation)
Ruiter, J.C.C. de, 1982. Incipient motion and pick-up of sediment as function of local variables. Delft Hydraulics Laboratory, Report R 657-XI (in preparation)
Sutherland, A.J. 1967. Proposed mechanism for sediment entrainment by turbulent flows. J. Geophys. Res.72(No. 24): 6183
White, C.M. 1940. The equilibrium of grains on the bed of a stream. Proc. Roy. Soc. London, A174 (no. 958): 322
Yalin, M.S. 1977. Mechanics of sediment transport (2nd ed.) Oxford: Pergamon Press
Yalin, M.S. & Karahan, E. 1979. Inception of sediment transport. J. Hydr. Div. ASCE, 105 (HY11): 1433.

Euromech 156: Mechanics of Sediment Transport / Istanbul / 12-14 July 1982

An experimental study of bed-load transport with non-uniform sediment

J.S.RIBBERINK
Delft University of Technology, Netherlands

ABSTRACT: An experimental study was carried out in the framework of a researchproject concerning the development of a mathematical model for morphological computations in rivers in the case of non-uniform sediment. During two years (1979-1981) a series of laboratory experiments were carried out in a straight flume under steady, uniform (equilibrium) conditions with a restriction to bed-load transport and the dune regime. The flume was fed upstream by different mixtures of two very narrow sieved size fractions. The behaviour of these fractions was studied by extensive sampling of the bed (dunes). The overall results of the experiments (mean water depth, velocity, energy slope, dune height and mean probability of each fraction in the bed) were used to verify existing models for the transport per fraction, dune height and bed roughness.

1 INTRODUCTION

The prediction of morphological changes in rivers due to natural causes or human interference is often carried out by using a mathematical model in which the movement of water and sediment is decoupled. In the one-dimensional approach the following set of equations can be used:

water $$u \frac{\partial u}{\partial x} + g \frac{\partial a}{\partial x} + g \frac{\partial z}{\partial x} = R \qquad (1)$$

$$u \frac{\partial a}{\partial x} + a \frac{\partial u}{\partial x} = 0 \qquad (2)$$

sediment $$\frac{\partial z}{\partial t} + \frac{\partial s}{\partial x} = 0 \qquad (3)$$

$$s = f(u, C_b) \qquad (4)$$

$$C_b = C_b(u, a) \qquad (5)$$

in which
u = average flow velocity
a = average water depth
z = average bed level
s = sediment transport in volume (including pores) per unit time and width
C_b = Chézy roughness for the bed
R = friction term

The main restricting assumptions are:

1. The sediment transport is a function of the local hydraulic conditions (transport-formula). This assumption becomes more acceptable if the time-scale of the morphological changes is large and bed-load transport prevails.

2. The watermotion is supposed to be quasi-steady which is acceptable for low Froude numbers (de Vries 1965).

3. The sediment is supposed to be sufficiently uniform leading to a constant representative grain size both in time and place.

In some cases the last assumption is not fulfilled e.g.
- in many rivers the average grain size decreases in downstream direction; if the region of calculation is very long this must be taken into account.
- if the upstream sediment input is suddenly stopped (e.g. behind a dam) erosion will take place often going together with longitudinal sorting because of selective transport of size fractions.

In these cases an extra dependent variable has to be taken into account viz. the composition of the available sediment mixture in a bed-surface layer. This can be achieved by splitting of the sediment mixture in size fractions and describing the transport and continuity of every size fraction separately. In case of n size fractions 2n sediment equations will replace the Eqs.(3) and (4). Another variable which enters into the equations is the thickness of the bed-surface layer. It is assumed that this layer is formed by the bedforms and the bedform height can be taken as layer-thickness.

Some years ago at the Delft University of

Technology a researchproject was started concerning the development and experimental verification of a mathematical model as described above. Other researchers as Hirano (1971), Thomas (1977) and Bettess and White (1979) followed a similar approach concerning the development of the mathematical model. However, a general lack of experimental verification remains. This is especially true for the following more or less empirical components of the mathematical model :

(i) Sediment transport formula per size fraction.

(ii) Dune-height (surface-layer thickness) predictor.

(iii) Bed-roughness predictor.

ad (i) Despite the large number of transport-formulae for uniform sediment there are only a few concepts available for a transportformula per size fraction. The most simple one is

$$s_i = \overline{p}_i \cdot s_i' = \overline{p}_i \cdot f_i'(u, C_b) \qquad (6)$$

in which

s_i = sediment transport of size fraction i in bulk volume (i.e. including pores) per unit time and width

s_i' = $f_i'(u, C_b)$ = hypothetical sediment transport of size fraction i in case of uniform sediment in identical hydraulic conditions.

$\overline{p}_i$ = average probability of size fraction i in the bed-surface layer
$(0 \leqslant \overline{p}_i \leqslant 1)$
$(\sum_{i=1}^{n} \overline{p}_i = 1$, n size fractions)

In this formula the transport of each fraction is directly proportional to its availability in the surface layer ($\overline{p}_i$). In case of bed-load transport a formula like Meyer-Peter & Mueller (1948) can be substituted for s_i' in Eq.(6).

An alternative formula is given by Egiazaroff (1965) who combines a formula like the one of Meyer-Peter & Mueller with a theoretical expression for the dimensionless critical shear stress of a size fraction as part of a sediment mixture. According to Egiazaroff the critical shear stress of the coarse fractions is smaller (and of the fine fractions larger) than according to the Shields criterion. His reasoning is that the drag force working on a large grain as part of a mixture is larger (and of a small grain smaller) than in case of uniform material.

ad (ii) For the calculation of the surface-layer thickness several dune-height predictors are available; for example an expression of Gill (1971)

$$\frac{H}{a} = \alpha(1 - \frac{\tau_c}{\tau})\ (1 - Fr^2) \qquad (7)$$

H = dune height
Fr = Froude number
α = constant

ad (iii) A roughness predictor can be used for example Engelund (1967) or White et al (1979).

In order to be able to carry out an experimental verification of the complete mathematical model for non-uniform sediment, it was decided to obtain first reliable empirical components of the model. Therefore a series of laboratory experiments were carried out with non-uniform sediment in a straight flume under steady and uniform (equilibrium) conditions. The results of the experiments were used for verification and calibration of the abovementioned semi-empirical formulae (in this paper only the bed-load formula per size fraction). The behaviour of separate size fractions was studied in detail through extensive sampling of a bed-surface layer.

2 EXPERIMENTAL CONDITIONS AND EQUIPMENT

2.1 Experimental conditions

The experiments were restricted to bed-load transport and the dune regime. Further all size fractions of the sediment mixture should move (no armouring). In order to be able to simplify the complete mathematical model (i.e. quasi-steady approach, decoupling of water and sediment equations) the Froude number has to be taken $Fr \lesssim 0.7$. Consequently constraints concerning the grain diameters of the mixture and the hydraulic conditions can be formulated.

As a result two narrow size fractions were chosen and composed to various mixtures. The sieve diameter of the fine fraction is between $0.71 \leqslant D_1 \leqslant 0.85$ mm and of the coarse fraction is between $1.19 \leqslant D_2 \leqslant 1.40$ mm with a possible 5 % error on both sides (see Fig.1).

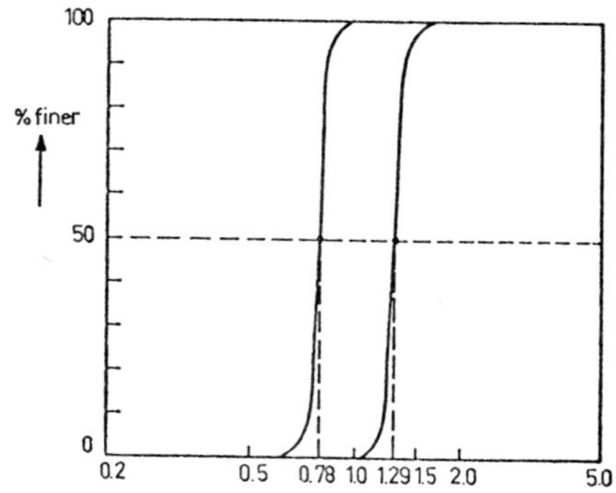

Fig. 1 Grain-size distribution of both size fractions

In every experiment the downstream waterlevel, the waterdischarge and the upstream sediment input (per size fraction) are controlled parameters which are kept at a constant level. When equilibrium is reached, i.e. when the amount and composition of the input and output mixture of the flume are equal, the measurements can take place.

Measured parameters are the waterlevel, bedlevel and the average composition of the bed-surface layer (dunes); these parameters are repeatedly measured with regular intervals in order to overcome statistical inaccuracy. The composition of the bed-surface layer is measured indirectly. Using a sampling method as described hereafter a vertical distribution of the mixture composition in the bed-surface layer can be determined at several positions in the flume ($p_i(z,z+\Delta z)$= probability of size fraction i between level z and $z+\Delta z$, $\sum_i p_i = 1$). In this way not only information about vertical sorting in the bed-surface layer is obtained but also the average composition of the bed-surface layer can be determined.

2.2 Experimental equipment

The experiments were carried out in a flume with a total length of 40 m and an inner width and height of 0.5 m. Over a length of 29 m the sand is transported; 11 metres are used for water in/outflow, sandfeeding and for a sandtrap. The flume width is reduced to 0.3 m in order to avoid large-scale bedforms (alternating bars) and produce bedforms as two-dimensional as possible.
The downstream water level is controlled with an adjustable weir in the last flume section. A sand elevator is used for the upstream feeding of the desired sediment mixture (Geldof and Groeneveld 1980). All the outflowing sediment is dried, sieved and weighed per fraction.

The measuring devices for water level and bed level are installed on a measuring carriage. A point-gauge measures the water level and electronic profile-indicators measure three longitudinal bed profiles. The analogue bed-profile signals are sampled, digitized and stored on digital tape.

Figure 2 shows an outline of the bed sampler as used for the sampling of the bed-surface layer. After stopping the flow thin-wall pipes (diameter 10 cm) are slowly pressed into the bed at distances of approximately 2 metres. The sampler consists of a hollow tube (inner diameter = 1 cm) which can be positioned exactly above the sampling pipes. By way of siphoning thin layers (thickness 0.5 cm) can be sampled separately out of the pipe. Every separate sample is caught in a sieve tray, dried in an oven and

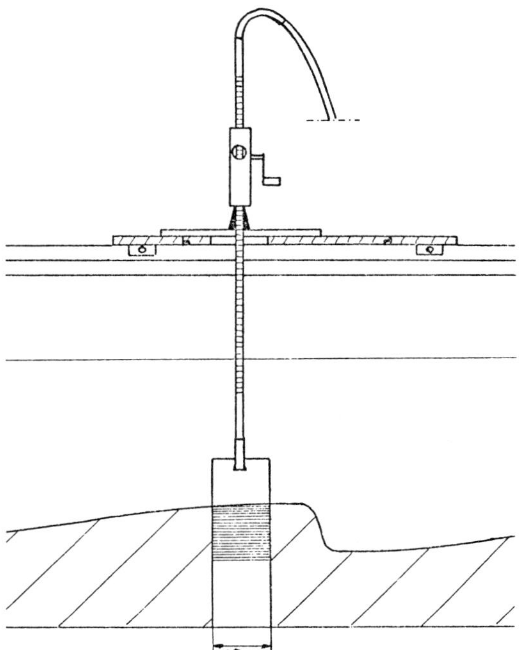

Fig. 2 Bed-sampler

next sieved and weighed per size fraction. After the grain-size analysis the samples are returned in the sampling pipes in their proper place in order to restore the surface layer as good as possible. The experiment can now be continued and new registrations can take place.

3 RESULTS

3.1 Overall-results

In the period 1979-1981 seven equilibrium experiments were carried out of which five were suitable for further analysis. The number of experiments is rather small because (i) reaching the equilibrium situation (especially reaching an equal composition of input and output mixture) takes much time (60-100 hours) and (ii) the sampling-procedure is rather time-consuming.

In these five experiments the water discharge was kept constant, only the total sediment transport and the composition of the transported mixture were varied.
Table 1 shows the main overall parameters. The Chézy roughness for the bed C_b was calculated via the hydraulic radius for the bed R_b (side-wall correction). Because a non-uniform situation existed in approximately the first 10 meters downstream of the

exp.	flume width (m)	effective measuring length (m)	water discharge ($m^3 . s^{-1}$)	water depth (cm)	water level slope (10^{-3})	bed level slope (10^{-3})	surface-layer thickness (cm)	% fine bed (-)	% fine transport (-)	total transport ($m^3 s^{-1} \cdot 10^{-6}$)	Chézy roughness bed ($m^{1/2} s^{-1}$)
3				16.8	1.65	1.53	4.4	41.0	50	2.95	36.8
4				16.0	1.79	1.82	3.5	16.8	20	2.95	35.8
5	0.3	19	0.0241	15.0	2.41	2.37	3.8	19.9	20	4.89	33.6
6				17.0	1.04	1.06	2.1	17.6	20	1.11	47.1
7				19.4	1.01	1.01	4.7	77.1	80	1.11	37.4

Table 1 Main experimental parameters (controlled and measured).

sand elevator (adaption section) an effective flume length of 19 metres remained.

3.2 Bed level, bed-surface layer

In each experiment the average bed level along the flume is determined through calculation of the linear regression line using the combined data of all bed-level registrations (effective measuring length). The number of bed-level registrations in each experiment varied from 5 to 20. The water-level registrations are treated in the same way and as a result the average water depth, water-level slope and bed-level slope are known (see Table 1).

Assuming that the bed-level fluctuations can be considered as a steady and ergodic stochastic process around the linear regression line, the bed-level data are further analysed by subtracting the linear regression line from the original registrations. The resulting data z_{ref} yield information about the bed-level fluctuations (bed forms) e.g. bedform length and height. The latter parameter has been put equal to the thickness of the bed-surface layer and is determined by gathering the bed-level fluctuations in a histogram. Figure 3 shows such a histogram belonging to experiment 7.

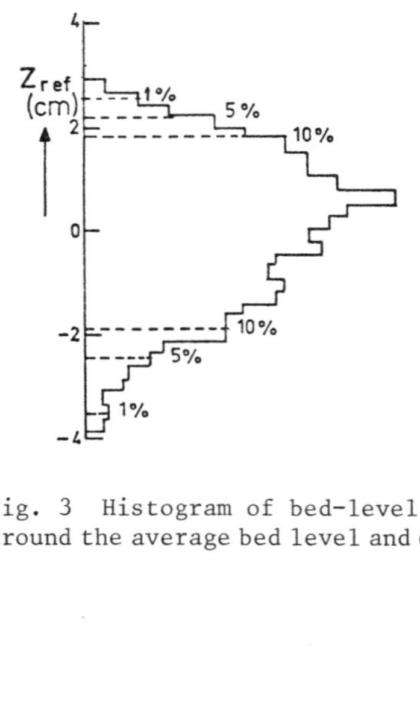

Fig. 3 Histogram of bed-level fluctuations around the average bed level and α-percentiles

It can be seen that the distribution is far from uniform which would result if the dunes would have been a series of triangles of equal shape and size. The tails of the distribution are caused by unfrequent occurrence of deep troughs and high crests. Consequently the definition of the thickness of the surface layer becomes problematic. It is assumed in the mathematical model that eroded and sedimentated material is directly mixed in the surface layer. However, in the deeper layers of the bed which are only reached occasionally the direct mixing cannot take place. Therefore an arbitrary definition of surface-layer thickness in which the long tails of the histogram are more or less excluded is chosen, viz. the distance between α-percentiles of the histogram of the bed-level fluctuations (see Fig. 3).
Three values of α are chosen: 1, 5 and 10 %. Table 2 shows the influence of α on the calculated surface-layer thickness in the different experiments.

α →	1 %		5 %		10 %	
exp. ↓	δ (cm)	$\bar{p}_1$ (-)	δ (cm)	$\bar{p}_1$ (-)	δ (cm)	$\bar{p}_1$ (-)
3	4.6	0.40	4.4	0.41	4.2	0.44
4	3.8	0.16	3.5	0.17	3.4	0.17
5	4.1	0.19	3.8	0.20	3.4	0.21
6	2.8	0.17	2.1	0.18	1.8	0.18
7	6.1	0.74	4.7	0.77	3.9	0.80

Table 2 Influence of α on the surface-layer thickness and the average probability of the fine fraction in the surface layer

The 5 % values are used for further analysis; they are included in Table 1.

The influence of the sediment-mixture composition on the surface-layer thickness can be seen by comparing the experiments 3 and 6 in which the average water depth and water velocity are almost equal. Figure 4 shows combined plots of all measured bed-level

registrations (sampling distance = 10 cm) in both experiments.
The transition from 20 % fine in transport (exp. 6) to 50 % fine in transport (exp. 3) causes a considerable increase of the surface-layer thickness (a factor 2 !) which accompagnies with a considerable change in bed roughness.

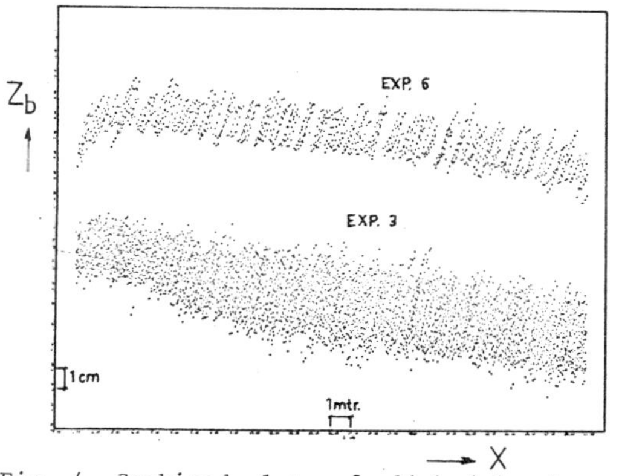

Fig. 4 Combined plots of all bed-level registrations in experiments 3 and 6

Remark: Especially in experiment 3 a large adaptation section of the bed in the first 10 metres of the flume can be seen; the dune height as well as the bed slope change gradually in that region.

3.3 Bed composition

In each experiment 1 to 4 sampling runs were carried out. In each sampling run 10 sampling pipes were positioned in the flume bed at one certain moment. The results of each run are gathered in one graph resulting in an average vertical probability distribution of one of the size fractions with respect to the average bed level (linear regression line).
Probability p_i = probability of size fraction i in a sample $(0 \leqslant p_i \leqslant 1)$ $p_1 + p_2 = 1$.
Figure 5a shows the average vertical probability distribution of the fine fraction p_1 in experiment 3 (as measured along the flume axis). It can be seen that going from the top layers into the deeper layers of the dunes a rather large gradient in mixture composition occurs; the coarse fraction shows a general preference to be deposited in the deeper layers of the dunes.

This may be explained by the difference in gravitational forces acting on the coarse and fine grains respectively. The coarse grains generally roll and slide further into the trough at the lee-side of the dunes than the fine grains. Although the variation in vertical distribution per sampling-pipe is rather large the shape of the vertical distribution as found in this experiment is typical for all sampling-pipes and all experiments.

Figure 5b shows the histogram of the bed-level fluctuations around the average bed level. Both figures in combination show the probability at level z_{ref} that a sediment mixture with composition $p_1(z_{ref})$ is eroded or sedimentated.

Generally the mixture composition along the flume axis is coarser than along the side-profiles. However, the shape of the vertical distribution is similar.
Figure 6a shows the average result of two different sampling runs (time-interval = 50 h) in experiment 7 along the side-profiles. The dotted lines indicate a region of variation of both runs. Figure 6b shows the respective histogram. Also here the preference of the coarse fraction for the lower layers of the dunes and the relatively low frequency that the lower layers are reached can be seen.

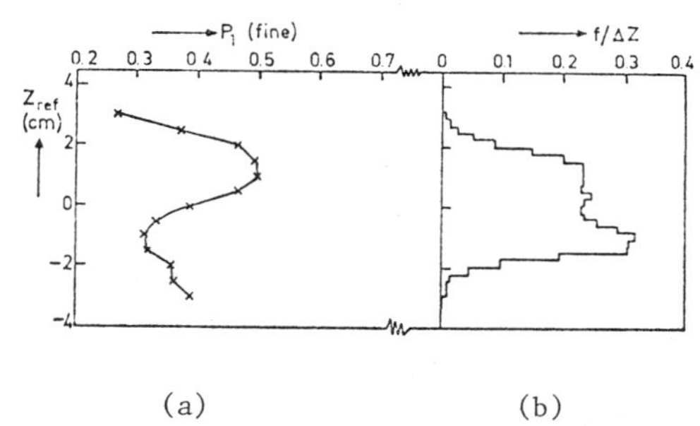

Fig. 5 Vertical probability distribution of the fine fraction in the surface layer around the average bed level (a) and histogram of bed-level fluctuations (b) in experiment 3.

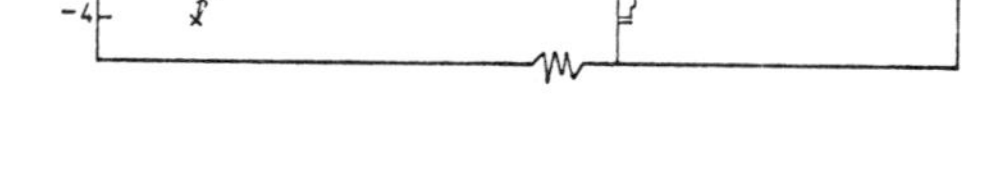

Fig. 6 Vertical probability distribution of the fine fraction in the surface layer around the average bed level (a) and histogram of the bed-level fluctuations (b) in two sampling-runs in experiment 7.

Because of the vertical sorting phenomenon the average percentage of the fine fraction in the surface layer ($\overline{p_1}$) is dependent of the choice of coefficient α in the definition of surface-layer thickness. Again three values of α were chosen: 1, 5 and 10 %. Direct vertical averaging of the $\overline{p_1}(z)$ distribution is not allowed because of the used sampling-technique. Material sampled from the lower layers of the bedform is present in a larger quantity (bedform shape !) and should therefore have a larger weight than material sampled from the upper layers of the bedform. The weighting is realised through multiplication of the $p_1(z)$ distribution with the cumulative histogram of the bed-level fluctuations (cumulated vertically downwards) between the α-percentiles.

Table 2 shows the influence of α on the average probability of the fine fraction $\overline{p_1}$ in all experiments. Similar to the surface-layer thickness the 5 %-values of $\overline{p_1}$ are included in Table 1.

The vertical sorting as was found in all experiments seems to be of large importance for the complete model for morphological computations in case of non-uniform sediment. It does not seem unrealistic that e.g. during sedimentation in stead of the whole transported mixture only a part of it (coarse part) is sedimentated.

The vertical sorting phenomenon is not limited to laboratory conditions and a mixture with only two size fractions. Zanke (1976) fiends similar vertical sorting of a 'large-range' sediment mixture in bedforms in the river Weser.

4 BED-LOAD TRANSPORT PER SIZE FRACTION

4.1 Verification and calibration

The overall results of the experiments in equilibrium conditions (Table 1) were used to verify and calibrate the two concepts for bed-load transport per size fraction as described in the Introduction. They can be written as :

$$\frac{s_i}{\sqrt{\Delta g D_i^3}} = 13.3\ \overline{p}_i\ (\tau_i - \tau_{c_i})^{3/2} \qquad (8)$$

with τ_{c_i} = dimensionless critical bed-shear stress of size fraction i

τ_i = dimensionless effective bed-shear stress of size fraction i

$$= \frac{\mu\ u^2}{C_b^2\ \Delta\ D_i} \qquad (9)$$

and D_i = grain diameter of size fraction i

Δ = relative density = $(\rho_s - \rho)/\rho$

μ = ripple factor = $(C_b/C_g)^{3/2}$ (Meyer-Peter & Mueller)

C_g = grain roughness = $18\ {}^{10}\log\ (12\ R_b/(D_{90} + 0.286\ \delta))$

δ = viscous sublayer thickness

According to the formula of Meyer-Peter & Mueller (1948) :

$$\tau_{c_i} = 0.047 \qquad (10)$$

According to Egiazaroff (1965) :

$$\tau_{c_i} = \frac{0.077}{({}^{10}\log\ 19\ \frac{D_i}{D_m})^2} \qquad (11)$$

with $D_m = \sum_{i=1}^{n} p_i D_i$ (n size fractions).

All parameters can be calculated from the results in Table 1.

Verification of both bed-load formulae per size fraction gave rather unsatisfactory results. In some experiments a zero transport for both size fractions was calculated. Therefore a calibration was carried out using a least-squares technique with the following model :

$$\frac{s_i}{\sqrt{\Delta g D_i^3}} = \alpha_i \cdot \overline{p}_i (\tau_i - \beta_i\ \tau_{c_i})^{3/2} \qquad (12)$$

Analogous to the formula of Meyer-Peter & Mueller the constants α_i and β_i were determined empirically.

A calibrated version of Eq.(12) including Egiazaroff's correction gave the best results. Figure 7 shows the agreement between measured and calculated transport per size fraction.

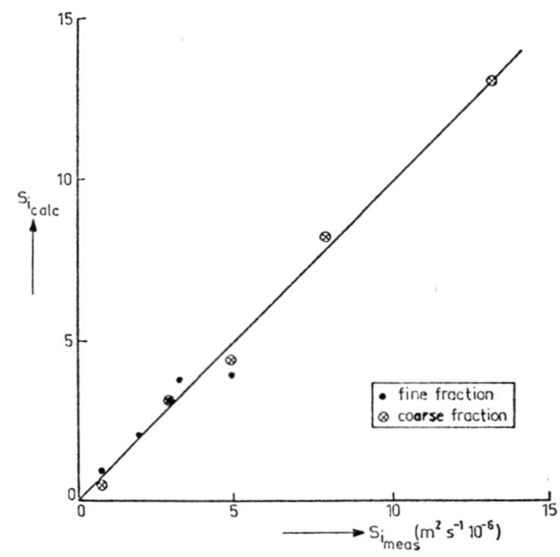

Fig. 7 Measured and calculated transport per size fraction according to Meyer-Peter & Mueller (1948) /Egiazaroff (1965).

The constants α_i and β_i hardly show any difference for the fine or the coarse fraction when Egiazaroff's correction is used : $\alpha_i \approx 15.9$ and $\beta_i \approx 0.65$.
It can be concluded that Egiazaroff's expression seems to be a useful correction to the bed-load formula of Meyer-Peter & Mueller in case of sediment mixtures.

4.2 Longitudinal sorting

The sorting phenomenon not only occurs in vertical direction but may also occur in longitudinal direction. In case of a sudden stop of the upstream sediment input (e.g. behind a dam in a river) the sediment bed will be eroded generally going together with longitudinal sorting. The latter is caused by the fact that the fine size fractions of the present sediment mixture are transported easier than the coarse size fractions which will stay behind. Eventually this may lead to armouring.

The sorting phenomenon will be more pronounced if the grain size distribution of the transported material differs more from the grain size distribution of the bed material. In case of a mixture with two size fractions this difference can be expressed by the probability of the fine fraction in the bed ($\bar{p}_1$) and the probability of the fine fraction in transport (p_{1_T}). The latter parameter is defined as :

$$p_{1_T} = \frac{S_1}{S_{total}} \qquad (13)$$

If the assumption is made that an (equilibrium) transport formula (per size fraction) can still be used during erosion or sedimentation, the difference between $\bar{p}_1$ and p_{1_T} can be calculated directly using a transport formula per size fraction. In case of the bed-load formula of Meyer-Peter & Mueller (per size fraction, Eqs.(8), (10)) this yields :

$$p_{1_T} = \frac{s_1}{s_1 + s_2} = \left\{1 + \left(\frac{D_2}{D_1}\right)^{3/2} \cdot \frac{(1-\bar{p}_1)}{\bar{p}_1} \cdot \left(\frac{\tau_2 - 0.047}{\tau_1 - 0.047}\right)^{3/2}\right\}^{-1} \qquad (14)$$

With $\tau_1 = \frac{D_2}{D_1} \cdot \tau_2$ (see Eq.(9)) Equation (14) yields p_{1_T} as a function of $\bar{p}_1, \tau_2$ and $\frac{D_2}{D_1}$.
Figure 8a shows this relation for $\frac{D_2}{D_1} = 1.5$. In all calculations both fractions are transported ($\tau_i > \tau_{c_i}$).
Figure 8b shows the same relation in case Egiazaroff's correction (Eqs.(8),(11)) is used. In all cases the percentage of the fine fraction in transport is larger than as present in the bed ($\bar{p}_1 < p_{1_T}$) or i.o.w. the bed (surface layer) has a relatively coarse composition. This fact is confirmed by the experimental results (see Fig. 8b and Table 1).

As the dimensionless bed-shear stress τ_2 decreases the difference between $\bar{p}_1$ and p_{1_T} increases; this indicates that the sorting phenomenon is more pronounced and the size fractions behave more differently close to 'initiation of motion'. Because of the small number of experiments this influence could not be confirmed by the experiments.

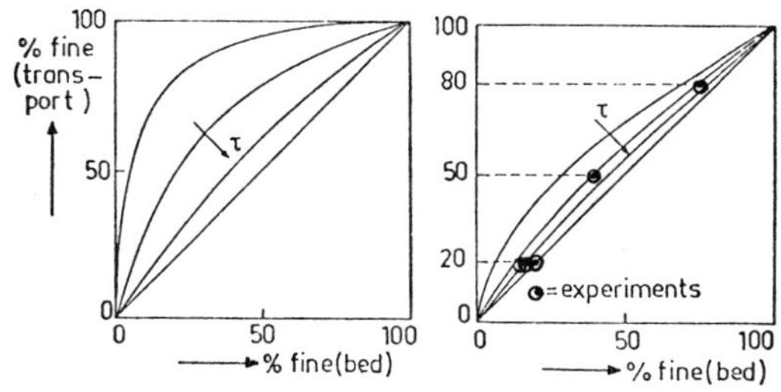

(a) (b)

Fig. 8 Difference in grain-size distribution of the material in transport and in the bed (surface-layer) according to Meyer-Peter & Mueller (a), according to Egiazaroff (b) and experimental results.

The influence of Egiazaroff's correction can clearly be seen in Fig. 8. A smaller difference between $\bar{p}_1$ and p_{1_T} or a more similar behaviour of both size fractions. This trend is confirmed by the experimental results; large differences between $\bar{p}_1$ and p_{1_T} as in Fig. 8a do not occur. Egiazaroff's expression is also confirmed by experiments of Pantelopoulos (1957), who finds that the critical shear stresses of the size fractions are much more equal than can be expected from the Shields-criterion.

The physical explanation for $\bar{p}_1 < p_{1_T}$ in case of bead load may be given by the fact that the vertical sorting occurs in combination with an irregular shape and size of the bed forms. In case of regular bed forms, e.g. represented by a series of triangles of equal shape and size, and no transport from one bed form to another, the grain-size distribution of the bed-surface layer material (in one bed form) must be equal to the grain-size distribution of the material in transport. Or, in case of a two-fraction mixture: $\bar{p}_1 = p_{1_T}$. However, the experiments learned that vertical sorting occurs in combination with bedforms of irregular shape and size (unfrequent occurrence of deep troughs and high crests). The coarse fraction has a general preference to be deposited in the lower layers where erosion will not take place for a relatively long period. Consequently the average rest period of the coarse fraction is relatively large and relatively more of the coarse material will be present in the bed-surface layer.

The same phenomenon appears in a more pronounced way in case (very fine) fractions go into suspension. Despite a major part of the total transport can be contributed to these fine fractions the concentration of these fine fractions in the bed-surface layer can be relatively small.

5 SUMMARY AND CONCLUSIONS

1. Laboratory experiments have been carried out with a sediment mixture of two size fractions in a straight flume in equilibrium conditions with a restriction to bed load and the dune regime. Overall data like water-depth, flow velocity, transport per size fraction, thickness and composition of the surface layer are available (Table 1).

2. Some semi-empirical models for bed load per size fraction were verified and calibrated. Egiazaroff's expression for the dimensionless critical shear per size fraction appears to be a useful correction of the bed load formulae per size fraction.

3. Because of unfrequent occurrence of high crests and deep troughs the surface-layer thickness is defined as the distance between the α-percentiles of the histogram of the bed-level fluctuations. For the determination of the average composition of the surface layer $\overline{p_1}$ (percentage fine) these α-% boundaries are used. The influence of α can be considerable (Table 2).

4. The used sampling-technique made it possible to determine vertical distributions of the probability of both size fractions in the surface layer. Vertical sorting occurred in all experiments. The coarse size fraction has a general preference for the lower layers and the fine size fraction for the upper layers of the dunes. This may be caused by the larger gravitational force working on the larger grains. The vertical sorting phenomenon seems to be of large importance for the one-dimensional morphological computations in case of non-uniform sediment.

5. The occurrence of longitudinal sorting is related to the difference in grain-size distribution of the material in transport and the material in the bed (surface layer). The sampling results showed that the surface layer always has a coarser composition than the transported mixture ($\overline{p_1} < p_{1_T}$). This may be caused by the vertical sorting in combination with the irregular shape and size of the bed forms. Again Egiazaroff's correction in the bed-load formula per size fraction appears to be useful in this respect.

REFERENCES

Bettess, R. & W.R. White 1979. A one-dimensional morphological river model. Wallingford Report nr. IT 194, Nov.

Egiazaroff, P.I. 1965. Calculation of non-uniform sediment concentrations. Proc. ASCE, Hydraulic Division, HY 4, July.

Engelund, F. & E. Hansen 1967. A monograph on sediment transport in alluvial streams. Teknisk Forlag, Copenhagen.

Geldof, H.J. & J. Groeneveld 1980. Sand feeding by elevator. Proc. ASCE, Hydraulics Division, HY 12, December.

Gill, M.A. 1971. Height of sand dunes in open channel flows. Proc. ASCE, Hydraulics Division, HY 12, December.

Hirano, M. 1971. River bed degradation with armoring. Transactions JSCE, Vol.3, part 2.

Meyer-Peter, E & R. Mueller 1948. Formulas for bed load transport. Proc. IAHR, Stockholm.

Pantalopoulos, J. 1957. Etude expérimentale du mouvement par charriage de fond d'un mélange de matériaux. Proc. IAHR, Lisbon.

Thomas, W.A. 1977. Mathematical modelling of scour and deposition. Proc. ASCE, Hydraulics Division, HY 8, August.

Vries, M. de 1965. Considerations on non-steady bed-load transport in open channels. Proc. IAHR, Leningrad.

White, W.R., E. Paris & R. Bettess 1979. A new general method for predicting the frictional characteristics of alluvial streams. Wallingford, Report nr. IT 187, July.

Zanke, U. 1976. Ueber den Einfluss von Kornmaterial, Strömungen und Wasserständen auf die Korngrössen von Transportkörpern in offenen Gerinnen. Mitt. Franzius Inst. Hannover, no. 44.

Euromech 156: Mechanics of Sediment Transport / Istanbul / 12-14 July 1982

Bedforms in relation to hydraulic roughness and unsteady flow in the Rhine branches (the Netherlands)

A.VAN URK
Ministry of Transport & Public Works, Arnhem, Netherlands

ABSTRACT: In order to improve existing methods for the computation of design levels of the dikes along the Rhine branches in the Netherlands the hydraulic roughness of the movable minor (alluvial) bed under unsteady floodwater conditions must be predicted. For this purpose soundings of longitudinal bed profiles along pre-programmed courses have been carried out. With the help of these measurements changes in the bedformpatterns during the rise and fall of a flood have been studied. Some preliminary results are presented about the hydraulic roughness as derived from dune dimensions and grainsize and about the changes to bedform dimensions and hydraulic roughness of the alluvial bed under unsteady flow conditions.

1 INTRODUCTION

The rivers Rhine and Meuse and their branches play a very important role in the economic and social life in the Netherlands. River management must weigh very carefully river engineering solutions for flood control, navigation, water management, landscape and nature and therefore prediction methods must have a high accuracy. There is an urgent need to model the river system mathematically so that the short and long term effects of human interventions can be quantified. Most of the research on river engineering is carried out within the cooperation program "Applied Research Waterstaat on Rivers" in which the Watercontrol and Public Works Deparment (Rijkswaterstaat), the Delft University of Technology and the Delft Hydraulics' Laboratory co-operate. One topic is the study of the bedforms, their development and their influence on the hydraulic roughness during very high discharges (Havinga, van Urk 1980). In addition to laboratory experiments soundings of longitudinal bedprofiles have been carried out during floods in the Rhine branches, in this connection, enabling changes in bedform patterns during rise and fall of floodflows to be studied. In this paper some preliminary results are given on the testing of predictors for resistance of bedforms to flow and about the change of bedforms dimensions and hydraulic roughness of the alluvial bed under unsteady flow condition.

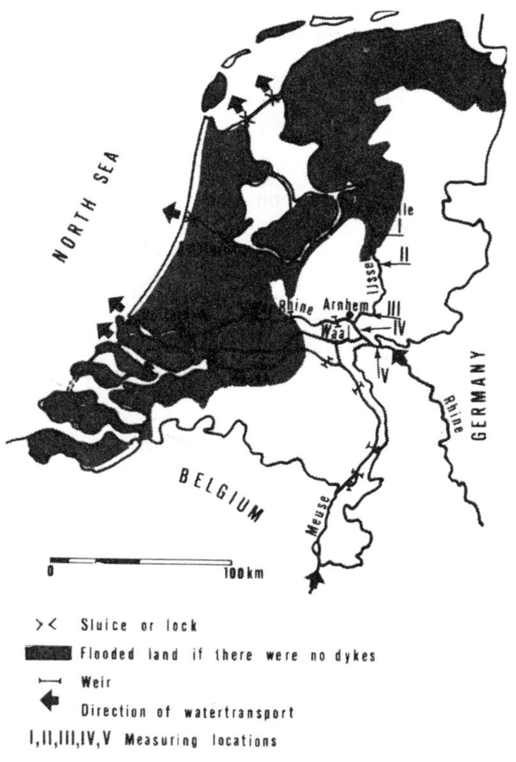

Fig.1 Rivers Rhine and Meuse in the Netherlands

2 PURPOSE

A mathematical flow model (2 dimensional, finite difference, implicit) has been used for the Rhine branches to derive the flood level along the dikes to be exceeded 1/1250 years i.e. once in 1250 years (8% per 100 years).

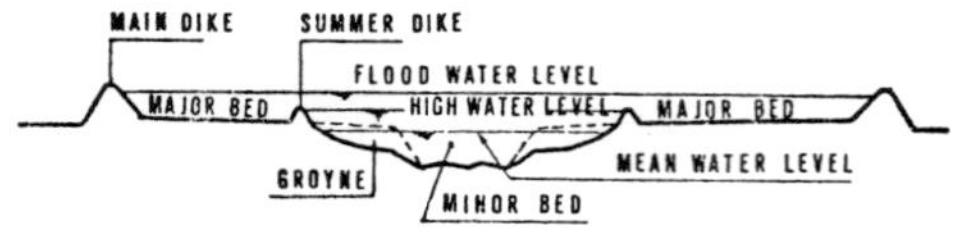

Fig. 2 Cross-section of the river bed.

For this model detailed information is needed about the hydraulic roughness of the major bed, minor bed, groynes, bends etc. In this context the terms "major" and "minor" mean: - the floodplain between summer dikes and main dikes and

- the bed between the summer dikes respectively.

In this paper some preliminary results are described related to the hydraulic roughness of the minor bed as derived from dune dimensions and the change of the hydraulic roughness under unsteady flow conditions. Usually a time lag occurs between dune formation and changes in flood flow and the difference between the hydraulic roughness under steady flow conditions and unsteady flow conditions is therefore of special importance in order to decide if this difference is significant.

The whole research topic consists of:

- Studies of bedforms and related hydraulic roughness under steady flow conditions both in a laboratory flume and in the prototype.
- Studies of existing predictors for the hydraulic roughness derived from dune dimensions.
- The development of mathematical formula to describe the changes in bedforms due to unsteady flow and their testing in laboratory flumes, e.g. a simple first order relation such as: (Wijbenga, Klaassen 1981)

$$\frac{dH(t)}{dt} = f\ (H(t),\ H_\infty\ .....) \qquad (1)$$

$H(t), H_\infty$ = Dune height at time t and at $t \to \infty$.

3 MEASURING PROCEDURES AND DATA PROCESSING

An automatic sounding system has been developed for the collection of reliable field data on sedimentary features like dunes in the Dutch rivers. This sounding system enables the bottom profiles along prefixed courses in all directions to be recorded. For this purpose the survey vessels are equipped with a digital reading echosounder and a two-dimensional horizontal positioning system, controlled by a desk-top computer. Deviations from the programmed course are computed and presented to the helmsman of the ship, to enable him to correct the course immediately. The actual position, together with the related water-depth data, is stored on tape and elaborated on a main frame computer at the office. Data processing then takes places including input of reference levels and correction of errors. With the help of mathematical filter techniques long waves (e.g. caused by bends) and short ripples are filtered out. The filter can be defined as a finite impulse response filter as described by (Rabiner, Gold 1975). Dune dimensions can then be calculated by considering the zerocrossings of the signal over the mean level (fig. 3). Distinction must be made between mean dune dimensions and weighted dune dimensions (weighted to the length of the dune). From other experiments it can be shown that the use of weighted dune dimensions in prototype is recommended. As can be seen in fig. 4 the bedforms do not have a very uniform shape in a given reach. However by using weighted dune dimensions the effects of small fluctuations around the mean

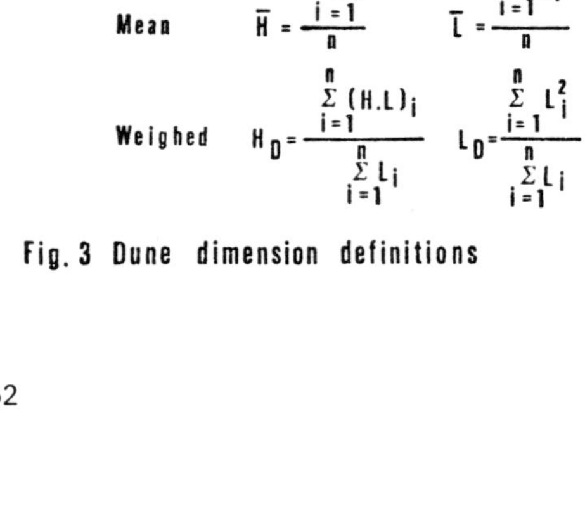

Fig. 3 Dune dimension definitions

level can be eliminated. The difference between mean dune dimensions and weighted dune dimensions is illustrated in fig. 3.

During some floods in the period 1979 to 1982 bedform patterns were measured at several locations. (fig. 1; locs. I to V). As an example the recordings at locs. I and V are shown in figs. 4 and 5 drawn to an arbitrary level datum.

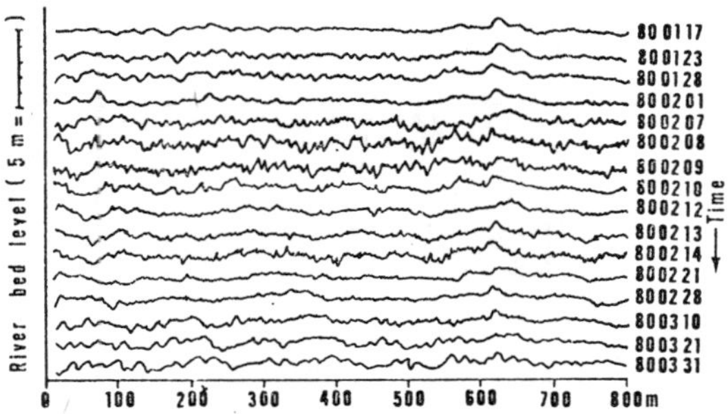

Fig.4 The longitudinal bed profile river IJssel in the period 80.01.17 to 80.03.31 (loc.I axis)

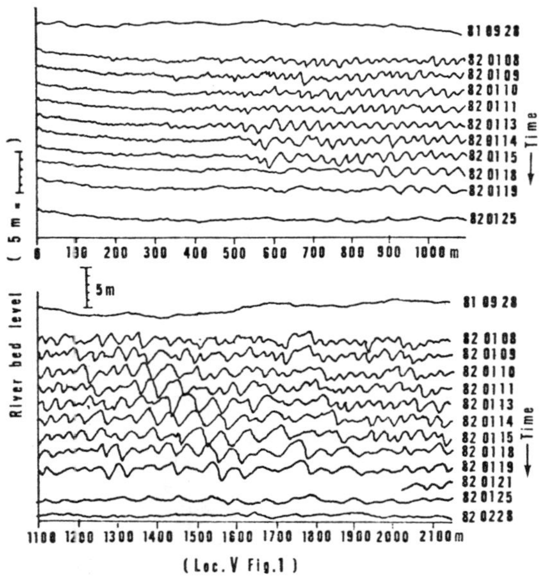

Fig. 5 The longitudinal bed profile Rhine in the period 81.02.28 to 82.02.28 (loc.V, axis)

At loc. I there was a flood in the period 80.02.01 to 80.02.28 with its crest passing this location at 80.02.12. Considerable changes in bedform dimensions are shown with a growth during the first stage of the rise and smoothing out of bed features just before the crest passed. At loc. V there was a flood in the period 82.01.01 to 82.01.28 with its crest passing at 82.01.10. At mean discharges the bed is flat in this location; at high discharges rather high dunes have been formed and after subsiding of the flood waters the dunes were washed out very rapidly (see 82.01.25). Other hydraulic parameters such as velocities, depths, enery slopes, grain sizes have also been recorded.

As an illustration some values are given in table A related to discharges which will be exceeded 90%,50%, and 10% of the time respectively, predicted values.

Table A Some hydraulic parameters at measuring locations on Dutch rivers

PLACE / PARAMETER		IJSSEL Veessen loc.I	Deventer loc.II	Rheden loc.III	PANNERD. CANAL loc. IV	RHINE loc. V
Discharge (m^3/s)	Q_{90}	221	221	221	246	1100
	Q_{50}	296	296	296	615	1950
	Q_{10}	467	467	467	1050	3200
Depth (m)	h_{90}	3.84	3.80	3.19	3.40	4.06
	h_{50}	4.40	4.56	3.95	4.67	5.49
	h_{10}	5.53	5.96	5.38	6.39	7.30
Velocity (m/s)	U_{90}	0.62	0.67	0.90	0.48	0.76
	U_{50}	0.71	0.74	0.96	0.94	1.03
	U_{10}	0.86	0.86	1.13	1.18	1.27
Slope ($*10^{-4}$)	i_{90}	0.45	0.90	1.27	0.24	0.50
	i_{50}	0.66	0.91	1.23	0.75	0.69
	i_{10}	0.86	0.95	1.20	0.99	0.84
Width (m)	B	103	92	76	135	340
Grain size (mm)	D_{10}	0.30	0.35	1.70	0.85	0.55
	D_{50}	0.45	0.55	4.80	3.20	3.60
	D_{90}	1.05	1.70	13.00	11.15	14.50
Roughness	-50% exceedance discharge only.					
($m^{½}/s$)	C_{50}	42	36	44	50	53
	f_{50}	0.045	0.060	0.041	0.031	0.028
($m^{-⅓}/s$)	n_{50}	0.031	0.036	0.029	0.026	0.025
(m)	k_{50}	0.25	0.53	0.18	0.09	0.076
Froude	F_{90}	0.10	0.11	0.16	0.08	0.12
	F_{50}	0.11	0.11	0.15	0.14	0.14
	F_{10}	0.12	0.11	0.16	0.15	0.15
Sed. Transp. (m^3/day)	S_{90}	60	140	0	0	0
	S_{50}	210	250	5	100	660
	S_{10}	680	540	60	500	2200

From soudings at these locations it is remarkable that the characteristics of the echo soundings do not change very much over the width of the river. However lengthwise they can change considerable from place to place.

4 VERIFICATION OF PREDICTORS OF HYDRAULIC ROUGHNESS BASED ON DUNE DIMENSIONS

Literature gives some predictors of the hydraulic roughness based on dune dimensions and grain size. Calculations with available

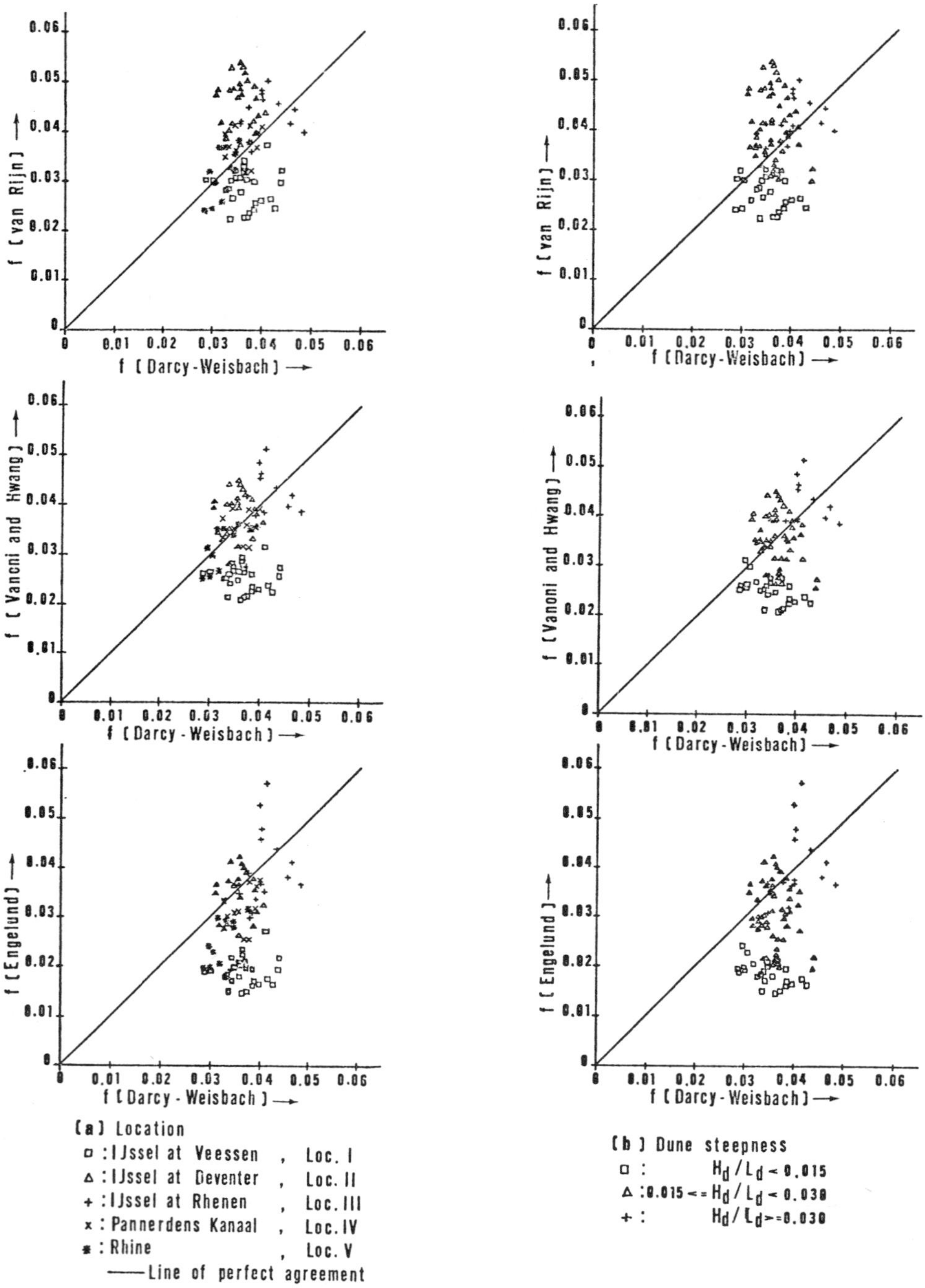

Fig. 6 Values of hydraulic roughness based on dune dimensions and grainsize compared with values calculated using the Darcy-Weisbach formula

data from laboratory flumes (World Flume Data) were carried out to select three of the most promising for use in the prototype (Klaassen, van Urk 1982/3) which are given in Table B.

Table B. Predictors for hydraulic roughness based on dune dimensions and grainsize.

References	Hydraulic roughness
Vanoni and Hwang (1967)	$f = f' + f''$
	$f' = 8g / (18 \log 12\frac{h}{D_{90}})^2$ ...(2)
	$\frac{1}{\sqrt{f''}} = 3,3 \log\frac{Lh}{H^2} - 2,3$ .....(3)
Engelund (1978)	$f = f' + f''$
	$f' = 8g / (6 + 2.5\ln\frac{h.^{f'}/_{f}}{D_{90}})^2$ ..(4)
	$f'' = 10e^{-2.5\frac{H}{h}} \cdot \frac{H^2}{h.L}$ ....(5)
van Rijn (1982)	$\frac{1}{\sqrt{f}} = \frac{18}{\sqrt{8g}} \log \frac{12h}{1.1(1-e^{-25\,H/L})H + 3D_{90}}$ .. (6)

From measured hydraulic parameters such as the water velocity above the alluvial bed, the waterdepth and the energy slope the hydraulic roughness can be calculated, expressed in C, f, k or n , with the help of Chezy and/or other formulas, as given below:

Chezy : $U = C.\sqrt{h.i}$ (m/s) .....(7)

Nikuradse : $C = 18\log\frac{12h}{k}$ ($m^{\frac{1}{2}}/s$)..(8)

Darcy Weisbach : $f = \frac{8.g.h.i.}{2}$ ....(9)

Mannings'n : $n = \frac{h^{1/6u}}{C}$ ($m^{-\frac{1}{3}}/s$) .....(10)

f, (Darcy Weisbach friction factor), calculated from (9), has been compared to f, calculated from (2) to (6), in fig. 6. In this figure data are distinguished by their location and by the dune steepness. This results in a grouping of the data. The predictors of van Rijn and of Vanoni and Hwang gives reasonable results; however the agreement is not entirely satisfactory. The dune dimensions and hydraulic roughness derived from both dune dimensions and hydraulic parameters for the lacations I, IV and V (fig. 1) have been plotted against time in figs. 7 and 8 respectively. The k-value (Nikuradse) has also been used for the hydraulic roughness because this value is directly comparable with the dune dimensions themselves. It is remarkable that the tendencies are more or less comparable. For the Rhine (fig. 8b) the agreement is particulary good. When dune dimensions are estimated from this figure for a steady flow of 8000 m^3/s, with help of an extrapolation of the given dune dimensions, it can be shown that the hydraulic roughness associated with a steady flow could be approximately 25% higher than the hydraulic roughness as given in fig. 8b because there is a considerable time lag in dune development, dune height, approximately 4 days. This will affect the waterlevels considerably.

It is also remarkable from figs. 7 and 8 that the time lag of the dune heights is much smaller than the time lag of the dune lengths. This is in accordance with results from laboratory experiments. (Wijbenga, Klaassen, 1981).

From this it can be seen that the possible predictors can be used in a relative way; they do not give information about the absolute values of the hydraulic roughness but they give some information about the changes in hydraulic roughness which occur when bedforms change.

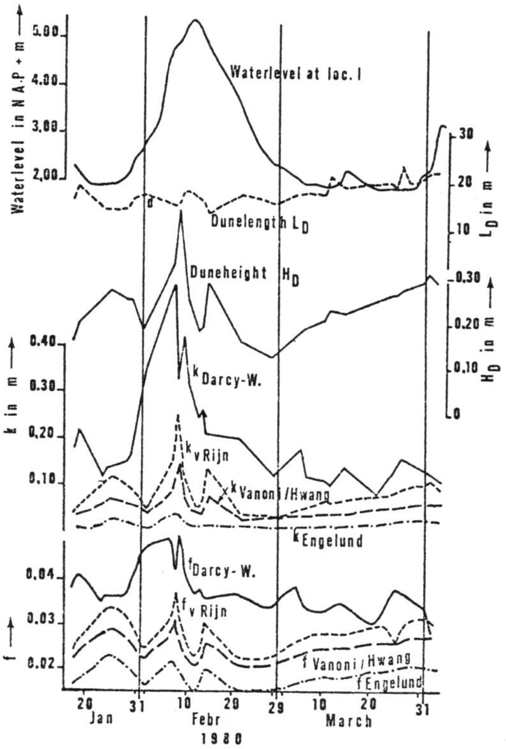

Fig. 7 Dune dimensions and hydraulic roughness during a flood on the river IJssel (loc. I)

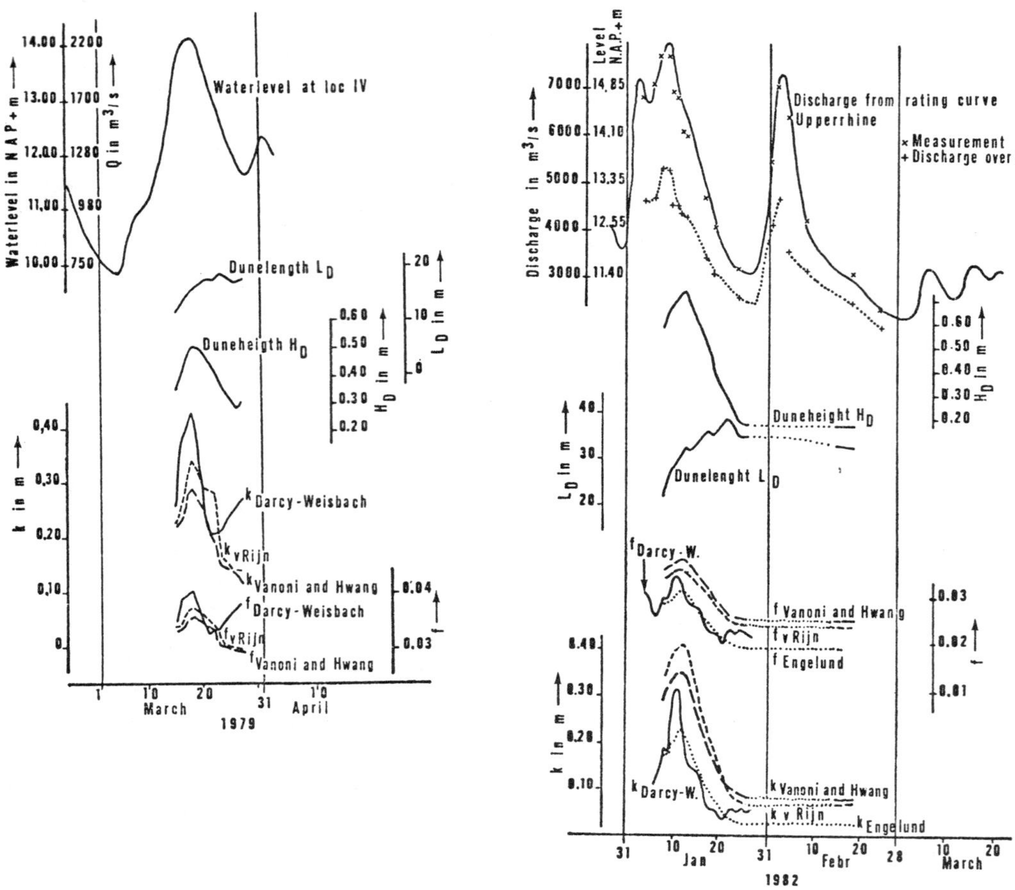

Fig. 8 Dune dimensions and hydraulic roughness during flood at the Pannerdens Canal (loc.IV) and the river Rhine (loc. V)

5 CONCLUDING REMARKS

a. Considerable and rapid changes in bedform dimensions have been measured on the Dutch rivers during unsteady flow conditions. These changes will affect the hydraulic roughness. The dune heights change more rapidly than the dune lengths.

b. There can be of the order of 25% difference between steady and unsteady roughness caused by a time lag in changes in bedform patterns.

c. The agreement between the hydraulic roughness derived from bedforms and from hydraulic parameters is not satisfactory. This is possibly caused by inaccuracies in measurements, by imperfections in the relationships themselves, or by local influences (e.g. bends).

However the relationships can be used in a relative way; they do not give information about the absolute value of the hydraulic roughness but they do give some information about the variation in hydraulic roughness when bedform dimensions change.

d. It is hoped that, with the help of a predictor for dune dimensions at extremely high discharges (steady flow), useful information about the hydraulic roughness to be expected at extreme high discharges (unsteady) can be derived (Wijbenga, Klaassen 1981) for use in flow models.

Acknowledgement

Measurements and data preparation have been carried out by employees River measurements and Research Sections respectively of the Southeast district, Rijkswaterstaat. Discussions about the progress have been held with mr. G.J. Klaassen of the Delft Hydraulics Laboratory.

REFERENCES

Engelund, F. (1978). "Hydraulic resistance for flow over dunes". Techn. Univ. of Danmark, ISVA, Progress Rep.no. 44, pp 16 - 20.

Havinga, H. and van Urk, A. (1980). "Solving river engineering problems in the Netherlands". IAHR Symposium on river engineering and its interaction with hydrological and hydraulic research. Belgrade May 1980.

Klaassen, G.J. and van Urk, A. (1982/3). "Testing predictors for resistance to flow of bedforms" (in preparation). Delft Hydraulics Laboratory, Rijkswaterstaat.

Rabiner, L.R. and Gold. B. (1975). "Theory and application of digital signal processing". Prentice Hall. U.S.

Van Rijn, L. (1982). "The prediction of bedforms and alluvial roughness". B.M.Sumer and A. Müller (eds.) Euromech 156: The mechanics of sediment transport, Rotterdam, Balkema.

Vanoni, V.A. and Hwang, L.S. (1967). "Relation between bedforms and friction in streams". Proc. Am. Soc. Civ. Engrs. Vol. 93, no. HY. 3.

Wijbenga, J.H.A. and Klaassen, G.J. (1981). "Changes in bedform dimensions under unsteady flow conditions in a straight flume". Sec. Int. Conf. on Fluvial Sediments. Univ. of Keele (U.K.), September 1981.

LIST OF SYMBOLS

C	= Chezy coefficient	$(m^{1/2}/s)$
$D_{10,50,90}$	= % of sample finer than given particle	(mm)
f	= Darcy Weisbach friction factor	(-)
f'	= Darcy Weisbach friction factor related to grain size	(-)
f''	= Darcy Weisbach friction factor related to bedforms	(-)
g	= Acceleration due to gravity	(m/s^2)
H	= Height of bedforms $(H(t), H_\infty)$	(m)
h	= Waterdepth	(m)
i	= Energy slope	(-)
k	= Nikuradse sand roughness	(m)
L	= Length of bedform	(m)
n	= Mannings roughness coefficient	$(m^{1/3}/s)$
U	= Flow velocity	(m/s)

4 Transport of sediment in suspension

Turbulent diffusion of solid particles in open channel flow

S.PETKOVIĆ & M.BOUVARD
Institut de Mécanique de Grenoble, France

ABSTRACT : The effect of the particle diameter upon the turbulent diffusion of solid particles in an open channel flow was examined. The results show the importance of the solid particle inertia effects and lift forces acting on a particle. The mean particle downward velocities are smaller than the settling velocities in still water.

1 INTRODUCTION

The results of several experimental studies (Coleman, 1970 ; Taggart & al. 1972 ; Barnard & Binnie, 1963) show that the coefficient of turbulent diffusion of solid particles, ε_s , is greater than the turbulent momentum transfer coefficient ε_m . This is rather surprising considering that the particles, due to their inertia, do not completely follow the high frequency turbulent velocity fluctuations. In fact this should result in $\varepsilon_s \leqslant \varepsilon_m$.

In order to clarify this problem an experimental study of the diffusion of spherical particles has been conducted in a turbulent two dimensional open channel flow. The basic idea was to deduce the concentration of solid particles from the probability density distributions of the particle position. The measurements of the mean and fluctuating components of solid particle velocities were carried out in order to verify the particle inertia effects. The fluctuating components of solid particle velocities are compared with those of the flow.

2 EXPERIMENTAL APPARATUS AND PROCEDURE

The experiments were carried out in a flume 0,35 m wide and 11,0 m long, with adjustable slope. Most of experiments were carried out with a flow depth h = 7,5 cm. The mean velocities were measured with a Pitot tube located at the centerline of the channel. The measured profiles agreed well with the logarithmic distribution proposed by Clauser $u = u_* \, (5{,}6 \log y^+ + 4{,}9)$.
Here, u_* is the shear velocity, $y^+ = y . u_*/\nu$, y is the distance from the wall, ν is kinematic viscosity. The shear velocity was predicted from the measured velocity profiles. This value was found to agree within 5 % with values obtained from the expression $u_* = (g.R.S)^{1/2}$ (g = acceleration due to gravity, R : hydraulic radius, S : channel slope).

Spherical plastic particles with different diameters (between 2 and 9 mm) and a density slightly greater than unity were used in the experiments.

An optical method, based on the interception of light sheet by the particles, was used for the detection of the particle position. When a particle intercepts a light sheet, correspondent photoelectric cell give its position. Probability density distributions were measured 6,0 m from the entrance of the channel. The probability density was defined as $P(y) = n/N.\Delta y$, where n was the number of particles in each strip of area Δy , N was the total number of particles in the cross-section. For each test about 500 particles were injected.

The particle velocities have been determined by a photographic technique. The system consisted of two units : a conventional 35 mm camera and a stroboscope. The flash frequency of the stroboscope was 10 Hz.

The co-ordinate data obtained from the photographic technique were processed to yield sequential pairs of instantaneous longitudinal and vertical velocities (U,V). The flow depth was divided into small increments Δy = 0,25 cm and 0,50 cm for the 2 and 9 mm particle diameter respectively. The mean and RMS values ($\langle U \rangle$, $\langle V \rangle$, U', V') of the particle velocities were computed from about 40 trajetories for each type of particle. The total number of sequential pairs (U , V) was about 800. The mean upward and downward vertical particle velocities were calculated separately.

An ultra-sonic velocimeter, with a frequency f = 2 MHz , was used for the measurements of turbulent longitudinal and vertical flow velocities.

3 RESULTS AND DISCUSSION

In Figure 1 three vertical probability density distributions for the particles of 2 ; 5 and 9 mm, having identical settling velocity in still water (W = 2,9 cm/s) are presented. The probability density distribution becomes more homogeneous with increasing particle diameter.

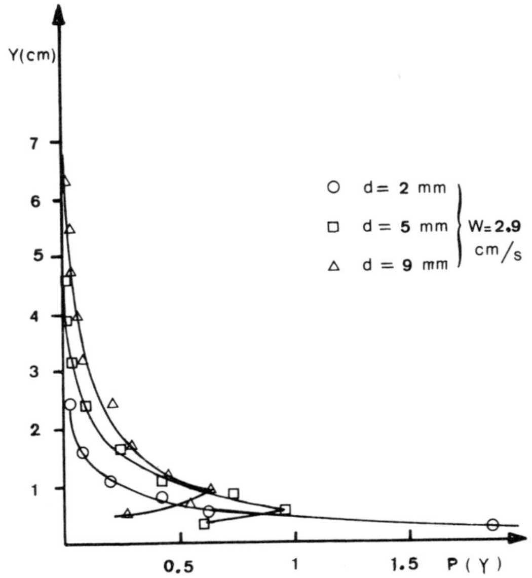

Fig.1. Distributions of vertical probability density. h = 7,5 cm ; U_* = 2,67 cm/s

The experimental results for 2 ; 5 and 9 mm particles are given in Figure 2 in terms of U_*/W and $(\bar{y} - d/2)/h$, ($\bar{y}$: mean particle position , d : particle diameter, h : flow depth). The mean particle position increases with increasing particle diameter.

Longitudinal velocity profiles for 2 and 9 mm particle diameters are given in Figure 3. It is seen that the particle velocities are smaller than the flow velocities. This can be attributed to the particle inertia.

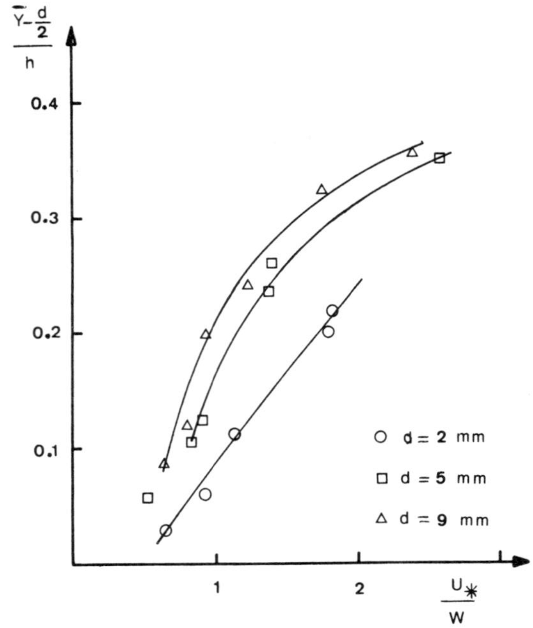

Fig.2. Effect of U_*/W upon the mean vertical position of the particle.

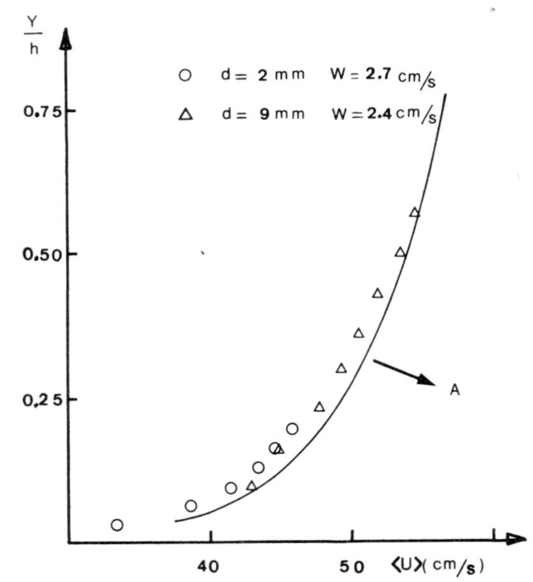

Fig.3. Mean longitudinal velocity profiles for 2 and 9 mm particles. A, logarithmic velocity profile $U = U_*(5,6 \log y^+ + 4,9)$ h = 7,5 cm ; U_* = 2,54 cm/s

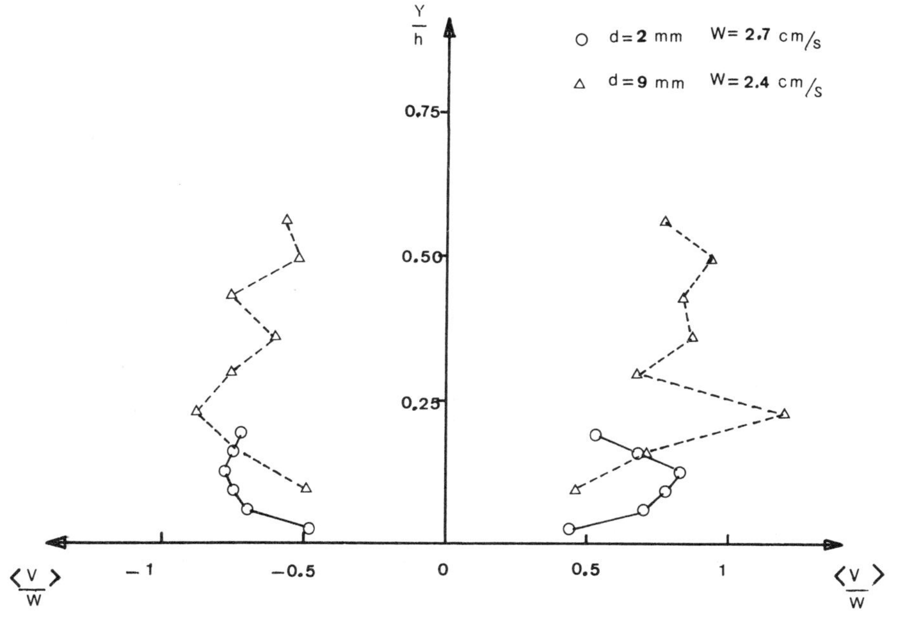

Fig.4. Mean vertical velocity data for 2 and 9 mm particles.
$-\langle V \rangle$, downward velocities ; $+\langle V \rangle$ upward velocities
$h = 7{,}5$ cm ; $U_* = 2{,}54$ cm/s

Vertical mean velocity data are given in Figure 4. The downward velocities of the particles are on the average 30 % smaller than the settling velocities in still water. This is in agreement with the experimental and theoretical results of Aydin (1979), Bouvard (1951) and Bagnold (1974).

The symmetry between upward and downward particle velocities (Figure 4) is a consequence of the effect of the lift forces which caused the upward motion of the particles. The expression for the lift on a sphere translated with velocity Vr relative to a flow possessing a shear dU/dy , proposed by Owen (1969) is
$L \sim \rho \, . d^3 . \, Vr . dU/dy$

As the particle velocities are on the average smaller than the flow velocities (Figure 3) it is expected that the lift force is on the average upward directed.

The drag on a particle and its submerged weight are
$F \sim \rho . d^2 . V^2$
$G \sim g \, (\rho_s - \rho) . d^3$
where V is vertical particle velocity, ρ_s and ρ particle and fluid density. Supposing that the relative velocity is proportional to the shear velocity, $Vr \sim U_*$, and that outside the viscous sublayer $dU/dy \sim U_*/y$, it follows that
$L \sim \rho . d^3 . U_*^2/y$

Considering the case where the solid particle is at the bottom of the channel with zero vertical velocity, the relation between lift force acting upward and gravity force acting downward is
$L/G \sim (d/y) \, \rho \, . U_*^2/g.d. (\rho_s - \rho)$
For example the values of the parameter $\rho . U_*^2/g.d. (\rho_s - \rho)$ for the three density probability distribution given in Figure 1 are 1,02 ; 0,66 and 0,54 for the 2 ; 5 and 9 mm particles respectively. For y = d/2 the values of L/G are 2,0 ; 3,0 and 3,7. Thus the solid particles are apt to be transported in the suspension.

When a particle has certain upward velocity, say $V \sim U_*$, a drag on the particle must be added to the gravity force.
It follows :
$(F+G)/L \sim (y/d) \, (1+g.d(\rho_s - \rho)/\rho . U_*^2)$
At y = 1,0 cm , L/(F+G) is 0,1 ; 0,3 and 0,6 for the 2 ; 5 and 9 mm particles respectively. This explains the differences between vertical probability distributions for the particles having the same settling velocity but different diameters. Lift force and particle inertia increase with increasing particle diameter.

In Figures 5 and 6 longitudinal and vertical fluctuations of the particle velocities are compared with turbulent fluctuations of the flow. It seems that for the particles of 2 mm the effects of particle

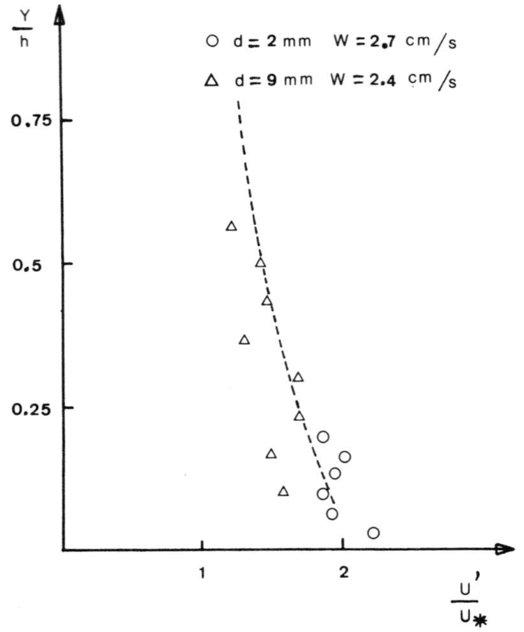

Fig.5. Longitudinal velocity fluctuations of particles and fluid (dashed line). h = 7,5 cm ; $U_* = 2,54$ cm/s

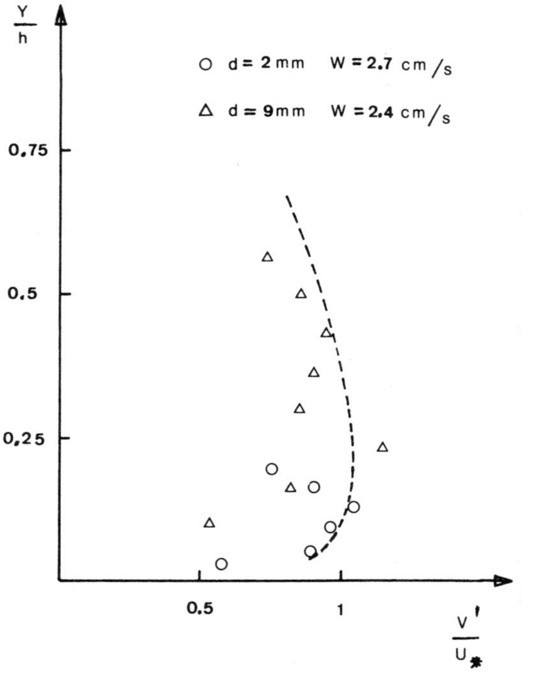

Fig.6. Vertical velocity fluctuations of particles and fluid (dashed line). h = 7,5 cm ; $U_* = 2,54$ cm/s

inertia are compensated by the effects of the lift forces. Thus the fluctuations of the particles agree on the average with those of flow. For the 9 mm particles effects of the particle inertia are more important since the fluctuations of the particle velocities are on the average smaller than those of flow, in spite of relatively large depth increment $\Delta y = 0,5$ cm.

4 CONCLUSIONS

From the "isolated particle experiments" better insight into the mechanisms governing particle diffusion in turbulent open channel flow has been obtained.

The effects of particle inertia and lift force are shown to be important. The lift increases with the particle diameter and near the bottom it can completely balance the particle weight.

The measured downward particle velocities are smaller than the settling velocities in steal water and this is consistent with preceding explanation.

REFERENCES

Aydin,M. 1979. The velocity of a solid particle settling in open channel flow. 18th Congress IAHR. Cagliari.

Bagnold,R.A. 1974. Fluid forces on a body in a shear-flow ; experimental use of stationary flow. Proc. R. Soc. London A 340.

Barnard,B.J.S. & Binnie,A.M. 1963. The vertical diffusivity and mean velocity of particles in a horizontal water pipe. J. Fluid Mech. Vol 15, part 1.

Bouvard,M. 1951 Etude de l'influence de la turbulence sur la chute des particules solides. La Houille Blanche. Nov-Dec.

Coleman,N.L. 1970. Flume studies of sediment transfer coefficient. Water Resources Research. Vol. 6, n° 3.

Owen,P.R. 1969. Pneumatic transport. J. Fluid Mech. Vol. 39, part 2

Taggart,W.C., Yermoly,C.A., Montes,S. & Ippen,A.T. 1972. Effects of sediment size and gradation on concentration profiles in turbulent flow. MIT, Repport n° 152.

Euromech 156: Mechanics of Sediment Transport / Istanbul / 12-14 July 1982

Stochastic model for particle movement in turbulent open channel flow

W.BECHTELER & K.FÄRBER
Hochschule der Bundeswehr, München, Germany

ABSTRACT: A stochastic model of suspended solid movement in turbulent open channel flow has been developed which can be used to calculate not only concentration distributions but also bed load transport rates. The stochastic model gives a statement of probability for the instant movement of a particle. By a Monte Carlo simulation many random walks are calculated to determine concentration, step lengths and resting periods distributions which determine bed load rates.

1. INTRODUCTION

The movement of sediment in streams or open channels is a very complex process and its calculation may never be completely possible because there is often a great uncertainity of knowledge of the parameters influencing the transport process. Turbulence and also the mechanism of dispersion and resuspension has a random character which can be corresponded in making stochastic models. These models describe the movement of solids by a statement of probability for the instant movement of a suspended particle which is based on measured turbulence and flow parameters. Stochastic models are described by Bayazit (1972), Li and Shen (1975), Chiu (1967), Bechteler and Färber (1981), reviews are given by Alonso (1981) and Bechteler (1980). With the help of stochastic models and the so called Monte-Carlo technique a physical process is simulated and in case of a dispersing or saltating particle many random walks have to be determined giving the probability of location or existence of a particle at a certain point. In mathematical theory it can be shown that by this probability an equivalent solution of a differential equation of second order describing transport can be achieved (Gnedenko, 1968).

2. STOCHASTIC MODEL FOR PARTICLE MOVEMENT

A two-dimensional, discrete stochastic model for the movement of a suspended solid can be made by deterministic and stochastic velocity components. A single time step Δt from i to i+1 can be modeled by

$$x_{i+1} = x_i + \Delta t \; (u(y) + u') \qquad (1)$$

$$y_{i+1} = y_i + \Delta t \; (-v_s + \phi(k) \; |v'|) \qquad (2)$$

The deterministic velocities are settling velocity v_s and flow velocity u(y) in direction x; u' and v' are fluctuating velocity components. The weighting function $\phi(k)$ is considered to depend on the difference of relative concentration around site k. k is denoting a location when flow depth is divided into M layers numbered k=1..M and spaced $\Delta y=h/M$.
The weighting function $\phi(k)$ randomly takes values +1 or -1, so that according to the concentration difference more or less up or down directed fluctuating velocities v' are chosen. In this model an idea of Prandtl is involved, who thought of the balancing mechanism of suspended sediment transport as of eddies carrying more or less solids from locations of different concentrations.
The effect of a boundary, for example at the channel bottom will increase the concentration at this site because particles will settle down. By a turbulent eddy more sediment is carried upwards from that layer of high concentration to an upper layer of low concentration than by an eddy acting in reverse direction.
The effect of different concentrations in different flow depth plays not only a

major role in suspended load transport but also the pickup rate determining bed load transport depend on the number of particles at channel bottom. This complex process of particle motions near the bottom in turbulent flow is determined in our stochastic model by the parameters of the weighting function $\phi(k)$. Fig. 1 shows model parameters and their relation to physical processes.

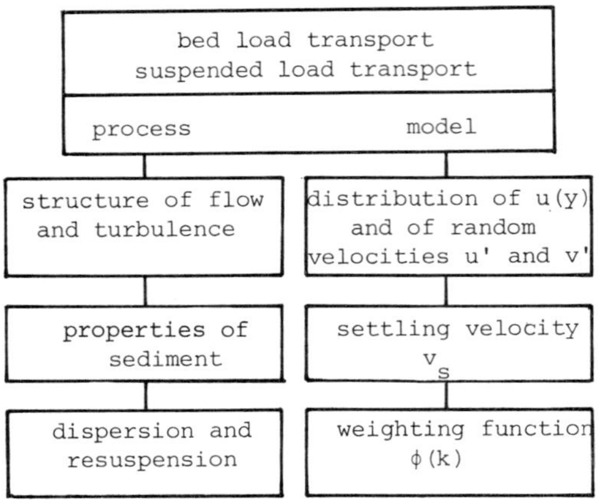

Fig 1 Stochastic model parameters

2.1 Turbulent velocity components

To simulate shear flow generated fluctuating velocity components, a simplified form of the Reynolds-stress equation is used to connect velocity correlations with shear stress. From these velocity correlations the parameters of the velocity distributions are determined. From these distributions single velocity components are randomly chosen to compute the random walk of a particle. For our stochastic model a turbulent model is used, which was proposed by Li and Shen (1975) to calculate the settlement of particles.
It is assumed that cross-streamwise fluctuating velocities v' are normally distributed:

$$p(v') = \frac{1}{\sqrt{2\pi}\,\sigma_{v'}} \exp\left(-\frac{1}{2}\left(\frac{v'}{\sigma_{v'}}\right)^2\right) \qquad (3)$$

The standard deviation $\sigma_{v'}$ is derived from the co-variance $\overline{u'v'}$, which is calculated by a simplified Reynolds-equation:

$$\overline{u'v'} = \nu \frac{du}{dy} - u_*^2 \left(1 - \frac{y}{h}\right) \qquad (4)$$

in which ν = kinematic viscosity of fluid, u_* = shear velocity refered to shear stress at channel bottom, and the definition of the correlation coefficient r:

$$r = \frac{\overline{u'v'}}{\sigma_{u'}\sigma_{v'}} \qquad (5)$$

and the assumption that

$$\sigma_{v'} = C \cdot \sigma_{u'} \qquad (6)$$

so that

$$\sigma_{v'} = \sqrt{\frac{C}{r}\,\overline{u'v'}} \qquad (7)$$

The values of correlation coefficient r and parameter C are determined according to measurements of McQuivey and Richardson (1969) and Laufer (1954), and are used for calculations in the same approximation as given by Li and Shen (1975).
For the longitudinal flow direction, a random variable u' is taken from a normal conditional distribution. This distribution considers the correlation between cross- and streamwise turbulent velocity fluctuations, while taking a random u' at a given v'

2.2 Weighting function

To determine the weighting function which depends on the concentration at different flow depths one has to know the concentration distribution, which however is to be determined.
To come along with this difficulty a modified Monte-Carlo technique is used, which is possible by the capabilities of modern computers. Instead of computing random walks one by one, many random walks are simultaneously calculated from an initial concentration distribution. So by one step the instant relative concentration c (k) of sediment at site k is known and therefore the weighting function ϕ (k) in Eq. 2 for the computations of the next step.
For present simulations the signs for up or down acting turbulent velocities are randomly chosen from a normal distribution. If a random variable s' of this normal distribution is below a certain value s_c given by the difference of relative concentrations:

$$s'_c = c(k-1) - c(k+1) \qquad (8)$$

an upwards directed velocity is used. Therefore the probability for this motion is

$$P(s' < s_c) = \tag{9}$$

$$\frac{1}{\sqrt{2\pi}\,\sigma_m} \int_{-\infty}^{s'_c} \exp\left(-\frac{1}{2}\left(\frac{s'}{\sigma_m}\right)^2\right) ds'$$

and for a downward directed motion:

$$P(s' \geq s_c) = 1 - P(s' < s_c) \tag{10}$$

as indicated in Fig. 2

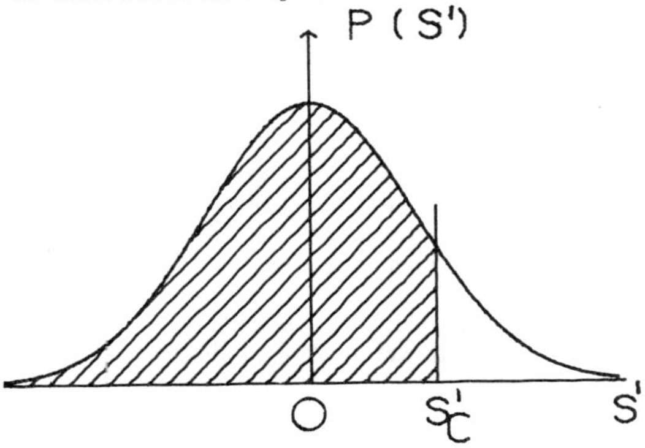

Fig. 2 Distribution of s'

The sensitivity of the effect of the weighting function is given by its standard deviation σ_m and is calculated from the settling velocity v_s and root-mean-values of cross-streamwise fluctuating velocities $\sigma_{v'}$:

$$\sigma_m = \frac{v_s}{\sigma_{v'}} m \tag{11}$$

m is a model parameter reflecting the influence of chosen computational parameters N and M, number of total walkers and number of layers.

2.3 Time step

A time step Δt is used for simulations which was derived by Bechteler and Färber (1982) from setting a dispersion coefficient of Rouse (1965) equal to a dispersion coefficient derived from a theoretical analysis of a dispersion process:

$$\Delta t = \frac{1.5\ \Delta y^2}{\kappa\ \beta . u_* (h-y) y/h} \tag{12}$$

κ, v.Kármán constant, v_* shear velocity, β parameter describing difference between mass and momentum transfer.

With the developed stochastic model it is possible to calculate steady state concentration profiles as concentration distributions in nonequilibrium conditions (Bechteler and Färber (1982)) by changing flow parameters during calculation procedure.

3. APPLICATION TO SEDIMENT MOTION ON FLAT BED

Most of stochastic models developed for bed load movement are based on the assumption that the movement of bed sediment particles is composed of a series of steps and resting periodes. Einstein (1937) was the first who developed a one-dimensional probabilistic model for bed load transport. Numerous investigators (see Cheong and Shen (1976) and Lee and Jobson (1977) have investigated the characteristics of bed load movement and determined the probability distribution for step length and rest periods of a sediment particle, which plays a major part in quantifying the bed sediment transport.

To calculate bed load transport with our discrete two-dimensional stochastic model the flow depth h is divided into M layers spaced Δy and of unit length, indexed k. N random walkers are distributed across these M layers. The mean number of walkers in a layer or cell is $\bar{n}_k = N/M$. At any time the number of walkers in a layer k may be more or less than this mean number. The relative concentration in one cell k at a discrete time step i is c(k,i) and is proportional to the number of random walkers n_k in this layer divided by the mean number $\bar{n}_k$.

As initial condition of bed load transport calculation all random walkers are in layer k=1, where they are considered to be immobile. The limiting variable s_c for the weighting function $\phi(k)$ at this site k is calculated by

$$s'_c = c(1)-c(2) \tag{13}$$

So, once the scour rate which depends on concentration of errodible bed is modelled with the weighting function. If c(1) is much smaller than c(2) the probability of pick-up or leaving layer k=1 for a random walker is vanishing, because the weighting function favours only transitions from higher layers to lower layers.

To determine bed load transport, step lengths and resting periods of particles leaving layer k=1 are recorded and analysed.

3.1 Bed load transport rate

The bed load transport rate q_B usually is given in terms of mean step length Λ and mean resting periods $\bar{t}$ or pick-up rates $p_s = 1/\bar{t}$. With correct geometric and shape parameters of sand grain A_2 and A_3 and with d diameter of sand q_B can be expressed (Nakagawa and Tsujimoto, 1980) as:

$$q_B = \frac{A_3}{A_{2d}} v_g \tag{14}$$

v_g group velocity of sand particles
$v_g = \Lambda/\bar{t}$.
Another expression using γ as the porosity of bed sand and $a_1 d$ as the moving layer thickness is also given by them:

$$q_B = (1-\gamma)\, v_g\, a_1\, d \tag{15}$$

In our stochastic model the bed load transport rate is expressed to be proportional to v_g and the mean number of random walkers in layer k=2 per unit time $n_{k=2}$:

$$q_B \approx n_{k=2} \cdot v_g \tag{16}$$

First the step length distributions and resting periods are determined.

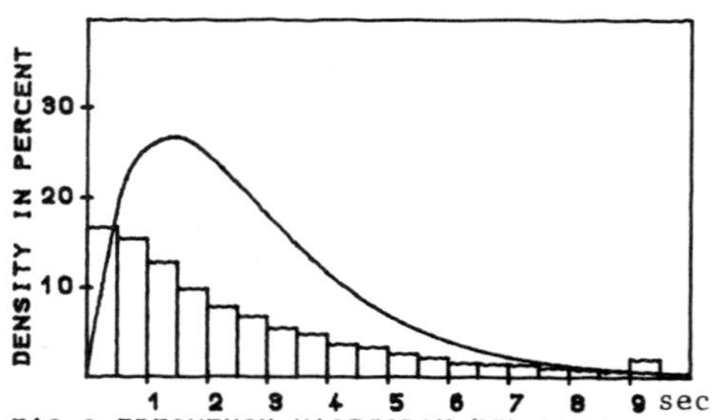

FIG.3 FREQUENCY HISTOGRAM FOR RESTING PERIODS WITH GAMMA FIT

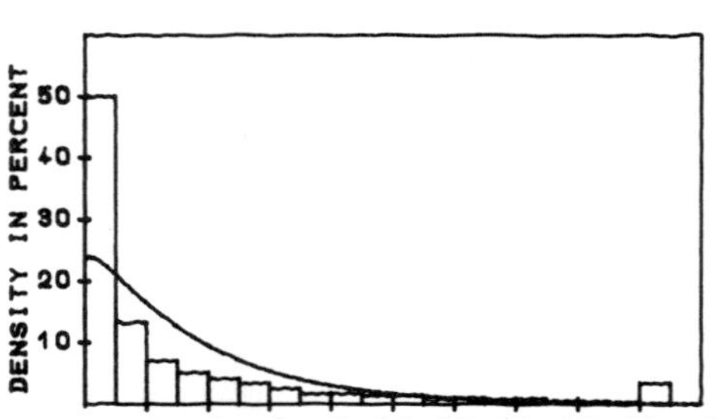

FIG.4 FREQUENCY HISTOGRAM FOR STEP LENGTHS WITH GAMMA FIT

In Fig 3 und 4 the results of a simulation are shown. Flow parameters were a logarithmic flow profile with mean flow velocity u_O = 0.9 m/sec, correlation coefficient $\bar{r}$ = 0.7 and factor C = 0.5, flow depth h = 0.5 m shear velocity u_* = 0.045 m/sec.
Sediment parameters were diameter d = 0.005 m, density ρ_s = 2380 kg/m^3 and settling velocity of a sphere v_s = 0.5cm/sec. Simulation model parameters were chosen as thickness Δy of a layer k equal diameter of settling particle d, so number of layers M was 100 and number of walkers N = 1000. Model parameter m was

$$m = 1/5\sqrt{\frac{M-1}{N}}$$

Hung and Shen (1971) found in their experiments that resting period has a distribution that cannot be represented by the conventional gamma or exponential distribution (Sayre and Hubbell, 1965). The stochastic model simulation results show a similar behaviour, because neither gamma nor exponential fit gave a satisfactory chi-square value.
The variation of bed load transport rate with flow intensity is shown in Fig. 5 in terms of dimensionless shear stress

$$T_o^* = To/(\rho_s - \rho_f)\, g\, d \tag{17}$$

and dimensionless transport rate q_B^*

$$q_B^* = c_s q_B / \sqrt{(\rho_s/\rho_f - 1)\, .\, g\, d^3} \tag{18}$$

Dots are simulation results and the drawn line shows Einsteins bed load function (Einstein 1950) ${}^E q_B^*$.

c_s is a scaling parameter ($c_s = 1.65 \cdot 10^{-5}$), so that both bed load rates $\bar{q}_B$ and ${}^E q_B^*$ could be on the same plot.
As a practial application of simulation the ratio between bed load and total load for different flow intensities can be calculated. This is simply done in setting this ratio R equal the number of walkers in layer K=2 divided by the number of total walkers:

$$R = \frac{n_{k=2}}{n_{k=2} + n_{k>2}} \tag{20}$$

The result is shown in Fig. 6. The plot shows a different shape as a result calculated by a model of Krishnappan (1977), but the transition from bed load transport can clearly be seen. In that transition domain the model assumptions for bed load transport will become invalid, because random walkers will not settle down, and so step lengths increase.

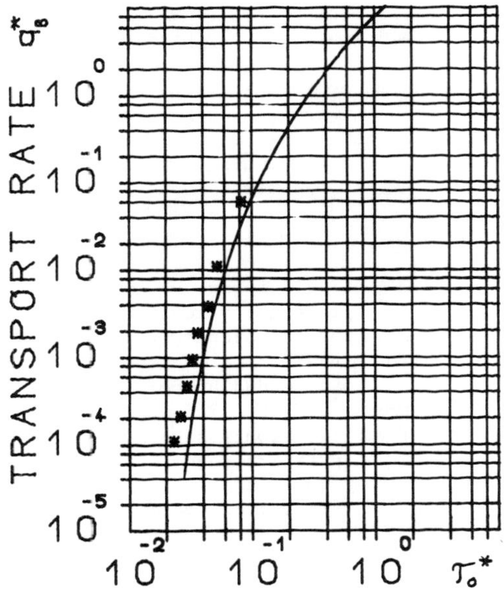

Fig. 5 Sediment transport rate

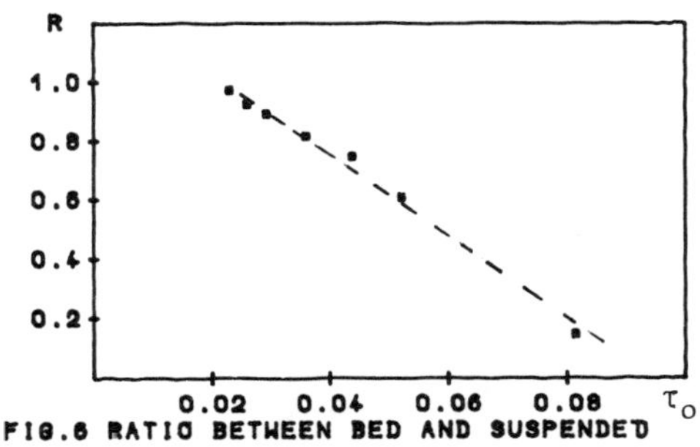

FIG.6 RATIO BETWEEN BED AND SUSPENDED LOAD

4. CONCLUSION

It could be shown that the stochastic model with the concept of the weighting function developed to calculate suspended load transport, can also be used to calculate bed load transport. Identifiing model parameters and calibrating these to practical cases is a task of future work.

REFERENCES

Alonso, C.V.: "Stochastic Models of Suspended Sediment Dispersion", Journal of the Hydraulic Division, ASCE, No HY 6, June 1981, pp. 733 - 757.

Bayazit, M.: "Random Walk Model for Motion of a Solid Particle in Turbulent Open-Channel Flow", Journal of Hydraulic Research 10 (1972), No. 1, pp. 1 - 14.

Bechteler, W.: "Stochastische Modelle zur Simulation des Transportes suspendierter Feststoffe", Wasserwirtschaft, Heft 6, 70. Jg. (1980).

Bechteler, W., Färber, K.: "Stochastische Modelle zur Beschreibung des Rückhalte- und Transportvorganges von suspendierten Sedimentpartikeln in turbulenten Gerinneströmungen", Mitt. Institut für Wasserwesen 5 (1981), S. 2 - 30.

Bechteler, W., Färber, K.: "Stochastic Model for Nonequilibrium Transport of Suspended Sediment", to be published (1982).

Cheong, H.F., Shen, H.W.: "On the Motion of Sand Particles in an Alluvial Channel", Proc. 2nd Int. IAHR Symp. on Stochastic Hydraulics, 1977, pp. 219 - 239.

Chiu, Ch.-L.: "Stochastic Model of Motion of Solid Particles", Proc. ASCE, 93 (1967), HY 5, pp. 203 - 218.

Einstein, H.A.: Der Geschiebetrieb als Wahrscheinlichkeitsproblem", Mitt. Versuchsanstalt für Wasserbau, Eidgenössische Technische Hochschule Zürich, Verlag Rascher und Co., 1937, S. 110.

Einstein, H.A.: "The bed-load function for sediment transportation in open channel flows", Techn.Bulletin 1026, 1950, Soil Conservation Service, U.S. Dept. of Agric.

Gnedenko, B.V.: "The Theory of Probability" Chelsea Publishing Company, New York, 1968.

Hung, C.S., Shen, H.W.: "Stochastic Models of Sediment Motion on Flat Bed", Jour. of Hyd.Div., Proc.ASCE, 1976, Vol. 102, No. HY 12, pp. 1745 - 1759.

Krishnappan, B.G.: "Simulation of Sediment Entrainment in Open channel Flows", Proc. of the 2nd Int.Symposium on Stochastic Hydraulics in Lund, 1976, pp. 336 - 350.

Laufer, J.: "The Structure of Turbulence in Fully Developed Pipe Flow", Report No. 1174, National Advisory Committee for Aeronautics, 1954.

Lee, B.K., Jobson, H.E.: "Stochastic analysis of dune bed profiles", Journ. of the Hydraulics Division, ASCE, 1974, Vol. 100, No. HY 7, Proc. Paper 10657, July, pp. 849 - 867.

Li, R.-M., Shen, H.W.: "Solid Particle Settlement in Open-Channel Flow", Proc. ASCE, 101, 1975, HY 7, pp. 917 - 931.

McQuivey, R.S., Richardson, E.V.: "Some Turbulence Measurements in Open-Channel Flow", Proc. ASCE, 95, 1969, HY 1, pp. 209 - 223.

Nakagawa, H., Tsujimoto, T.: "On Probabilistic Characteristics of Motion of Individual Sediment Particles on Stream Beds", Proceedings of the 2nd International Symposium on Stochastic Hydraulics, International Association of Hydraulic Research, Lund, Sweden, 1976, pp. 293 - 316.

Rouse, H.: "Modern Conceptions of Mechanics of Fluid Turbulence", Transactions ASCE, Vol. 102, 1937, pp.463.

Sayre, W.W., Hubbell, D.W.: "Transport and Dispersion of Labeled Bed Material, North Loup River, Nebraska", U.S.G.S., Prof.Paper 433-C, pp. 48.

NOTATION

The following symbols are used in this paper

A_2, A_3, a_1	shape parameters of sand particles
C	proportionality factor between turbulent velocity components
c	relative concentration
c_s	scaling parameters
c(k,i)	concentration in layer k at time step i
g	gravity constant
i	index denoting time step
k	index denoting layer or cell
M	number of layers across flow depth
m	model, parameter
N	total number of random walkers
n_k	number of walkers in one layer k
$\bar{n}_k$	mean number of walkers in one layer k
P(k/k+1)	probability for transition from k to k+1
p(v')	probability density of turbulent velocity component v'
q_B	bed load transport rate
q_B^*	dimensionless bed load transport rate
r	correlation coefficient of turbulent velocity components
R	ratio of bed load to total load
s'	random variable of weighting function ϕ
u'	random velocity component in flow direction
u(y)	deterministic velocity distribution across flow depth
u_o	mean flow velocity
$\overline{u'v'}$	covariance of u' and v'
u_*	shear velocity
v'	random velocity component in cross-streamwise direction
v_g	group velocity
v_s	settling velocity
x	streamwise direction
x_i	position in streamwise direction at step i

y_i	position in cross-streamwise direction at step i
β	proportionality factor between mass and momentum transfer, set to one
γ	porosity of sand
Δt	time step
$\bar{t}$	mean resting period
κ	v. Kármán`s constant
Λ	mean step length
ν	kinematic viscosity
ρ_s, ρ_f	density of solid, fluid
σ_m	standard deviation of weighting function distribution
$\sigma_{u'}$	standard deviation of random velocity distribution of u'
$\sigma_{v'}$	standard deviation of random velocity distribution of v'
τ_o, τ_o^*	shear stress at channel bottom, dimensionless form
$\phi(k)$	weighting function

Euromech 156: Mechanics of Sediment Transport / Istanbul / 12-14 July 1982

Numerical modelling of sediment transport in open channel flows

I.ÇELIK
University of Karlsruhe, Germany

ABSTRACT: Sediment concentration distributions are predicted in a fully developed open channel flow with a fixed, rough bottom for various types of sediment material. Calculations are done using the k-ε turbulence model and relating the mass transfer coefficient to the turbulent eddy-viscosity distribution which is also calculated as a part of the solution. Deposition and entrainment of the sediment particles are considered. Influence of various hydraulic parameters on the predicted sediment concentration are investigated. Predictions are compared with experiments. The results show that concentration distributions can be predicted adequately with the calculated velocity and eddy-viscosity distributions.

1 INTRODUCTION

The presence of a free surface in open channel flows, similar to a solid wall boundary, causes a damping in the vertical turbulent motion due to surface tension and geometrical restrictions, thus, reducing the intensity of the vertical velocity fluctuations and the characteristic turbulent length scale in the vicinity of the surface. The surface damping of the vertical fluctuations and the reduction of the turbulent length scale, both, inhibit the vertical momentum transport by the turbulent motion so that the eddy viscosity is reduced near a free surface (Figure 1). The experimental work reported by Nakagawa et al. (1975) and Ueda et al. (1976) for fully developed two-dimensional turbulent open-channel flows strongly support the foregoing arguments. Under the guidance of these experiments Çelik et al. (1982) have recently presented a modified version of the standard k-ε turbulence model which is capable of simulating the above mentioned free surface-surface-proximity effects on turbulence. In particular, their model produces the measured parabolic eddy-viscosity distribution in open channel flows quite satisfactorily.

In this paper the turbulence model presented by Çelik et al. is applied to sediment laden fully developed open-channel flows with a fixed, rough bed. The governing equations are solved numerically using a finite-difference procedure suited especially for boundary-layer-type flows. The predicted velocity and eddy-viscosity distributions are used in calculating suspended sediment-concentration distribution by assuming that the Reynolds' analogy holds. The experiments reported by Jobson and Sayre (1970a) are used to compare predictions with measurements. The relative influence of the turbulent Schmidt number, fall velocity and the boundary condition at the bed on the predicted concentration distribution, are investigated. Although, in the past many other investigators have attempted to predict the suspended sediment concentrations numerically, the present method seems to be one of the first attempts to use the predicted velocity and eddy-viscosity distributions with a second order turbulence-closure model in calculating sediment concentration distributions. The work presented here constitutes the initial stage of a more general analysis, currenty under investigation, of numerical modelling of suspended sediment transport in non-uniform, unsteady free-surface flows with mobile beds by using the k-ε turbulence model.

2 THE MATHEMATICAL MODEL

The mean velocity field of steady two-dimensional turbulent open channel flow is governed by the thin-shear layer form of the x-momentum equation

$$U\frac{\partial U}{\partial x} + V\frac{\partial U}{\partial y} = - g\frac{dh}{dx} + g\sin\theta - \frac{\partial \overline{uv}}{\partial y} \qquad (1)$$

and by the continuity equation

$$\frac{\partial U}{\partial x} + \frac{\partial V}{\partial y} = 0 \qquad (2)$$

where U and V are the mean flow velocity, u and v are the fluctuating velocity components in the x- and y-directions, respectively, g is the acceleration of gravity, h is the flow depth and θ is the channel inclination.

The turbulent shear stress uv appearing in Eq. (1) is determined from a modified version of the k-ε turbulence model obtained by simplifying $\overline{u_i u_j}$-transport equations as presented by Çelik et al. (1982). In this model $\overline{uv}$ is determined from the eddy-viscosity relation

$$-\overline{uv} = \nu_t \frac{\partial U}{\partial y} \qquad (3)$$

and the eddy-viscosity ν_t is related to the turbulent kinetic energy k and the dissipation rate ε by

$$\nu_t = c_\mu k^2/\varepsilon \qquad (4)$$

where c_μ is a function of production P and dissipation rate of k, and the vertical distance from the free surface. k and ε are obtained by solving the thin shear layer forms of the semi-empirical transport equations for these quantities:

$$U\frac{\partial k}{\partial x} + V\frac{\partial k}{\partial y} = \frac{\partial}{\partial y}\left(\frac{\nu_t}{\sigma_k}\frac{\partial k}{\partial y}\right) + \nu_t\left(\frac{\partial U}{\partial y}\right)^2 - \varepsilon \qquad (5)$$

$$U\frac{\partial \varepsilon}{\partial x} + V\frac{\partial \varepsilon}{\partial y} = \frac{\partial}{\partial y}\left(\frac{\nu_t}{\sigma_\varepsilon}\frac{\partial \varepsilon}{\partial y}\right) + c_{\varepsilon 1}\frac{\varepsilon}{k} P - c_{\varepsilon 2}\,\varepsilon^2/k \qquad (6)$$

where σ_k and σ_ε are empirical functions given by $c_\mu/c_k\overline{v^2}/k$ and $c_\mu/c_\varepsilon\overline{v^2}/k$; c_k, c_ε, $c_{\varepsilon 1}$ and $c_{\varepsilon 2}$ are empirical constants the values of which are taken 0.18, 0.18, 1.44 and 1.92, respectively, as in Çelik et al. (1982). For more details of the model the reader may refer to Çelik et al. (1982).

The basic mass-transfer equation which can be derived by applying the principle of mass conservation for the steady state condition can be simplified, see for example Jobson and Sayre (1970b), to obtain the two-dimensional thin shear-layer form

$$U\frac{\partial C}{\partial x} + V\frac{\partial C}{\partial y} = \frac{\partial}{\partial y}\left(\frac{\nu_t}{\sigma_c}\frac{\partial C}{\partial y}\right) + \frac{\partial}{\partial y}(V_s C) \qquad (7)$$

where C is the concentration, σ_c is the turbulent Schmidt number and V_s is the mean fall velocity of a particle in a turbulent flow and is positive in the downward direction. The longitudinal component of the sediment velocity, U_s, is taken equal to the fluid velocity, U. In Eq. (7) the Reynolds' analogy (i.e. the turbulent transfer coefficient for C is proportional to the turbulent momentum transfer coefficient ν_t), and that the longitudinal diffusion term is negligible in comparison with the other terms are assumed. Furthermore, it is assumed that the concentration C is sufficiently small, such that the grouping effects and the consequences of particle collisions are not important.

2.1 Boundary conditions

The boundary conditions on U, k and ε are not applied at the channel bed itself but at a point $y = y_c$ near the bed where the logarithmic law of the wall prevails. As a result (see for example Rodi, 1980) one gets

$$U_c = \frac{U_*}{\kappa}\ell n\left(E\frac{y_c U_*}{\nu}\right),\; k_c = \frac{U_*^2}{\sqrt{c_{\mu c}}},\; \varepsilon_c = \frac{U_*^3}{\kappa y_c} \qquad (8)$$

where ν is the kinematic viscosity, U_* is the friction velocity, E is a wall roughness parameter, κ is the von Kármán constant which is taken to be 0.42, and the subscript c denotes values at the point $y = y_c$.

At the free surface symmetry conditions (i.e. zero gradient) applied for U and k but for ε an empirical surface boundary condition suggested by Çelik et al. (1982) as

$$\varepsilon = 5.54\, k^{3/2}/h \quad \text{at } y = h \tag{9}$$

is used. This boundary condition is based on the assumption that the macro-length scale of turbulence, $\ell (\propto k^{3/2}/\varepsilon)$ is reduced by the presence of a free surface.

The boundary conditions for the concentration C are: at the water surface there is no transport across the water surface, hence

$$\frac{\nu_t}{\sigma_c}\frac{\partial C}{\partial y} + V_s C = 0 \tag{10}$$

and at the bed, when there is no entrainment from the bed load,

$$\frac{\nu_t}{\sigma_c}\frac{\partial C}{\partial y} + (1-A)\, V_s C = 0 \tag{11}$$

wherein A is the probability that a particle settling to the bed is deposited there; thus, AV_sC is the rate of transfer from suspension to the bed. The measured concentration distribution at the beginning value of x is taken as the additional boundary condition that is needed in order to integrate Eq. (7).

3 METHOD OF SOLUTION AND APPLICATION

The governing differential equations (1), (2), (5), (6) and (7) are numerically integrated using the forward-marching finite difference procedure for two-dimensional boundary layer type flows described by Spalding (1977). In this procedure the equations are first transformed to x and ω coordinates, the latter being the stream function normalised to unity at the edge of the flow (the free surface in the present case). The resulting equations have a general form

$$\frac{\partial \phi}{\partial x} + (a+b\omega)\frac{\partial \phi}{\partial \omega} = \frac{\partial}{\partial \omega}(c\frac{\partial \phi}{\partial \omega}) + S_\phi \tag{12}$$

where a and b are arbitrary functions of x, c and S_ϕ are functions of any dependent or independent variable, S_ϕ being the source term associated with the variable ϕ which stands for anyone of the set of dependent variables U, k, ε and C. The finite difference equations are obtained by integrating Eq. (12) over the ith control volume of the grid, resulting in fully implicit equations in "conservation" form. Coefficients of the quasi-linear difference equations are evaluated at the upstream station whenever not known at the downstream station. Finally, each equation is solved at a given forward step inverting a tri-diagonal matrix by successive substitution.

The procedure discussed above was applied to an open channel flow, first, to determine the fully developed velocity and eddy-viscosity distributions. The calculations were started with an uniform velocity profile and with step profiles for k and ε (i.e. $k = U_*^2/\sqrt{c_\mu}$, $\varepsilon = U_*^3/\kappa y$ at $y = y_c$ and $k = \varepsilon = 0$ otherwise), and continued downstream until the flow became fully developed (that is no further changes in the longitudinal direction).

Only two cases of Jobson and Sayre's (1970a)experiments defined as Flow B and Flow C were simulated. The discharge per unit width, q, the water depth, h, and the velocity ratio U_*/U_m, U_m being the mean velocity equal to q/h, are matched with the experimental values. An approximate correction is applied to the measured h and U_* such that $h = h_{exp} - 2k_r/3$ and $U = \sqrt{gRS}$ where h_{exp} is the measured flow depth,

Table 1. Hydraulic data for calculated cases

Flow	q (m^3/s-m)	h (m)	k_r (m)	E	U_m (m/s)	U_*/U_m	$Re = \frac{U_m h}{\nu}$ $\times 10^{-5}$
B	0.232	0.387	0.027	0.0015	0.60	0.134	2.32
C	0.356	0.387	0.027	0.0010	0.92	0.134	3.56

k_r is the wall-roughness height, S is the channel slope and R is the hydraulic radius based on h. The hydraulic parameters describing Flow B and Flow C used in calculations are given in Table 1. The given friction velocities were matched by changing the value of the roughness parameter E (Eq. 8) until the right U_* was obtained. Typical values of E are given in Tabel 1 (E = 9.0 for a smooth wall).

The results of flow-field calculations are shown in Fig. 1. Two additional flow cases with $U_*/U_m = 0.083$ and $U_*/U_m = 0.039$, the latter being for a smooth wall, are also calculated and included in Fig. 1 for comparison.

The fully developed velocity and eddy-viscosity distributions calculated as described above, were then used in calculation of the concentration distributions. The concentration distributions at the smallest value of x/h (Figs. 2, 3, 4 and 5) were used as initial boundary conditions to predict the others in each figure. The initial profiles were obtained by drawing smooth curves through the data points. The measured distributions sufficiently far from the actual line source at the free surface were used as initial boundary conditions instead of the source itself, because in the vicinity of the source the assumptions made in driving Eq. (7) would not be valid. Rather than using the boundary conditions (10) and (11) directly, finite difference form of these were obtained by applying the principle of mass balance at the boundary control volumes. The fall velocity term in Eq. (7) was treated as a source term in Eq. (12).

Grid independence of the numerical solution was ensured by taking sufficiently large number of grid points in both x- and y-directions, the grid being denser near the wall and the free surface boundaries. Since the numerical method used is a very widely used and tested one (see Spalding, 1977), no comparison of the accuracy of the numerical scheme is presented in this paper.

The results of the concentration predictions are depicted in Figs. 2 - 5. The concentration C was normalised by a reference mean concentration defined as

$$C_r = \frac{1}{\rho q} \int_0^h \rho U C \, dy \tag{13}$$

where ρ is the fluid density.

The experimental data points shown in these figures were reported by Jobson and Sayre (1970a). The experiments were performed with a continuous line source of dispersant supplied at the water surface. Three different dispersant were used: dye (Rhodamine WT), fine sediment (spherical glass beads with a median sieve diameter of 0.123 mm, specific gravity of 2.42 and a median standard fall velocity of 1.09 cm/s), and coarse sediment (natural river sand with a median sieve diameter of 0.39 mm, specific gravity of 2.65 and median standard fall velocity of 6.30 cm/s).

4 DISCUSSION OF THE RESULTS

4.1 Mean flow field

The comparison of the predicted velocity profile ($U_*/U_m = 0.134$) with experiments (Figure 1) shows considerable disagreement. Better agreement is obtained when a smaller U_*/U_m (= 0.083) than the measured one is used. This could be attributed mostly to the three-dimensional effects encountered in the experiments. It was reported in the original work (Jobson, 1968) that, in most cases the maximum velocity accured below the free surface and secondary currents were present, which are known to be usual in three-dimensional channel flows. Furthermore, every data point on the measured velocity profile in Fig. 1 was obtained by averaging several other measurements at different longitudinal locations in the test section. Hence, the experimental velocity profiles cannot be considered a truly two-dimensional channel flow profile, and thus it could not be predicted well by a two-dimensional analysis. Nevertheless, the predicted velocity profile with $U_*/U_m = 0.134$ (i.e. the experimental value) is believed to be more representative of the flow field, since there could be very little error in the experimental value of U_*/U_m, where $U_* = \sqrt{gRS}$. The other velocity profile with $U_*/U_m = 0.083$ was used in calculations to illustrate the effects of such a change in the velocity profile on the predicted concentration profiles.

It is seen from Figure 1 that the predicted eddy-viscosity distribution is in reasonable agreement with measurements of Ueda et al.

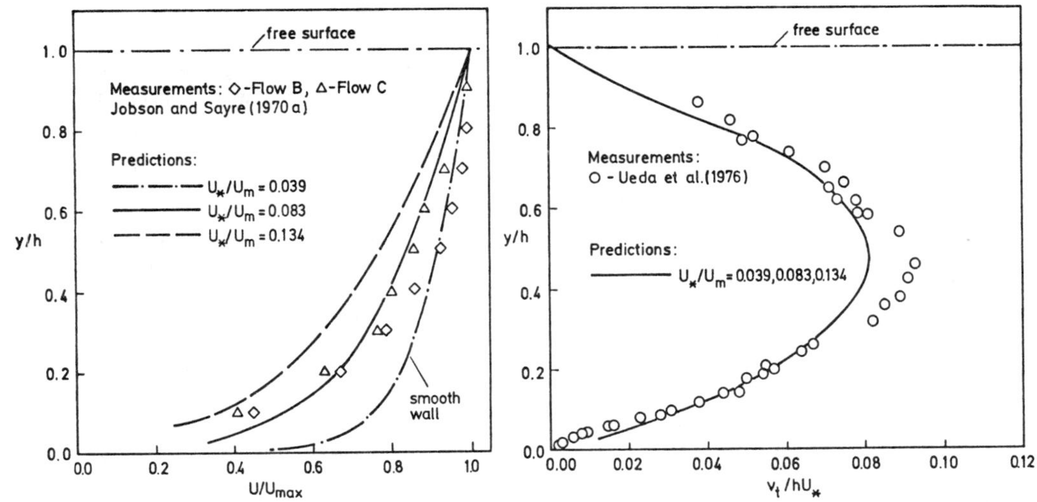

Figure 1. Velocity and eddy-viscosity distributions; Flow B and Flow C.

(1976). The slightly lower predicted values (< 10 %) of ν_t/hU_* away from the wall region were attributed (Çelik et al., 1982) to the prediction of higher velocity gradients compared to measurements. However, since ν_t was obtained by using numerical derivatives of the measured velocity profile, dU/dy, which is small in the outer half of the channel, great accuracy should not be expected in the experimental values of ν_t. The present predictions do not indicate any change in the relative eddy-viscosity distribution with increasing U_*/U_m (i.e. with increasing roughness), the flow depth and the discharge being fixed. This seems to be reasonable since the velocity profiles reamin similar with increasing roughness when the velocity is made nondimensional with the friction velocity, U_* .

4.2 Dye concentration

Predicted dye concentration profiles are compared with experiments in Fig. 2. Fig. 2 depcts the strikingly similar influence of larger mass transfer coefficient (i.e. smaller σ_c) with that of larger U_*/U_m. The dispersant is more uniformly distributed both with larger U_*/U_m and smaller σ_c. Larger U_*/U_m correspond to larger wall roughness in this case, which is known to increase mixing. The best agreement between predictions and measurements was obtained with $\sigma_c = 1.0$ and $U_*/U_m = 0.134$, which supports the argument (Jobson and Sayre, 1970a) that for dye the turbulent Schmidt number, σ_c, is approximately one. The difference between predictions and experiments near the free surface in case of $U_*/U_m = 0.134$ and $\sigma_c = 1$ could be due to the discrepancy between measured and calculated velocity profile (Figure 1). If it is noted that there was up to 40 % scatter in the data (see Jobson and Sayre, 1970a) the predictions can be considered quite satisfactory.

4.3 Sediment concentration

The mechanics of sediment transport is more complicated than that of a dye, since it involves two more hydraulic parameters; the fall velocity, V_s, and A, the parameter controlling the deposition on the bed, in addition to velocity distribution and the turbulent Schmidt number, σ_c. First, the effects of V_s, σ_c and A on the predicted concentrations were investigated, keeping the velocity profile fixed. The results of the predictions with Flow C and $U_*/U_m = 0.083$ for fine sediment are given in Fig. 3. It is seen in Fig. 3 that large changes in σ_c brings about minor changes in the concentration profiles.

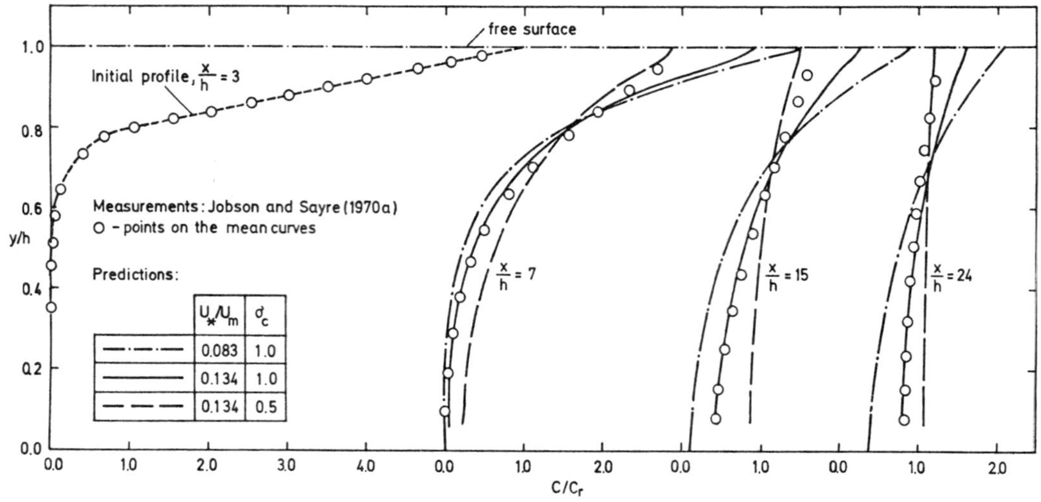

Figure 2. Dye concentration; Flow B and Flow C.

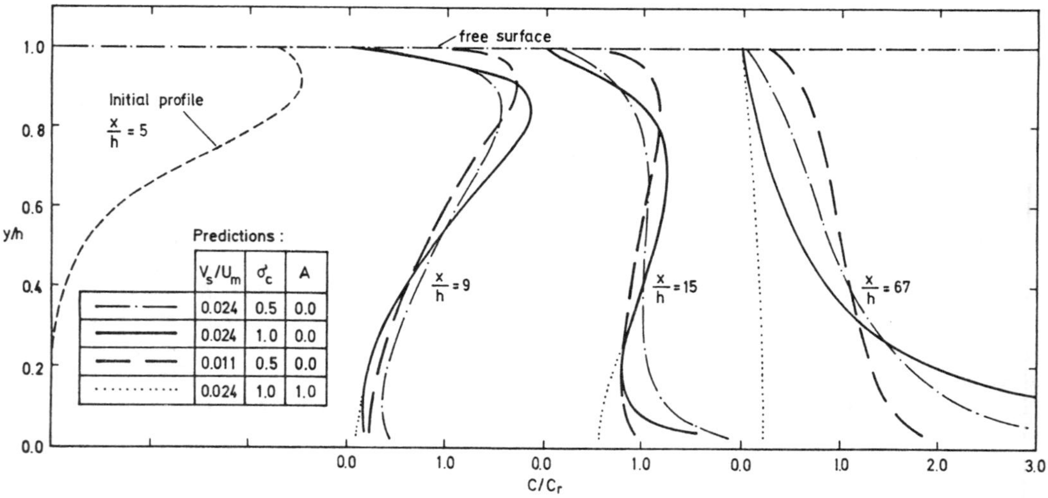

Figure 3. Influence of various parameters on concentration (fine sediment with Flow C and $U_*/U_m = 0.083$).

The shapes of the profiles remain similar, but the peak concentrations do change considerably, decreasing near the surface and increasing near the wall and becoming more uniformly distributed with larger mass-transfer coefficient (smaller σ_c) as it was observed in case of dye concentration. The fall velocity, V_s, has quite different influence than σ_c. With increasing V_s the rate of decent of the peak concentration is faster, more sediment being collected near the wall (no deposition occurs since A = 0). Thus, one sees that σ_c controls the rate of spread, whereas V_s controls the rate of decent of the dispersant mass as it was

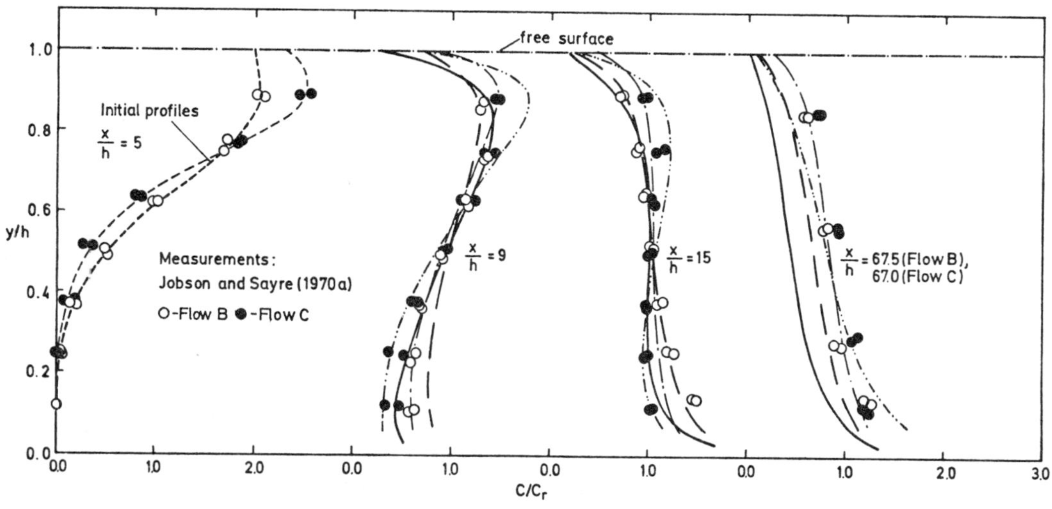

Figure 4. Fine sediment concentration. Predictions: Flow B, A = 0.2, V_s/U_m = 0.025, σ_c = 0.5, ——— U_*/U_m = 0.083, — — U_*/U_m = 0.134; Flow C, A = 0.2, V_s/U_m = 0.017, U_*/U_m = 0.134, —·— σ_c = 0.5, —··— σ_c = 1.0

observed in Jobson and Sayre's (1970b) predictions. The effect of the probability of deposit, A, is minor and restricted to the near wall region at small distances downstream of the source. However, at large distances the influence becomes pronounced when A = 1, since most of the material deposits on the bed. Both A = 0 (no deposition) and A = 1 (100 % deposition) give unrealistic solutions; A = 0 giving very large C and $\partial C/\partial y$ near the bed and with A = 1 almost no material being left in suspension at large distances (Fig. 3, x/h = 67). Therefore, reasonable prediction of the concentration at large distances from the source could only be achieved with sufficiently accurate values of A.

The predictions are compared with experiments for fine sediment with Flows B and C in Fig. 4, and for coarse sediment only with Flow C in Fig. 5. Since the measured fall velocity of sediment particles varied significantly (Jobson and Sayre, 1970a) from the standard fall velocity in quiescent fluid, the mean value of measured V_s/U_m was used in the calculations. As for A, a value of 0.2 was used. This was estimated by Jobson and Sayre (1970b) from the distribution of fine sediment deposited on the bed after each run. Although, it is believed that the velocity profile with the experimental U_*/U_m is more representative of the flow field, calculations with U_*/U_m = 0.083 were also included in Figs. 4 and 5 for comparison. Since the value of σ_c is not known, two extreme values were used; 0.5 and 1.0 for fine sediment, and 0.5 and 1.5 for coarse sediment. The value of A estimated for fine sediment was also used for coarse sediment and another value, A = 0.05, was used to check the sensitivity of predictions to the changes in A.

The comparisons depicted in Figs. 4 and 5 both show that the predicted concentration profiles are not very sensitive to the changes in the hydraulic parameters mentioned in the proceeding paragraph. The changes produced in the predicted concentration distributions as a result of large changes in the above parameters are in the order of magnitude of the scatter in the experimental data. It is seen that sufficiently good agreement between predictions and experiments is obtained with σ_c = 0.5 and A = 0.2, in both fine and coarse sediment concentration

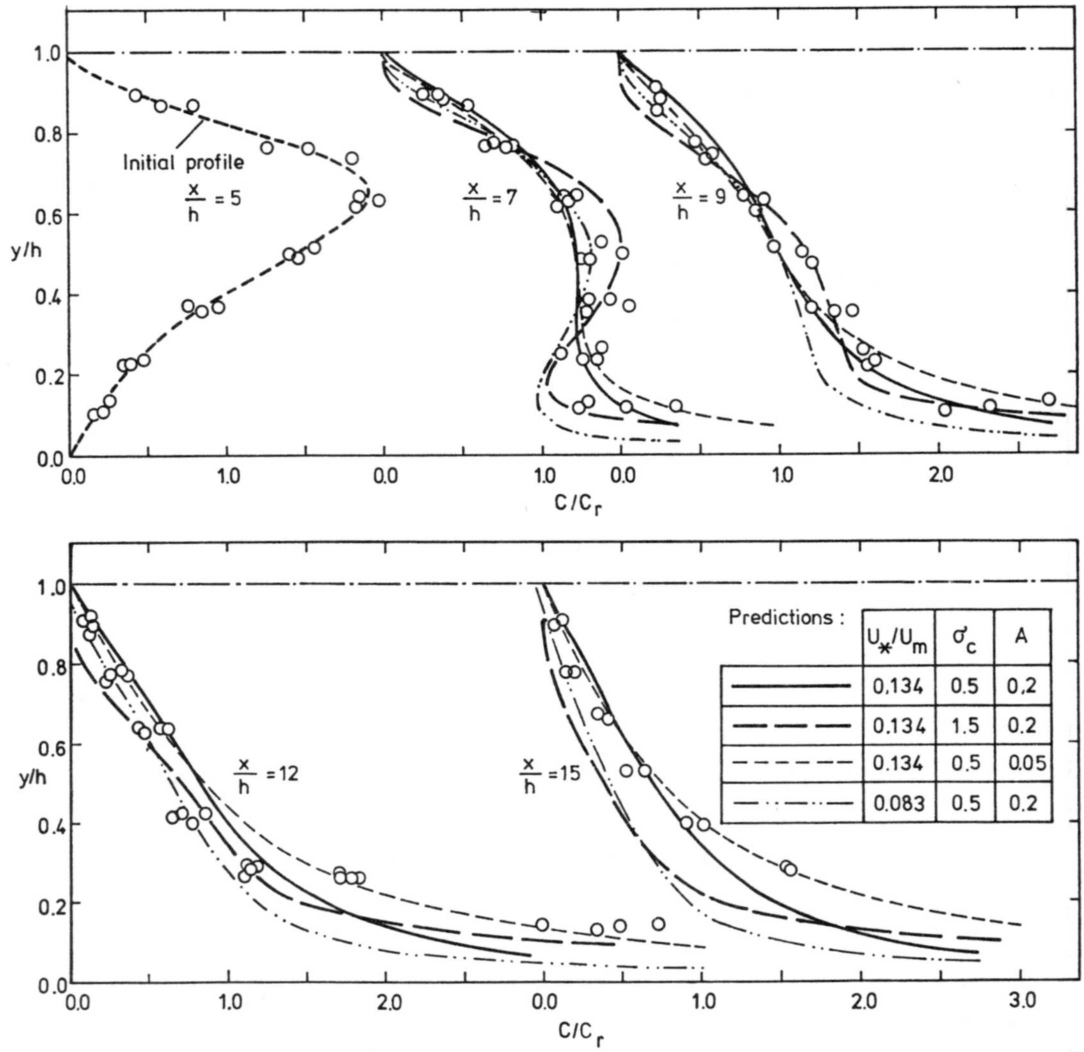

Figure 5. Coarse sediment concentration. Predictions: Flow C, $V_s/U_m = 0.072$; ○ -Measurements, Jobson and Sayre (1970 a)

calculations at least for the cases investigated.

5 CONCLUSIONS

It was shown that suspended sediment concentration distributions in turbulent open-channel flows can be predicted adequately using the calculated velocity and eddy-viscosity distributions with the k-ε turbulence model with a turbulent Schmidt number $0.5 \leq \sigma_c \leq 1.0$ provided that an approximately correct value of A, the probability of deposition, and the turbulent fall velocity of sediment particles are given. The results of the calculations indicate that $\sigma_c \cong 1$ for dye and $\sigma_c \leq 1$ for sediment.

REFERENCES

Çelik, I., Hossain, M.S. & Rodi, W. 1982. Modelling of free-surface-proximity effects on turbulence. Proc. Int. Symposium on Refined Modelling of Flows, Laboratoire National d'Hydraulique, Chatou, France, Sept. 7-10.

Jobson, H.E. 1968. Vertical mass transfer in open channel flows. Openfile Report, U.S. Geological Survey, Fort Collins, Colorado.

Jobson, H.E. & Sayre, W.W. 1970a. Vertical transfer in open channel flow. Proc. ASCE, Jnl. Hydraulic Div., Vol. 96, No. HY3, pp. 703-724.

Jobson, H.E. & Sayre, W.W. 1970b. Predicting concentration profiles in open channels. Proc. ASCE, Jnl. Hydraulic Div., Vol. 96, No. HY10, pp. 1983-1996.

Nakagawa, H., Nezu, I. & Ueda, H. 1975. Turbulence of open channel flow over smooth and rough beds. Proc. of JSCE No. 241, Sept.

Rodi, W. 1980. Turbulence models and their application in hydraulics. International Association for Hydraulic Research, Delft, The Netherlands.

Spalding, D.B. 1977. Genmix: A general computer program for two dimensional parabolic phenomena. Pergamon Press, New York.

Ueda, H., Möller, R., Komori, S. & Mizushina, T. 1976. Eddy diffusivity near the free surface of open channel flow. Int. Jnl. Heat Mass Transfer, Vol 20, pp. 1127-1136.

Euromech 156: Mechanics of Sediment Transport / Istanbul / 12-14 July 1982

Improved numerical calculation of sedimentation for different bed-roughness and various turbulence models

W.BECHTELER & W.SCHRIMPF
Hochschule der Bundeswehr, München, Germany

ABSTRACT: In order to evaluate the sediment transport in rivers and open channels an improved explicit difference scheme was developed to provide solutions for the diffusion-convection equation. Numerical diffusion can be avoided when the continuity of mass is taken into account. Different assumptions for the diffusion coefficient and the initial concentration are taken to demonstrate their influence on the removal ratio, vertical concentration profile and the deposition of solid matter at the bottom. Experiments with fine sediment are run in a laboratory flume in order to check the applicability of the numerical model.

INTRODUCTION

The sedimentation process of solid particles in turbulent flowing water is of fundamental interest for many hydraulic problems. Most often the behavior of the suspended sediment is described by a diffusion-convection equation. An analytical solution of this equation is given by Dobbins (1944). However for simplification constant coefficients are taken for the solution of the concerned partial differential equation. But for practical cases these coefficients (velocity and turbulent diffusion coefficient over flow depth) cannot generally assumed to be constant. An analytical solution for one case of non-constant flow velocity and non-constant diffusion coefficients is given by Sumer (1977). Although analytical solutions should be preferred when they are available, the numerical methods offer greater applicabilities in regard to different boundary conditions and various functions for the coefficients.

An explicit difference scheme is chosen to provide solutions for the diffusion-convection equation. Comparing this numerical procedure with that one given by Sarikaya (1977) numerical diffusion can be prevented if the conservation of mass is considered as an additional condition. This improved numerical scheme is used to calculate concentration profiles, removal ratios and the distribution of solid matter at the bed for various assumptions of the initial solid concentration, the turbulent diffusion coefficient and the bed-roughness. Some theoretical and experimental results are presented in the following.

BASIC EQUATIONS

As already mentioned suspended sediment transport can be treated as a convection-diffusion process. Reducing the resulting differential equation to steady state conditions and two dimensions where diffusion in longitudinal direction is neglected the dimensionless Eq. 1 is obtained:

$$\frac{\partial C}{\partial X} = \frac{v_s}{u}\frac{\partial C}{\partial Y} + \frac{\varepsilon_s}{hu}\frac{\partial^2 C}{\partial Y^2} + \frac{1}{hu}\frac{\partial \varepsilon_s}{\partial Y}\frac{\partial C}{\partial Y} \qquad (1)$$

in which C - the dimensionless local sediment concentration; X = x/h - the distance in direction of main flow; Y = y/h - the height above bed; u - the local mean flow velocity of the fluid and the particles; v_s - the mean fall velocity of the particles; h - the flow depth; ε_s - the vertical sediment diffusion coefficient. The boundary conditions at the bottom and at the water surface are defined by Eqs. 2 and 3:

$$\frac{\varepsilon_s}{hu}\frac{\partial C}{\partial Y} + (1-A)\frac{v_s}{u}C = 0 \quad \text{when} \quad Y = 0 \qquad (2)$$

and

$$\frac{\varepsilon_s}{hu}\frac{\partial C}{\partial Y} + \frac{v_s}{u}C = 0 \quad \text{when} \quad Y = 1 \qquad (3)$$

There is no net transport across the water surface. The factor (1-A) can be interpreted as a resuspension coefficient in which A is the probability that a particle settling to bed is deposited there. Referring to Gessler (1971) the probability of grains to rest can be calculated by means of the ratio τ_c/τ, where τ is the shear stress at the bottom and τ_c is the critical shear stress at incipient motion of the grains.

Taking a steady function for $C = C(X = 0, Y)$ the area where a solution $C = C(X, Y)$ has to be found is well defined in the mathematical sense by Eqs. 1 to 3.

NUMERICAL SOLUTION

Eq. 1 represents a partial differential equation of the parabolic type. Therefore it is convenient to choose an explicit difference scheme to derive suitable solutions. The computer process time is reduced considerably in comparison with an implicit method. However difficulties can arise as far as stability and convergency are concerned.

Forward differential quotients are used to approximate the first derivatives for the numerical solution. The second derivatives in Eq. 1 are described by symmetric expressions. Because the mathematical solution area is identical with the longitudinal cross section of the flume, the notation for computation and discretization is shown in Fig. 1.

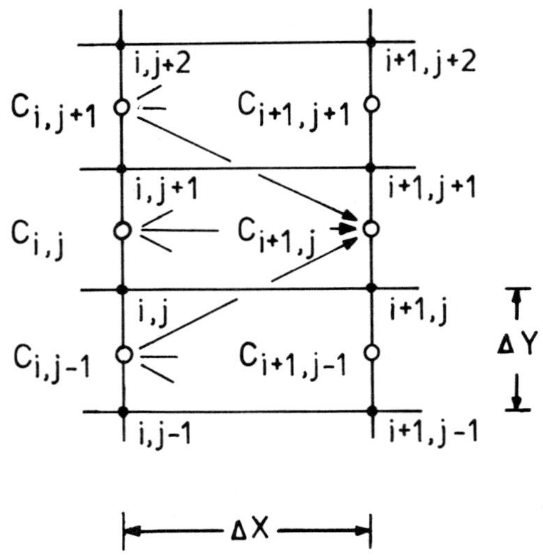

Fig. 1. Grid system and notation for the numerical solution.

The desired concentration $C_{i,j}$ will not be computed at the knot i,j but in the center between i,j and i,j+1. According to the statements of Marsal (1976) this method should be used if the boundary conditions are defined by means of differential equations. Therefore the resulting grid system is a mixture of a centered grid and a corner grid. The concentration $C_{i,j}$ is taken to be constant between i,j and i,j+1.

Remembering that Eq. 1 is derived for conservation of solid mass in turbulent flowing water, this important fact must not be forgotten when partial derivatives are approximated by differential quotients. Thus the continuity of mass is reached if the concentration $C_{i,j}$ is combined with the corresponding velocity u_j in each term containing the local flow velocity u; analogously u_{j+1} with $C_{i,j+1}$ and u_{j-1} with $C_{i,j-1}$. Taking these conditions into account the following set of equations offers the mathematical basis for the numerical solution. The total number of mesh lines in X- and Y-direction are set to I and J, respectively.

$j = 1$ $(Y = 0)$:

$$C_{i+1,j} = C_{i,j} + \frac{\Delta X}{\Delta Y}\frac{v_s}{u_{j+1}} C_{i,j+1} - \frac{\Delta X}{\Delta Y}\frac{v_s}{u_j} C_{i,j} + \frac{\Delta X}{\Delta Y^2}\frac{\varepsilon_{j+1}}{hu_{j+1}} C_{i,j+1} - \frac{\Delta X}{\Delta Y^2}\frac{\varepsilon_{j+1}}{hu_j} C_{i,j} + (1-A)\frac{\Delta X}{\Delta Y}\frac{v_s}{u_j} C_{i,j} \qquad (4)$$

$1 < j < J-1$:

$$C_{i+1,j} = C_{i,j} + \frac{\Delta X}{\Delta Y}\frac{v_s}{u_{j+1}} C_{i,j+1} - \frac{\Delta X}{\Delta Y}\frac{v_s}{u_j} C_{i,j} + \frac{\Delta X}{\Delta Y^2}\frac{\varepsilon_{j+1}}{hu_{j+1}} C_{i,j+1} - \frac{\Delta X}{\Delta Y^2}\frac{\varepsilon_{j+1}}{hu_j} C_{i,j} + \frac{\Delta X}{\Delta Y^2}\frac{\varepsilon_j}{hu_{j-1}} C_{i,j-1} - \frac{\Delta X}{\Delta Y^2}\frac{\varepsilon_j}{hu_j} C_{i,j} \qquad (5)$$

$j = J-1$ $(Y = 1)$:

$$C_{i+1,j} = C_{i,j} - \frac{\Delta X}{\Delta Y}\frac{v_s}{u_j} C_{i,j} + \frac{\Delta X}{\Delta Y^2}\frac{\varepsilon_j}{hu_{j-1}} C_{i,j-1} - \frac{\Delta X}{\Delta Y^2}\frac{\varepsilon_j}{hu_j} C_{i,j} \qquad (6)$$

STABILITY, CONVERGENCY, CONSISTENCY

If Taylor series are applied for the approximation of the differential equation it can be shown (Keller (1972)) that the discretization error can be evaluated for this case by

$$e = O\ (\Delta Y^2) + O\ (\Delta X) \tag{7}$$

If the functions for $u = u(Y)$ and $\varepsilon_s = \varepsilon_s\ (Y)$ are limited in the mathematical sense, the discretization error disappears with ΔX and ΔY approaching zero. The consistency is certain then. As far as stability is concerned, Keller derives three conditions from the so-called Maximum Principle. These stability conditions are reduced to the Eqs. 8 and 9 for the explicit difference scheme used for the diffusion-convection equation:

$$\frac{\varepsilon_s(Y)}{hu(Y)} - \Delta Y \left| 0.5\left(\frac{v_s}{u(Y)} + \frac{1}{hu(Y)}\ \frac{\partial \varepsilon_s(Y)}{\partial Y}\right) \right| \geq 0 \tag{8}$$

$$1 - 2\ \frac{\Delta X}{\Delta Y^2}\ \frac{\varepsilon_s(Y)}{hu(Y)} \geq 0 \tag{9}$$

In each grid which satisfies Eqs. 8 and 9 stability is achieved. It has to be stated that for some special cases the first condition cannot be attained by realistic step sizes ΔY. Experience has shown that stability can also be received if the left hand side of Eq. 8 is of magnitude greater $O(-10^{-4})$. Following the conclusions of Keller stability involves convergency if the consistency of the approximations is ensured.

APPLICATIONS

Solutions of Eq. 1 are often desired to calculate removal ratios and settling lengths for suspended solids in horizontal flow tanks. When the diffusion coefficient for the sediment is equal the coefficient for turbulent transfer of momentum, $\varepsilon_s = \varepsilon$ is given by Eq. 10 for logarithmic velocity profile:

$$\frac{\varepsilon_s}{hu_*} = \frac{\varepsilon}{hu_*} = \kappa Y\ (1-Y) \tag{10}$$

in which κ - the v. Kármán constant and u_* - the bottom shear velocity. Jobson and Sayre (1970) derived from their experimental data Eq. 11:

$$\frac{\varepsilon_s}{hu_*} = \beta_1\kappa\ (1-Y)Y + \beta_2\ \left(\frac{1-Y}{0.9}\right)^3;\ Y \geq 0.1$$

$$\frac{\varepsilon_s}{hu_*} = \beta_1\kappa\ (1-Y)Y + \beta_2\ \left(\frac{Y}{0.1}\right)^3;\ Y \leq 0.1 \tag{11}$$

The constants are different for fine ($\beta_1 = 0.98$, $\beta_2 = 0.038$) and coarse sediment ($\beta_1 = 0.49$, $\beta_2 = 0.10$). Experimental data of Coleman (1970) were used by Kerssens, v. Rijn and v. Wijngaarden (1977) to derive the following expression for ε_s:

$$\frac{\varepsilon_s}{hu_*} = \frac{\varepsilon_{smax}}{hu_*} = \gamma_1 + \gamma_2\ \left(\frac{v_s}{u_*}\right)^{\gamma_3}\ ;\ Y \geq 0.5$$

$$\frac{\varepsilon_s}{hu_*} = 4Y\ (1-Y)\ \frac{\varepsilon_{smax}}{hu_*}\ ;\qquad Y \leq 0.5 \tag{12}$$

The constants are calculated to $\gamma_1 = 0.1$, $\gamma_2 = 0.38$, $\gamma_3 = 4.31$ for flumes and to $\gamma_1 = 0.13$, $\gamma_2 = 0.20$, $\gamma_3 = 2.12$ for natural channels. When logarithmic velocity distribution is assumed and ε_s is given by an expression like Eqs. 10 to 12, solutions of Eq. 1 can be provided in terms of v_s/u_* and $\bar{u}/u_*$, where $\bar{u}$ is the mean flow velocity. Fig. 2 shows for two examples the calculated longitudinal deposition of fine sediment in a flume.

As expected, the distribution of the sediment at the bottom tends to flatten with decreasing ratio v_s/u_*. The different relations for ε_s lead to no significant deviations of the computed curves for these two numerical examples. However, the differences increase for calculations with Eq. 12 with increasing values for v_s/u_*. Then the relatively great value for ε_s in the upper half of the flume in combination with the concentration gradient prevents the settling of the solid particles in the numerical computation.

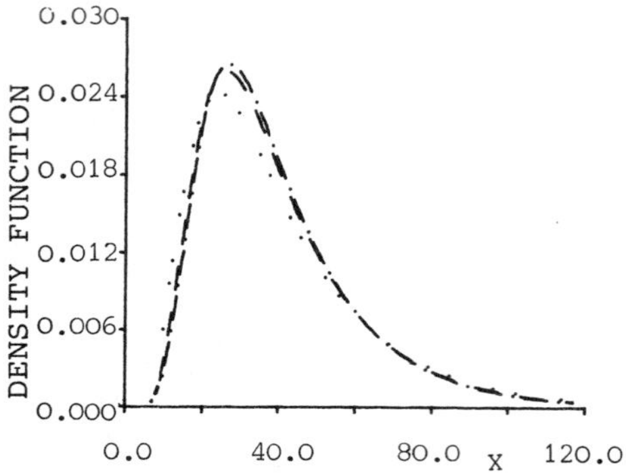

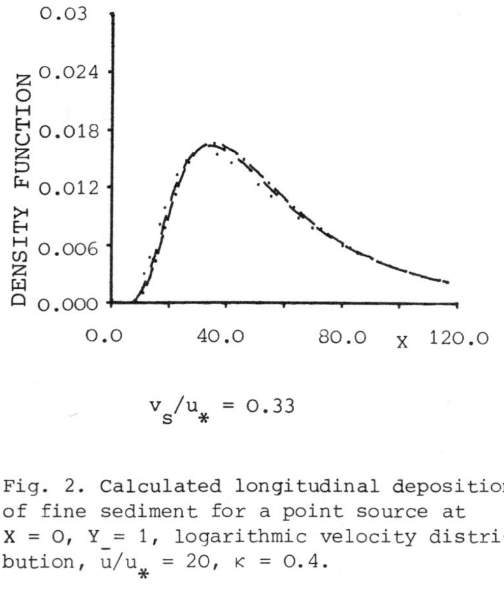

Fig. 2. Calculated longitudinal deposition of fine sediment for a point source at $X = 0$, $Y = 1$, logarithmic velocity distribution, $\bar{u}/u_* = 20$, $\kappa = 0.4$.

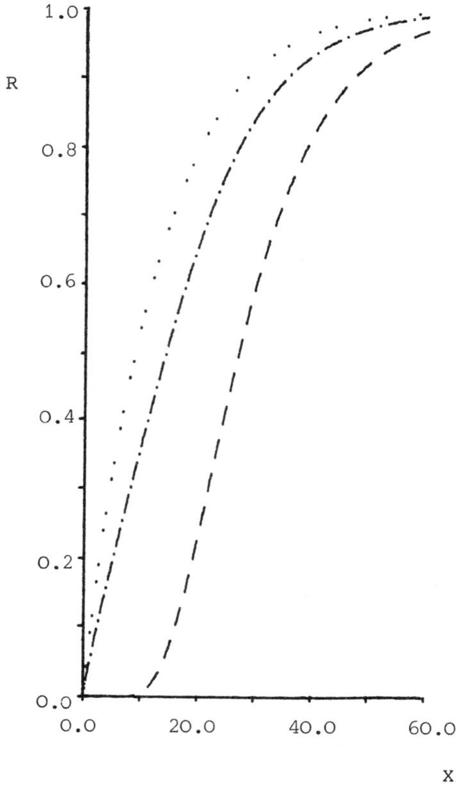

Fig. 3. Calculated removal ratios for three different initial concentration profiles; $v_s/u_* = 0.77$, $\bar{u}/u_* = 23.1$, logarithmic velocity distribution, $\kappa = 0.4$.

When the removal ratios in horizontal settling tanks have to be predicted it is important to know the influence of the initial concentration profile on the separation effect. Three different initial concentrations are assumed to calculate the removal ratios $R = R(X) = 1. - \bar{c}(X)/\bar{c}_0$, shown in Fig. 3.

The vertical concentration profiles at $X = 6.0$ of the three cases in Fig. 3 are shown in Fig. 4.

Settling tests are performed in a tilting flume of 33 m length in order to check the described mathematical-physical model. In Fig. 5 the calculated settling lengths X_{90} for a removal ratio of 90 % are com-

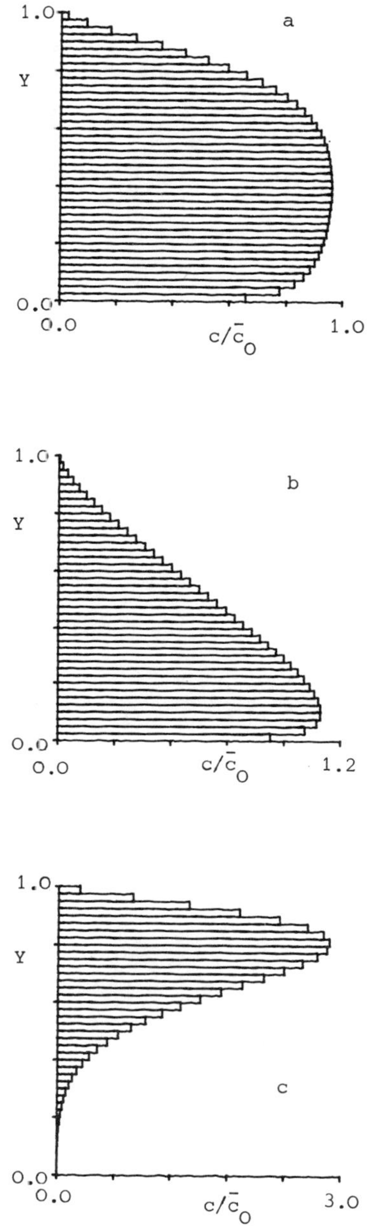

Fig. 4. Vertical concentration profiles at X = 6.; concentration at X = 0.: constant (a), triangular (b), point source (c).

pared with some experimental data of a test series. The median standard fall velocity of the sand sample was measured to be v_s = 1.25 cm/s. Therefore it can be assumed that the relation for ε_s given by Jobson and Sayre is valid for this case. Eq. 12 cannot be used because the ratio v_s/u_* exceeds the value of 1.0. The third curve in Fig. 5 results from Camp's diagram (1946) computed for constant flow velocity and constant diffusion coefficient. In all cases the resuspension coefficient (1-A) in Eq. 2 is set to zero.

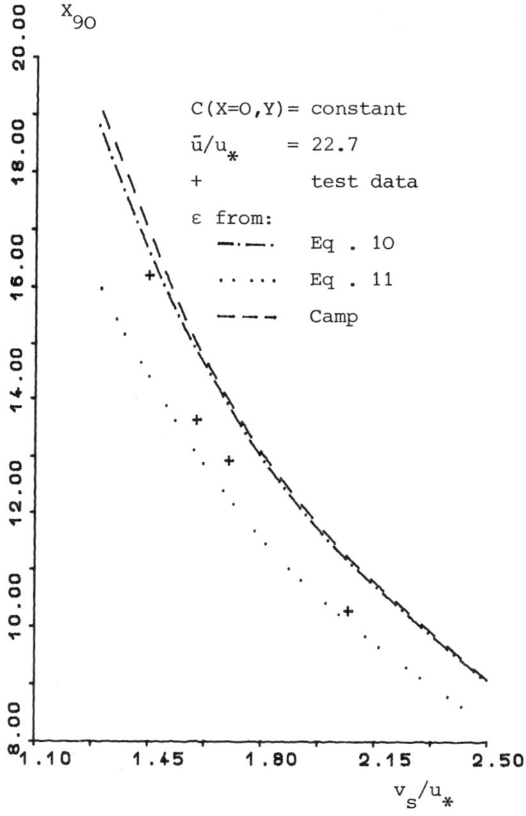

Fig. 5. Settling lengths for R = 0.9; theoretical and test data.

The test data are in relatively good agreement with the range of the theoretical values. The influence of fluid turbulence on the sedimentation of the solid particles seems to be reduced with increasing ratio v_s/u_*. The fall velocity distribution of the sand sample could be determined by means of a computerized sedimentation balance. However no significant differences were stated for this case when the fall velocity distribution was taken into account. Furthermore the test data and the results from the numerical model are compared with stochastic calculations Bechteler (1981)). Both are in good agreement and prove the applicability of the improved numerical model for practical sedimentation engineering.

CONCLUSIONS

An improved explicit difference method was developed to provide numerical solutions of the diffusion-convection equation for steady-state sediment transport in two dimensions. The numerical scheme can be applied to calculate removal ratios, settling lengths and concentration profiles for various cases of the turbulent diffusion coefficient, the initial concentration profile and the bed-roughness. The theoretical results are in good agreement with test data. However the presented examples also reveal the great importance of predicting precise boundary conditions for the mathematical-physical model.

REFERENCES

Bechteler, W. 1981, Stochastische Modelle zur Simulation des Transportes suspendierter Feststoffe. Wasserwirtschaft 70; 191-196.

Camp, T.R. 1946, Sedimentation and the Design of Settling Tanks. Transactions of the ASCE, 111, Paper No. 2285: 895-958.

Coleman, N.L. 1970, Flume Studies of the Sediment Transfer Coefficient. Water Resources Research, 6, No. 3: 801-809.

Dobbins, W.E. 1944, Effect of Turbulence on Sedimentation. Transactions of the ASCE, 109, Paper No. 2218: 629-656.

Gessler, G. 1971, Beginning and Ceasing of Sediment Motion. River Mechanics, Vol. 1, Editor: H.W. Shen.

Jobson, H.E., Sayre, W.W. 1970, Vertical Transfer in Open Channel Flow. Proceedings of the ASCE, Jnl of the Hydraulics Division, 96, No. HY3: 703-724.

Keller, H.B. 1972, Numerische Integration von partiellen parabolischen Differentialgleichungen. Mathematische Methoden für Digitalrechner I, Editor: A. Ralston, H.S. Wilf, R. Oldenbourg Verlag München Wien: 243-281.

Kerssens, P.J.M., Rijn, L.C.v., Wijngaarden, N.J.v. 1977, Model for Non-Steady Suspended Sediment Transport. Proceedings of the Seventeenth Congress of the IAHR, Aug. 15-19, Baden-Baden, Federal Republic of West Germany: 113-120.

Marsal, D. 1976, Die numerische Lösung partieller Differentialgleichungen. BI- Wissenschaftsverlag, Mannheim/Wien/Zürich.

Sarikaya, H.Z. 1977, Numerical Model for Discrete Settling. Proceedings of the ASCE, Jnl of the Hydraulics Division, 103, No. HY8: 865-876.

Sumer, B.M. 1977, Settlement of Solid Particles in Open-Channel Flow. Proceedings of the ASCE, Jnl of the Hydraulics Division, 103, No. HY11: 1323-1337.

NOTATION

A	probability that a particle which reaches the bottom is deposited there;
$\underline{C}$	dimensionless concentration;
c	mean solid concentration;
$\bar{c}_0$	mean solid concentration at X = 0;
h	flow depth;
i	subscript in the numerical scheme (X-direction);
I	number of mesh lines in the X-direction;
j	subscript in the numerical scheme (Y-direction);
J	number of mesh lines in the Y-direction;
R	removal ratio, $R = 1.-\bar{c}/\bar{c}_0$;
u	local mean flow velocity in X-direction;
u_*	shear velocity;
v_s	settling velocity of the particles;
x	distance in direction of main flow;
X	= x/h;
X_{90}	settling length for R = 0.9;
y	height above bed;
Y	= y/h;
$\beta_1 \beta_2$	constants in Eq. 11;
$\gamma_1 \gamma_2, \gamma_3$	constants in Eq. 12;
ΔX	mesh size in X-direction;
ΔY	mesh size in Y-direction;
ε	turbulent diffusion coefficient;
ε_s	turbulent diffusion coefficient for the sediment;
ε_{smax}	maximum diffusion coefficient according to Eq. 12;
κ	v. Kármán constant;
τ	shear stress at the bottom;
τ_c	critical shear stress at incipient motion of the grains

Euromech 156: Mechanics of Sediment Transport / Istanbul / 12-14 July 1982

Some phenomena associated with hyperconcentrated flow

ZHAOHUI WAN
Technical University of Denmark, Copenhagen

ABSTRACT: Some unusual phenomena associated with the hyperconcentrated flow and experiments relating to it are briefly described. In the experiments the threshold condition, the transport rate, the bed configurations and the flow resistance have been measured and discussed.

1 A SURVEY OF THE YELLOW RIVER

The Yellow River is famous for its tremendous amount of sediment. The annual sediment load is 1.6×10^9 tons giving an average concentration of 1.4 pct. (in volume). Most sediment is carried along during the flood season in the form of hyperconcentrated flood. Examples taken from Longmen gauging station located in the middle reach of the Yellow River are shown in Tab. 1.

Nearly half a millard tons of sediment or a quarter to one third of annual load was transported in a flood within 2-3 days. Samples with concentrations as high as 35 pct. have been measured in the Yellow River and 56 pct. in its tributaries.

In recent years hyperconcentrated floods happen more often because of the development cf irrigation, the construction of reservoirs, besides, the destruction of plant cover in some regions.

2 SOME PHENOMENA AND CHARACTERISTICS OF THE HYPERCONCENTRATED FLOW

Extremely high concentration can not be simply explained as wash load. As a common rule, the higher the concentration, the coarser the particles (Chien, N. et al. 1979). Fig. 1, for instance, is taken from a gauging station in a tributary of the Yellow River.

Table 1. Examples of hyperconcentrated floods measured at longmen gauging station.

date	time (hrs.)	maximum discharge (m^3/sec)	maximum concentration (%)	total sediment load (10^8 tons)	percent (%) *	eroded bed depth in this period (m)
1966 18/7~20/7	58	7460	35.2	4.53	26.5	7.0
1970 2/8 ~ 4/8	72	13800	31.2	4.97	35.3	8.8

$$* \text{ percent} = \frac{\text{total sediment load}}{\text{annual sediment load}}$$

Sometimes serious siltation associates with the hyperconcentrated flow. In some small tributaries the phenomenon called

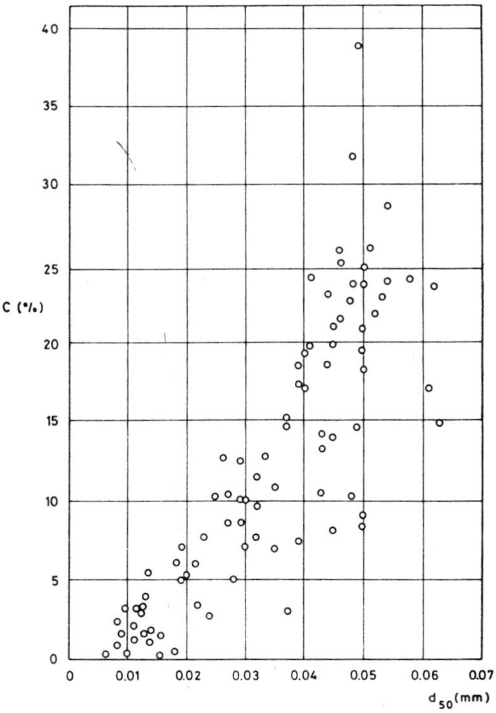

Fig. 1 The correlation between the median diameter of sediment and the concentration at Zhuangtou gauging station in the Luo River.

"clogging" happens during the decline of a hyperconcentrated flood, while the discharge becomes smaller and smaller, but the concentration is still high. The river stops flowing for a time, but the water level rises gradually until the river starts flowing again. After a while it stops again and the process repeats. Fig. 2 is a record taken from a small tributary. Such unsteady flow phenomenon has been reproduced under laboratory conditions. (Wan, Z.H. et al. 1979)

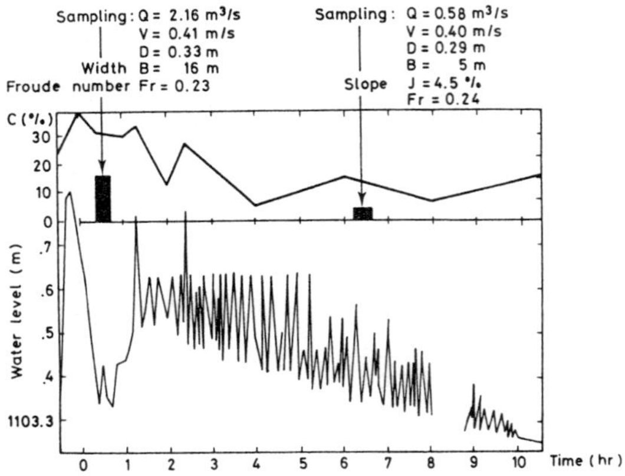

Fig. 2 The unsteady flow phenomenon associated with the hyperconcentrated flow in the Pu River.

A hyperconcentrated flood is not always associated with serious siltation. On the contrary, sometimes vigorous erosion happens. The flood listed in Tab. 1 scoured the river bed for nearly 9 meters while passing through the Rongmeng gauging station, see Fig. 3. The witnesses describe such vigorous erosion as follows: As the flood passes, a block of bed about 1 m thick, is torn off by the flow, turns around and stands up towering over the water surface for several seconds. Then it collapses with a great noise. After several minutes another block of the river bed towers over the water surface and collapses again, ...

Because of the reduction of particle settling velocity the hyperconcentrated flow has large transport capacity, which can be utilized.

Hyperconcentration irrigation has been largely developed in recent years. Not only the water, but also the sediment is considered as resources and utilized. Flow with concentration as high as 36 pct. has been diverted and utilized. Flow with concentration of more than 30 pct. has been conveyed as much as 67 km in irrigation canals. In three large irrigation districts of the province Shaangxi more than 40 mill. tons of sediment were utilized in this way in a year (1977).

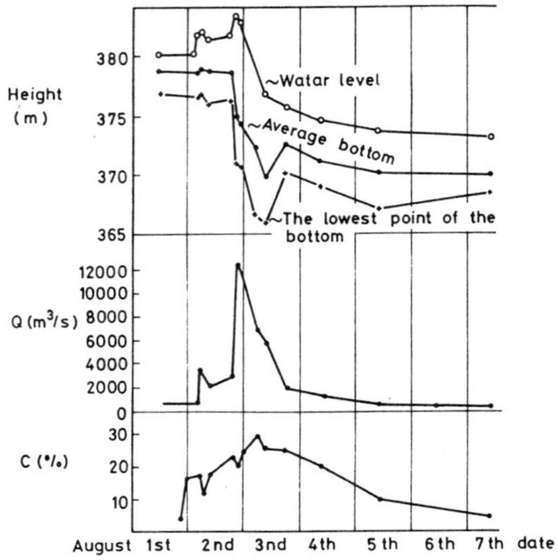

Fig. 3 The vigorous scour associated with the hyperconcentrated flood at Longmen gauging station.

In some cases, a hyperconcentrated flood passes rather easily through a reservoir. Table 2 is data taken from the Sanmenxia reservoir. Although in front of the dam the depth was over 30 m and the slope of the water surface was less than 10^{-4}, most sediment was released from the reservoir.

Table 2. Hyperconcentrated floods passing through the Sangménxia Reservoir.

Time	conditions at the inlet of reservoir			maximum concentration at the outlet (%)	releasing rate (%)
	maximum discharge (m^3/sec)	maximum concentration (%)	average concentration (%)		
1977 6/July	13,600	23.2	13.8	22.2	91
" 3/Aug.	12,000	8.98	6.72	12.1	115
" 6/Aug.	15,400	34.4	10.4	34.4	113

$$\text{releasing rate} = \frac{\text{total amount of sediment released from the reservoir}}{\text{total amount of sediment entering the reservoir}}$$

In some small reservoirs a sub-reservoir of mud forms underneath the clear water. Particles settle very slowly in mud and the mud keeps its fluidity for a long time. The mud can be released in a couple of days from the subreservoir formed by a flood, lasting only a couple of hours.

The hyperconcentrated flow easily forms

turbidity currents, not only in reservoirs, but also in rivers. Sometimes the hyperconcentrated flow from the Wei River - a tributary of the Yellow River - is submerged underneath the relatively clear flow from the Yellow River. Fig. 4 is taken from a gauging station located just below the confluence. A distinct interface can be detected.

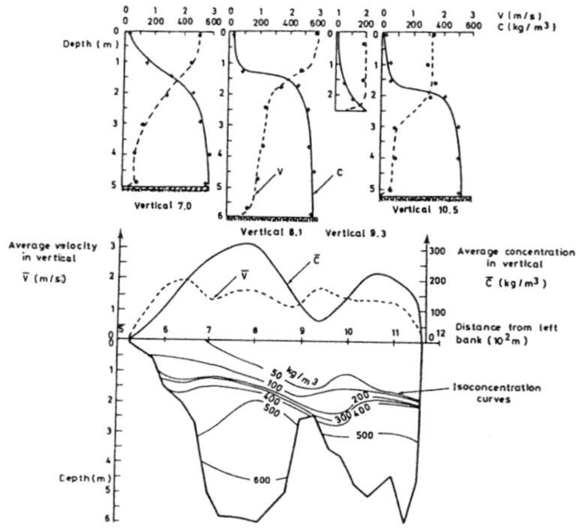

Fig. 4 The turbidity current in river at Tongguan gauging station.

3 A BRIEF DESCRIPTION OF THE FLUME EXPERIMENTS

The behaviour of the hyperconcentrated flow seems different from that of an ordinary sediment-laden flow, due to the existence of the large amount of clay particles. Some experiments have been carried out at ISVA in a 10 m long closed conduit with a rectangular cross section (30 cm wide). Bentonite is used as clay particles to form the suspension. Plastic beads of volumetric diameter (d) 3.45 mm with density (s) 1.29 are used as coarse particles. Both bentonite suspension and plastic beads are recirculated. The total height of the conduit is 20 cm and about half of it is occupied by the granular bed.

4 SOME INTRODUCTORY EXPERIMENTS

Before the main experiments, some introductory experiments have been done.

4.1 Rheological property of the bentonite suspension

On the basis of the measurement in a rotating viscometer, the rheological property of the bentonite suspension can be approximately characterized as a Bingham fluid:

$$\tau = \tau_B + \eta \frac{du}{dy} \qquad (1)$$

where τ and $\frac{du}{dy}$ are shear stress and shear rate, respectively, the yield stress τ_B and the rigidity η vary with the concentration C.

$$\tau_B (\text{dyne/cm}^2) = 4.15 \times 10^6 \cdot C^3 \quad (2)$$

4.2 The resistance properties of the bentonite suspension in a rigid conduit

The hydraulic parameters were measured when the bentonite suspension flowed in the rigid conduit without a granular bed.

If a relevant universal Reynolds number Re_1, consisting of an effective viscosity μ_e, is adopted, the f (friction factor) $\sim Re_1$ curve of the suspension is the same as that of the clear water, as Fig. 5 (Wan Zhaohui et. al. 1979), where

$$Re_1 = 4R \cdot V \cdot \rho / \mu_e \qquad (3)$$

$$\mu_e = \eta + \frac{1}{2}\frac{\tau_B R}{V} \qquad (4)$$

ρ is the density of the bentonite, V and R are the average velocity and the hydraulic radius, respectively.

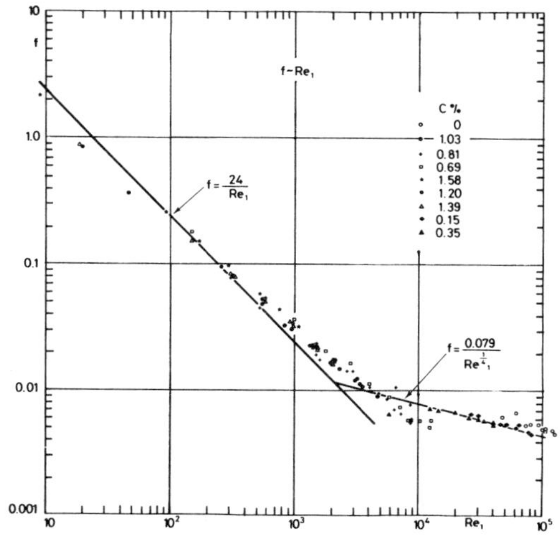

Fig. 5 The flow resistance property of the bentonite suspension.

4.3 The settling velocity of particles in Bingham fluid

An upward flow with nearly uniform velocity distribution was created in a specially designed box, see Fig. 6. Natural sand as well as plastic beads of different densities and sizes were put into the center of the upward flow to see if they were carried away by the flow or settled in the bottom. In this way the settling velocity was determined.

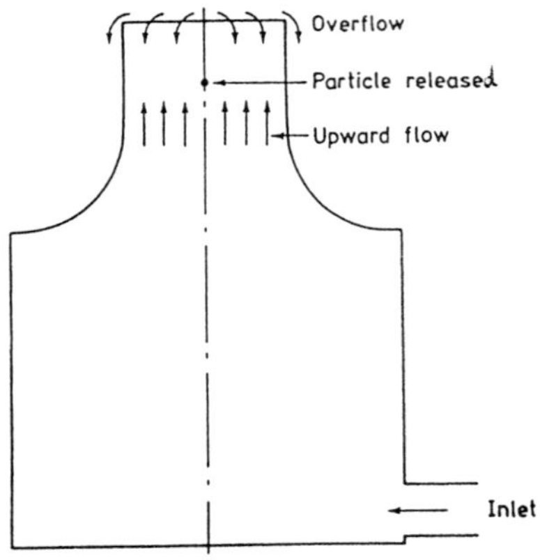

Fig. 6 The experimental equipment for settling-velocity measurement.

The drag exerting on a particle in Bingham fluid is caused not only by η, but also by τ_B. If a universal Reynolds number Re_2 suggested by R.W. Ansley and T.N. Smith is adopted (Ansley, R.W. et al, 1967), the experimental points properly follow the curve suggested by E. Navntoft (Navntoft, E., 1968) for natural particles settling in clear water, see Fig. 7, in which C_D is the drag coefficient and:

$$Re_2 = \frac{\rho W^2}{\eta \frac{W}{d} + \frac{7\pi}{24} \tau_B} \qquad (5)$$

W is the particle settling velocity in Bingham fluid.

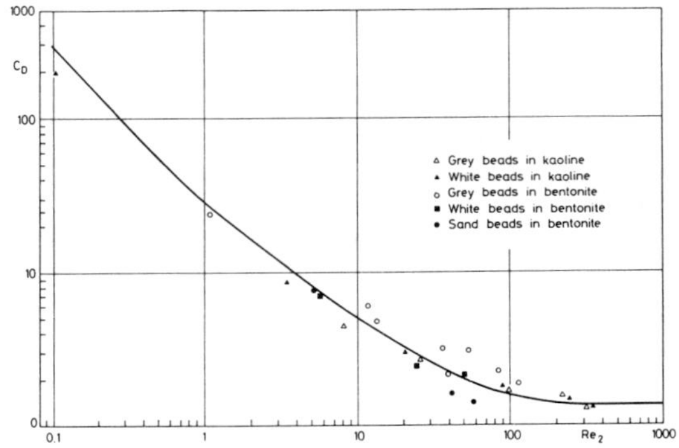

Fig. 7 $C_D \sim Re_2$

5 SOME OBSERVED PHENOMENA

In the main experiments, at first the clear water, later the bentonite suspension of different C flowed over a movable granular bed in the conduit.

The coarse particles started moving at a larger discharge (or velocity) in bentonite suspension than in clear water. Dunes in bentonite suspension were lower (Fig. 8) and smoother in shape. The bed with dunes transformed to flat bed at a smaller discharge in bentonite suspension than in clear water. When C was larger than 1 pct., no dune formed at any discharge and the bed stayed plane.

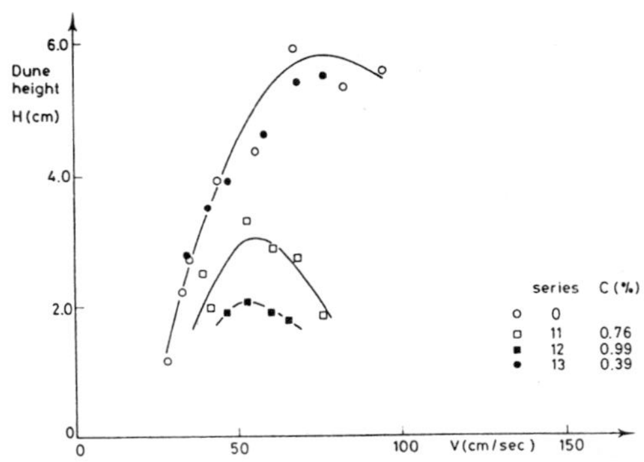

Fig. 8 The dune heights under different bentonite concentrations.

In most cases coarse particles moved as bed load. In high flow intensity region with high concentration, considerable coarse particles were suspended. Compared with that in clear water, the transport rate of coarse particles in bentonite is smaller at low flow intensity and larger at high flow intensity, see Fig. 9, in which

$$\theta' = \frac{R_b' J}{(s-1)d} \qquad (6)$$

$$\phi = \frac{q_T}{\sqrt{(s-1)gd^3}} \qquad (7)$$

where J is the slope, q_T is the transport rate in unit width, R_b^1 is the hydraulic radius with respect to bed surface friction.

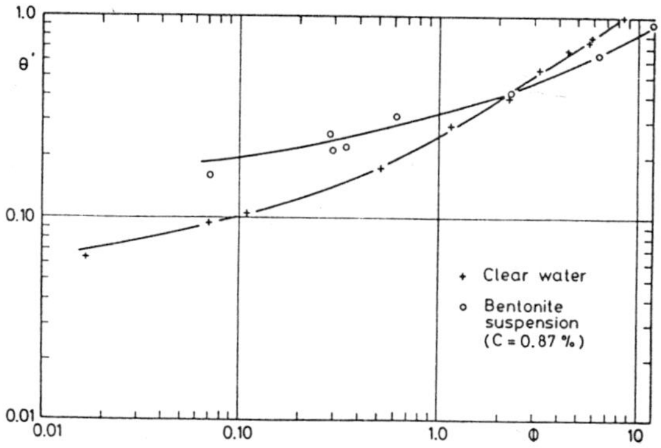

Fig. 9 The comparison of $\phi \sim \theta'$ correlation in bentonite and that in clear water.

6 CONDITIONS FOR INCIPIENT MOVEMENT AND THE TRANSPORT RATE

An empirical threshold condition has been obtained, as shown in Fig. 10.

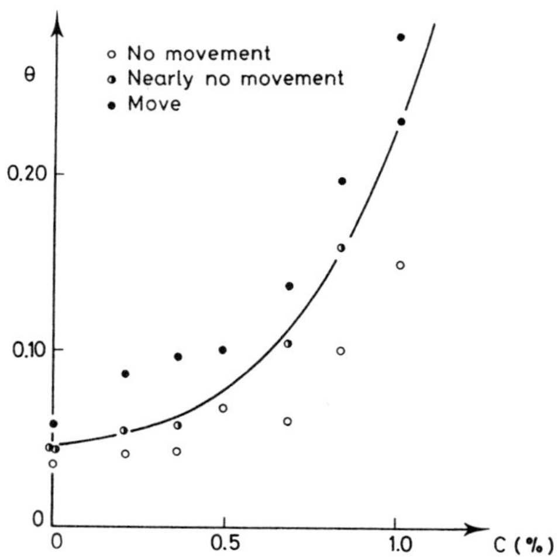

Fig. 10 The threshold condition $\theta_c \sim c$.

For clear water the transport rate can be well described by Engelund-Fredsøe bed load formula (see Fig. 9) (Engelund et al, 1976).

$$\phi_b = 10(\theta' - \theta_c)(\sqrt{\theta'} - \sqrt{\theta_0}) \qquad (8)$$

For bentonite suspension it is shown that this equation is still applicable, only some adjustments should be made.

First, θ_c should be determined according to Fig. 10.

Secondly, θ_0 should be adjusted.

Considering the equilibrium of a particle moving near the bed, they got:

$$C_D \cdot \frac{1}{2} \rho (\alpha U_f' - U_B)^2 = \rho g (s-1) \frac{\pi}{6} d^3 \beta \qquad (9)$$

in which C_D is the drag coefficient, β is the dynamic friction coefficient, U_B is the migrating velocity of the particle and $\alpha U_f'$ is the flow velocity near the bed. From Eq. (9) the relative velocity between the flow and the particle can be written as:

$$\alpha U_f' - U_B = \frac{\theta_0}{\theta'} \qquad (10)$$

where

$$\theta_0 = \frac{4\beta}{3\alpha^2 C_D} \qquad (11)$$

Such a relative motion is analogous to a particle settling in bentonite. So in bentonite we can calculate C_D, and consequently θ_0, by means of the curve in Fig. 7.

After these two adjustments, ϕ_b can be calculated, and it is compared with measured values in Fig. 11. The experimental points coincide with the calculated curves except at high flow intensities with high C. Under such conditions, suspended load is considerable and the bed load formula can not describe it. The presence of suspended load can be judged by the criterion $\frac{W}{U_f'} = 0.8$. In Fig. 11 points corresponding to $\frac{W}{U_f'} < 0.8$ are marked. The ratio between the suspended load and the bed load ζ can be calculated based on Einstein's suggestion. And the bed load is only part of the measured load ϕ.

$$\phi_b = \frac{1}{1+\zeta} \phi \qquad (10)$$

After such a correction, the measured point coincide with the calculated curves again.

7 THE FLOW RESISTANCE

The flow resistance in alluvial channel consists of two parts: surface friction and formal resistance due to bed configuration. In dimensionless form it reads:

$$\theta = \theta' + \theta'' \qquad (11)$$

Because the dunes are lower and smoother in shape, in bentonite the formal resistance is smaller than in clear water, see Fig.12.

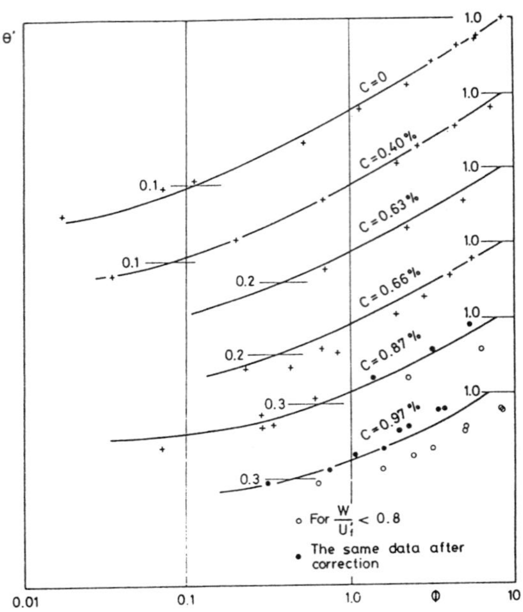

Fig. 11 The comparison of measured and calculated $\phi \sim \theta'$ correlation.

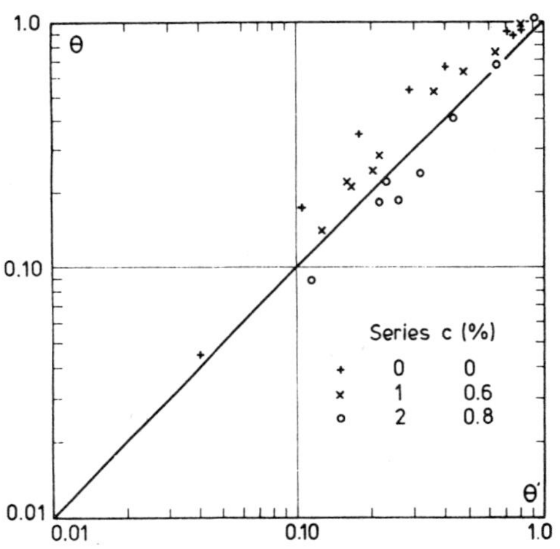

Fig. 12 $\theta \sim \theta'$ for different concentrations.

REFERENCES

Ansley, R.W. and Smith, T.N. 1967, Motion of spherical particles in a Bingham plastic. AICHE J. vol. 13, No. 6; 1193-1196.

Chien, N., Wan, Z.H. and Qian, Y.Y. 1979. The flow with heavy sediment concentration in the Yellow River. Jnl. of Qinghua University 19(2): 1-17 (in Chinese).

Engelund, F. and Fredsøe, J., 1976. A sediment transport model for straight alluvial channels. Nordic Hydrology, 7: 293 - 306.

Navntoft, E., 1968. Et experimentelt studium i materialvandring og om opslemmet transport. Licentiatrapport, Laboratoriet for hydraulik, Danmarks tekniske Højskole, Bulletin No. 16 (in Danish).

Wan, Z.H., Qian, Y.Y., Yang, W.H. and Chao, W.L., 1979. A laboratorial investigation of hyperconcentrated flow. Jnl of the Yellow River 1: 53-65 (in Chinese).

5 Sediment transport in steep channels

Euromech 156: Mechanics of Sediment Transport / Istanbul / 12-14 July 1982

Flow structure and sediment transport mechanics in steep channels*

M.BAYAZIT
Technical University, Istanbul, Turkey

ABSTRACT: In steep channels significant changes occur in the flow structure due to the effect of roughness elements that cover a substantial part of the flow depth. These are observed in the friction factor, velocity distribution and turbulence intensities. The modification of flow structure influences the initiation of motion of bed particles and rate of bed-load transport.

1 INTRODUCTION

Steep channels in nature (mountain streams, torrents) are characterized by their following properties: they have high gradients (usually a few percent), a coarse mixture on the bed which is nonuniform in size and shape (rock, boulder, gravel, sand), water depths that may be of the same order of magnitude as the roughness height, rates of flow and sediment transport that are widely fluctuating, and Froude numbers that are usually rather high. It is necessary to investigate whether the existing methods of sediment transport computation are applicable under these circumstances. For this purpose flow structure in such channels should be first examined and then its effect upon sediment transport should be clarified. This paper aims at presenting a review of the state-of-the-art on this subject.

2 FLOW STRUCTURE NEAR LARGE ROUGHNESS ELEMENTS

In recent years flow structure (velocity distribution, turbulent fluctuations, energy loss) near large roughnesses has been investigated in a number of laboratory and field studies. These aim at examining the complicated flow pattern that is caused by the separation due to roughness elements. The effect of these elements on the flow is felt even in the region away from the boundary and it is most intense when the roughness size is a large fraction of the flow depth.

Special problems arise in investigating the flow structure near large roughnesses. These are to be considered in the following discussion.

2.1 Theoretical bed level (reference datum)

Flow along a rough boundary is not uniform in the exact sense as the depth varies in the flow direction. In order to apply the approximation of uniform flow conditions, it has to be decided at which height the bed level should be chosen with respect to which the flow depth is to be measured. This problem, negligible in small depths, gains significance in the case of large relative roughness. Various ideas have been proposed for the determination of the reference datum, which are based on the following criteria :

1. Schlichting (1936) introduced the concept of geometric bed level, which is the level that would arise if all the roughnesses were melted to form a smooth surface. By using this level as the reference datum

* This paper was presented by invitation at the colloquium as a survey lecture.

the weight of fluid is distributed evenly along the flow direction to produce an average bed shear stress. O'Loughlin, MacDonald (1964) suggested to use a modified version of this concept where the separation zones behind the roughness elements are included in the computation of the volume of roughnesses.

2. Another approach is to determine the theoretical bed level such that the logarithmic velocity law and the resulting energy loss equation give the best fit with the experimental data at large depths (Bayazit 1976).

Results obtained by various authors for different types of roughness geometry are given in Table 1. It is seen that the distance of the theoretical bed level below the tops of the roughness elements varies in the range of (0.15-0.35) k, where k is the physical roughness height.

2.2 Equivalent sand roughness

Another problem concerns with the determination of the equivalent sand roughness k_s, defined by Nikuradse, that will produce the observed energy loss when inserted into the equation for rough open channels obtained from the logarithmic velocity law:

$$(8/f)^{1/2} = (1/\varkappa)\ \ln(h/k_s) + 6 \qquad (1)$$

where f is the Darcy friction factor, $\varkappa$ Karman's constant (usually assumed to be equal to 0.40), h flow depth. k_s is not a measure of the physical roughness height but rather of the effect on the flow of the roughnesses, determined experimentally. For large roughness elements k_s depends on the size, shape and spacing of roughnesses as well as the relative roughness (ratio of roughness size to flow depth).

Experiments by Schlichting (1936) and O'Loughlin, MacDonald (1964) have shown that k_s assumes a maximum value for a certain concentration in the case of discrete roughness elements. This maximum value is equal to 3.5 times the roughness height for cubes and 4.0 times the diameter for spheres. In the case of natural gravel and closely packed roughness elements k_s is found to be equal to a few times the physical roughness size as shown in Table 2. The increase of k_s over the actual roughness size can be attributed to the wake interference losses in the lee of roughness elements (Morris 1955).

For a natural mixture composed of various sizes most authors prefer to express k_s as a multiple of a particle size larger than the median, such as D_{84} or D_{90}, assuming that larger particles are likely to influence the resistance more than the smaller particles. This is, however, not a very important point since the ratio of a certain percentile to any other percentile does not vary significantly for most natural sediments.

2.3 Friction factor of a flow with large relative roughness

When h/k is smaller than a certain value f can no longer be computed by Eq.(1). It has been found that the friction factor takes larger values than those predicted by Eq.(1). Scheuerlein (1973) pointed out that this was the case for many mountain

Table 1. Distance ak of theoretical bed level below the tops of roughness elements

Author	Roughness shape	a
Schlichting (1936)	Sphere	0.16
Einstein, El-Samni (1949)	Hemisphere	0.20
O'Loughlin, MacDonald (1964)	Sand	0.27
Goma, Gelhar (1968)	Sphere	0.23
Blinco, Parthenaides (1971)	Irregular Particles	0.27
Grass (1971)	Gravel	0.18
Cheng, Clyde (1972)	Sphere	0.15
Kamphuis (1974)	Gravel	0.30
Bayazit (1976)	Sphere	0.35
Denker (1980)	Cylinder	0.35
Denker (1981)	Triangular	0.35

Table 2. Equivalent sand roughness k_s for various boundaries

Author	Roughness shape	k_s
Leopold, Wolman, Miller (1964)	Gravel	3.5 D_{84}
Limerinos (1970	Gravel	3 D_{84}
Kamphuis (1974)	Gravel	2 D_{90}
Bayazit (1976)	Closely packed spheres	2.5 D
Charlton et al. (1978)	Gravel	3.5 D_{90}
Hey (1979)	Gravel	3.5 D_{84}
Thompson, Campbell (1979)	Gravel	4.5 D_{90}
Gladki (1979)	Gravel	2.5 D_{80}
Denker (1980)	Closely packed cylinders	2 D
Bray (1980)	Gravel	3.5 D_{84} (3.1 D_{90})
Griffths (1981)	Gravel	5 D_{50}
Denker (1981)	Triangular	8.5 k

streams. Roberson (1963),O'Loughlin, MacDonald (1964) observed this in the laboratory experiments with separated roughnesses, Bayazit (1976) found that the friction factor increases significantly when $h/k < 3$ for a flow over closely packed spheres (k is the radius of the spheres). Bathurst (1978) observed the same phenomenon for river data when $h/D_{84} < 2.5-3$. Day (1977) showed that radical departures from the logarithmic resistance equation occurs when $h/D_{90} < 3$ in flume experiments over coarse alluvial beds. Charlton et al. (1978) reached the same conclusion. Thompson, Campbell (1979) proposed the following equation for natural channel data:

$$(8/f)^{1/2}=(1-0.1\,\frac{k_s}{h})\,(\frac{1}{\varkappa}\ln\frac{h}{k_s}+6) \qquad (2)$$

This equation is in qualitative agreement with the finding mentioned above.

Fig. 1 shows the variation of the friction with relative roughness in experiments over two-dimensional regular triangular roughness elements. It is seen that Eq. (1) is valid

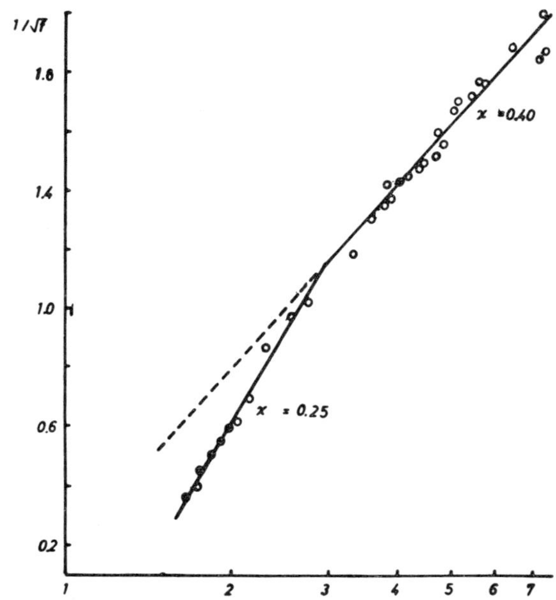

Fig. 1. Darcy friction factor as a function of relative roughness (flume experiments with triangular roughness).

only when $h/k > 3$. For smaller depths Eq. (1) underpredicts the friction factor. In this region a steeper straight line can be approximately fitted, which corresponds to an average value of $\varkappa = 0.25$.

Bathurst (1978) investigated an extreme case where some roughness elements project through the flow. In this case flow resistance is determined mainly by the combined form drag of the individual elements and is related to the Froude number, as well. f increases with the increase of relative roughness as more and more elements protrude through the free surface. Bathurst et al. (1981) proposed a rather complicated resistance equation for the flume data with large-scale roughness. However, a simpler equation in the following form agrees with his data just as well :

$$(8/f)^{1/2} = 2.18 \ln \frac{h}{S_{84}} + 2.95 \qquad (3)$$

where S_{84} is the size of the short axis of the bed material which is bigger than 84 % of short axes. Eq. (3) differs little from the equation given by Hey (1979).

It can be concluded that additional energy losses occur as the relative roughness increases above a certain value, which depends on the geometry of roughness elements. This is due to the fact that the zone where wake eddies shed by the roughness elements interact with the main flow covers a substantial part of the flow depth when relative roughness is high. Exact amount of increase in the friction factor is a function of the roughness geometry and therefore no single equation can be given.

2.4 Velocity distribution near a rough boundary

Various authors argued that logarithmic law does not hold for the velocity distribution in the vicinty of roughness elements and proposed certain modifications. Schlichting (1936) found that this was the case in the region $y/k < 2.5\text{-}4$. Rotta (1950) proposed a slightly modified law of velocity distribution assuming $l = l_0 + \varkappa y$ for the mixing length in the vicinity of the roughnesses where l_0 was expected to be related to the shape and pattern of roughness, Christensen (1972) assumed $l_0 \propto k$.

O'Loughlin, Annambhotla (1969), in their measurements of the velocity distribution in a flow over a boundary covered with spaced cubes, found a departure from the logarithmic law in the region $y/k < 1\text{-}2$, where y is the distance from the smooth boundary between the cubes. Larger velocities were measured in this region than those which would be obtained from the logarithmic law. This was explained by defining a modified eddy diffusion coefficient that is larger because of the more efficient mixing in the wake layer. This is equivalent to an increase of the mixing length as discussed above. It was assumed that the eddy diffusivity remained at a constant value in the region near the roughnesses because of an approximately uniform layer of highly intense turbulence in this region. Outside the wake region ($y/k > 1\text{-}2$), it was expected that the turbulence characteristics would be independent of the detailed shape of the roughness elements, and would exhibit the same properties as flow past a sand-roughened boundary of equivalent roughness. Similar results were obtained by Chen, Roberson (1974) in experiments where hemispherical roughness elements were glued on the wall of a smooth pipe at regular intervals. It was found that two logarithmic profiles with different slopes could be fitted to velocities in the regions $y/k < 1$ and $y/k > 1$. Near the roughnesses $\varkappa$ values were larger than 0.40 and increased with $u_* k/\nu$ where u_* is shear-stress velocity, ν kinematic viscosity.

Pyle, Novak (1981) are of the opinion that the apparent increase of velocities near the roughnesses can be eliminated by shifting the reference datum. They obtained a single velocity profile with a slope corresponding to $\varkappa = 0.285$ by adjusting the datum as a function of the concentration of roughness elements.

In all the above experiments individual roughness elements were attached to the wall of the conduit or flume at intervals corresponding to a certain concentration. In order to investigate the case of closely packed roughness elements, which is more representative of the beds of natural channels, a series of experiments have been carried out in our laboratory. In these experiments regular two-dimensional triangular roughnesses (k = 25 mm) were attached to the bottom of a flume so that no smooth boundary was present between them. Quasiuniform flows were then established at various depths.

Velocity measurements were performed along several verticals in the flow direction by means of a hot-film anemometer. It was found that in each run velocity profiles measured at different verticals were essentially the same in the region $y/k > 1$ (y is the distance from the bottom of the flume, k height of the triangle). This implies that the eddies shed by the roughnesses are confined in the pockets between the consecutive elements and that the flow can be considered uniform outside of these pockets. Velocities measured in the region $y/k < 1$ decreased very slowly as the physical boundary was approached as a consequence of the intense mixing in this zone.

Velocity profiles in the region $y/k > 1$ exhibit two zones: a transition zone where the velocity gradient is larger than the value corresponding to $\varkappa = 0.40$ and decreases gradually with the distance from the boundary, and a zone where the conventional logarithmic law with $\varkappa = 0.40$ holds. The transition zone extends throughout the flow depth for high values of relative roughness ($h/k \lesssim 5$) (Fig. 2). $\varkappa$ values computed from the slopes of straight lines fitted to measured velocities in the transition zone are always smaller than 0.40. In the inner part of the transition zone, where the effect of roughness elements is directly observed, $\varkappa$ adopts a constant value of about 0.18, independent of the value of relative roughness. When the relative roughness is not large ($h/k \gtrsim 5$), the two zones can be distinguished. In the run with $h/k = 8.5$, the transition zone is confined to $y/k \lesssim 4$.

2.5 Turbulence characteristics in a flow with large relative roughness

Chen, Roberson (1974) found that in pipe flows turbulence intensity decreased with increase of relative roughness in the region near the wall. Bayazit (1976) found similar results in open channel flows over beds with spherical roughnesses. The decrease

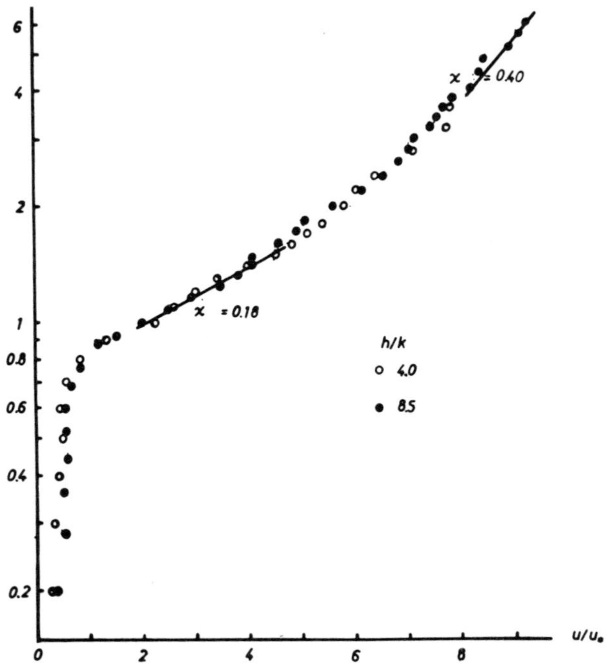

Fig. 2. Velocity distribution in flows with large relative roughness (flume experiments with triangular roughness).

observed in the intensity of turbulence fluctuations can be explained by the fact that a substantial part of the energy of the mean flow is converted into turbulence in the separation zones between the roughness elements in the case of large-scale roughness. No information is available at present on the bursting phenomenon near large roughness elements.

2.6 Resistance due to bed forms

Bed forms in streams with coarse bed material are generally less dominant than for the case of sand-bed rivers. Ripples and dunes that are observed for flows over sand beds have not been found for flows over gravel beds. According to Parker, Peterson (1980) alternating diagonal bars are formed with the accompanying pool and riffle formation, which causes an additional resistance. Friction factors, however, are at least an order of magnitude lower than those of dune beds. These bars are removed at flood stages. Cunha (1967), Gladki (1979) and Griffiths (1981) investigated the resistance due to bed forms in gravel-bed rivers. Simons et al. (1979) argued that there is a significant decrease in resistance when sand and gravel size sediment are deposited between rocks and cobbles.

3 SEDIMENT TRANSPORT IN STEEP CHANNELS

Sediment is transported mainly as bed-load in steep channels with coarse bed material. Sediment transport occurs only at relatively high discharges. Among the various bed-load formulae, Meyer-Peter, Müller's formula and Einstein's bed-load function are most commonly used. Parker (1979), using gravel-bed river data, proposed a bed-load equation in terms of Einstein's parameters which is almost identical with Einstein's relation in the range of $0.032 < S < 0.1$, where S is Shields' parameter. Parker's equation predicts slightly smaller rates of transport.

The above discussion showed that flow structure will be changed when relative roughness of the bed is very large. It may be expected that this will have an effect on the initiation of motion of particles on the bed and on the rate of bed-load transport

3.1 Initiation of motion of particles on the bed of a very rough channe

Ashida, Bayazit (1973), in flume experiments with natural gravel, found that the critical shear stress for initiation of motion increased significantly with the relative roughness. Shields' parameter

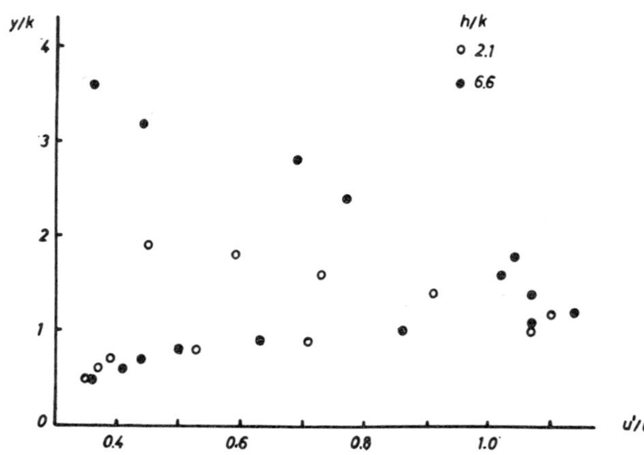

Fig.3. Distribution of turbulence intensity in flows with large relative roughness (flume experiments with triangular roughness).

has to be modified in steep channels as

$$S_c = \frac{\tau_0}{D(\gamma_s - \gamma)(\cos\theta \cdot \tan\phi - \frac{\gamma_s}{\gamma_s - \gamma}\sin\theta)}$$

where τ_0 is bed shear stress, γ_s and γ specific weight of sediment and fluid, respectively, θ channel slope and ϕ angle of internal friction of bed material. S_c increased from 0.04 up to 0.12 when the relative roughness k/h exceeded the value of 0.3, approximately (Fig. 4).

Bayazit (1978) investigated scour of a sphere in a flume with a bed covered by hemispheres in the most compact configuration. In these experiments values of Shields' parameter were recorded at the stage of initiation of vibration of the sphere and at the critical stage when the sphere was moved by the flow. These values increased about 4 times for very large values of relative roughness. This was explained by the reduction of the time-average velocity and the turbulence intensity in the vicinity of the sphere as the relative roughness increases. As a result, critical velocity occurs at higher bed shear stress as compared with flows at mild slopes.

These findings are supported by the observation of Thompson, Campbell (1979) in a natural channel on a 0.052 slope that "boulders of the size that would be expected to move by Shields' criterion did not move".

3.2 Bed-load transport in very rough channels

Bayazit, Denker (1980) studied the motion of a spherical particle on a bed lined with fixed cylinders of uniform size for shallow depths of flow. The motion was treated as a stochastic process composed of subsequent resting periods, steps and moving periods. It was found that the average velocity, V, of the particle decreased significantly as the relative roughness increased, mainly due to the increase of resting period, T. As a result bed-load transport rate was reduced by approximately 30 times when relative roughness increased from 0.20 to 0.35 for a constant value of bed shear stress (Fig. 5).

In almost all the models of bed-load transport bed shear stress is considered to be the only parameter that controls the phenomenon. Experiments described above have shown that this is not correct when the bed roughness is of the same order of magnitude as the flow depth. For such high values of relative roughness resting periods of the particles increase substantially. This was explained as follows. Rest of the particle comes to an end when the fluctuating force of the turbulent

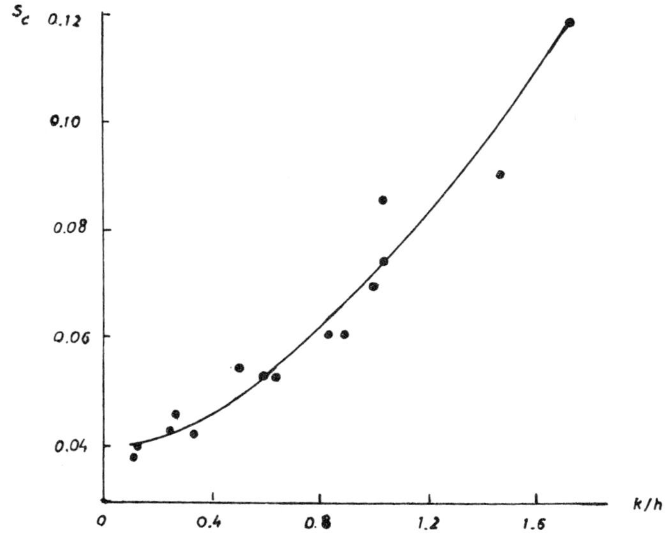

Fig. 4. Critical value of Shields' parameter as a function of relative roughness (Ashida, Bayazit, 1973).

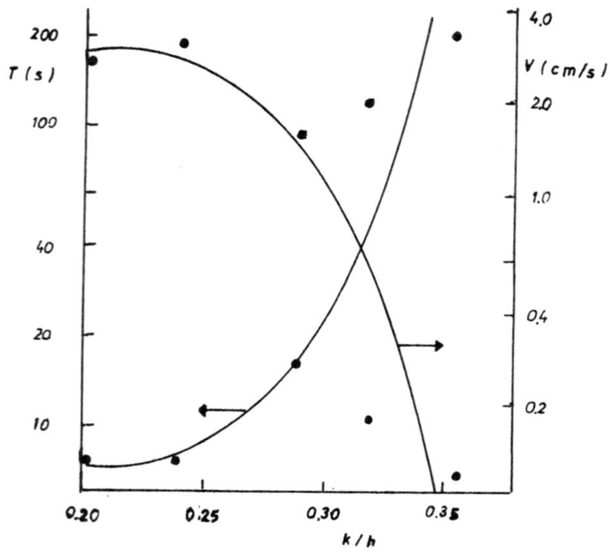

Fig.5. Mean resting period and mean particle velocity as function of relative roughness (Bayazit, Denker, 1980).

flow acting on the particle reaches a critical magnitude which is sufficient to destroy the stability. As turbulence intensity diminishes with increasing relative roughness for a constant value of bed shear stress, fluctuating force exceeds the critical value less frequently for large values of relative roughness. Magnitudes of the step length and moving period, on the other hand, are related to the initial momentum applied by the flow to the particle when it is eroded. As this momentum is approximately proportional to the critical value of the disturbing force it is not much affected by the change of the flow structure as the relative roughness increases.

CONCLUSIONS

The above review shows that flow and bed-load transport in steep channels has certain characteristics which make it significantly different from the corresponding phenomenon in mild gradients. Altough we know the basic reasons for these differences to a certain extent, more extensive laboratory and field data are needed in order to be able to set up equations that can be used in predicting the energy loss, critical stage of bed-load and rate of sediment transport in mountain streams. The main difficulty arises from the nonuniformity of the various flow and sediment properties that have an effect on this phenomenon. Therefore it looks unlikely that universal equations can be derived for various aspects of this interesting subject in the near future.

ACKNOWLEDGEMENT

Experiments described in this paper were performed by Dr. H. Denker.

REFERENCES

Ashida, K. and M. Bayazit. 1973. Initiation of motion and roughness of flows in steep channels. Proc. of the 15 th Congress of the IAHR, 1: 475-484.

Bathurst, J.C. Flow resistance of large scale roughness. J. of the Hyd. Div., ASCE, 104 (HY 12): 1587-1603.

Bathurst, J.C.,R.-M. Li and D.B. Simons. 1981. Resistance equation for large-scale roughness. J. of the Hyd. Div., ASCE, 107 (HY12): 1593-1613.

Bayazit,M. 1976. Free surface flow in a channel of large relative roughness, J. of Hyd. Res., IAHR, 14(2): 115-126.

Bayazit, M. 1978. Scour of bed material in very rough channels. J. of the Hyd. Div., 104(HY9): 1345-1349.

Bayazit, M. and H. Denker. 1980. Stochastic analysis of bed-load motion in very rough channels. 3rd Int. Symp. on Stochastic Hyd.: 347-357.

Blinco, P.H. and E. Parthenaides. 1971. Turbulence characteristics in free surface flows over smooth and rough boundaries. J. of Hyd. Res., IAHR, 9(1): 43-72 .

Bray, D.I. 1980. Evaluation of effective boundary roughness for gravel-bed rivers. Canadian J. of Civ. Eng., 7(2): 392-397.

Charlton, F.G., P.M. Brown and R.W. Benson. 1978. The Hydraulic geometry of some gravel rivers in Britain. Rep. No. IT 180, Hyd. Res. St., Wallingford, UK.

Chen, C.K. and J.A. Roberson. 1974. Turbulence in wakes of roughness elements. J. of the Hyd. Div., ASCE, 100 (HY1): 53-67.

Cheng, E.D.H. and C.G. Clyde. 1972. Instantaneous hydrodynamic lift and drag forces on large roughness elements in turbulent open channel flow. Sedimentation (Einstein), ed. H.S. Shen, Chap. 3.

Chistensen, B.A. 1972. Incipient motion on cohesionless channel banks. Sedimentation (Einstein), ed. H.S. Shen, Chap. 4.

Cunha, L.V.D. 1967. About the roughness in alluvial channels with comparatively coarse bed materials. Proc. 12th Congress of the IAHR, 1: 76-84.

Day, T.J. 1977. Discussion of Resistance equation for alluvial channel flow, by D.E. Burkham and D.R. Dawdy. J. of the Hyd. Div., ASCE, 103(HY5): 582-584.

Denker, H. 1980. Icing in open channels and structure of turbulent flow between two parallel plates of infinite length. Ph.D. thesis, Tech. Univ. of Istanbul (in Turkish).

Denker, H. 1981. Personal communication.

Einstein, H.A. and E.S. El-Samni. 1949. Hydrodynamic forces on a rough wall. Rev. Mod. Phys., 21(3): 520-524.

Gladki, H. 1979. Resistance to flow in alluvial channels with coarse bed materials. J. of Hyd. Res., IAHR, 17(2): 121-128.

Goma, M.R.J. and L.W. Gelhar. 1968. Turbulent pipe flow with rough and porous walls. Rep. No. 109, Hydrodynamics Lab., Dept.of Civ. Eng., M.I.T.

Grass, A.J. 1971. Structural features of turbulent flow over smooth and rough boundaries. J.of Fluid Mech., 50(2): 233-255.

Griffiths, G.A. 1981. Flow resistance in coarse gravel bed rivers, J.of Hyd. Div., ASCE, 107 (HY7): 899-918.

Hey, R.D. 1979. Flow resistance in gravel-bed rivers, J. of Hyd. Div., ASCE, 105 (HY4): 365-379,

Kamphuis, J.W. 1974. Determination of sand roughness for fixed beds. J.of Hyd. Res., IAHR, 12(2): 193-203.

Leopold, L.B., M.G. Wolman and J.F. Miller. 1964. Fluvial processes in geomorphology. San Francisco: Freeman and Co.

Limerinos, J.T.1970. Determination of the Manning coefficient from measured bed roughness in natural channels. Water Supply Paper 1898-B, USGS.

Morris, H.M. 1955. Flow in rough conduits. Trans. ASCE, 120: 373-398.

O'Loughlin, E.M. and V.S.S. Annambhotla. 1969. Flow phenomena near rough boundaries. J. of Hyd. Res., IAHR , 7(2): 231-250.

O'Loughlin, E.M. and E.C. MacDonald. 1964. Some roughness concentration effects on boundary resistance. La Houille Blanche. 19(7): 773-783.

Parker, G. 1979. Hydraulic geometry of active gravel rivers. J. of Hyd. Div., ASCE, 105 (HY9): 1185-1201.

Parker, G. and A.W. Peterson. 1980. Bar resistance of gravel-bed streams, J. of Hyd. Div., ASCE, 106(HY10): 1559-1575.

Pyle, R. and P. Novak. 1981. Coefficient of friction in conduits with large roughness. J.of Hyd. Res., IAHR, 19(2): 119-140.

Roberson, J.A. 1963. Discussion of rougness spacing in rigid open channels, by W.W. Sayre and M.L. Albertson. Trans. ASCE, 128(1): 400-401.

Rotta, J. 1950. Das in Wandnaehe gültige Geschwindigkeitsgetz turbulenter strömungen. Ing.Arch., 18: 277-280.

Scheuerlein, H. 1973. The mechanics of flow in steep, rough open channels. Proc. of the 15th Cong. of the IAHR, 1: 457-465.

Schlichting, H. 1936. Experimentelle untersuchungen zum Rauhigkeitsproblem. Ing. Arch., 7: 1-34.

Simons, D.B., K.S. Al-Shaikh-Ali and R.Li. 1979. Flow resistance in cobble and boulder river beds. J. of Hyd. Div., ASCE. 105 (HY5): 477-486.

Thompson, S.M. and P.L. Campbell. 1979. Hydraulics of a large channel paved with boulders, J. of Hyd. Res., IAHR, 17(4): 341-354.

Initiation of sediment transport in steep channels with coarse bed material

J.C.BATHURST
Institute of Hydrology, Wallingford, Oxon, UK
W.H.GRAF & H.H.CAO
Laboratoire d'Hydraulique, Ecole Polytechnique Fédérale de Lausanne, Switzerland

ABSTRACT: Measurements in a flume (0.6 m wide) with a gravel of median size 22 mm, a range of water discharges 0.01-0.1 m^3/s and slopes of 3%,5%,7% and 9% are used, along with other flume and field data, to investigate the criteria which determine initiation of sediment transport in steep channels. The most common method for predicting initiation of transport is the Shields diagram which plots the Shields dimensionless shear stress against shear Reynolds number. However, this may not be appropriate to steep channels with small ratios of depth to sediment size because (a) the processes of sediment transport are not adequately described by shear stress considerations; and (b) the Shields diagram is poorly designed mathematically. By contrast the Schoklitsch approach, giving critical flow discharge as a function of slope, has a more suitable, process-related basis. The flume data support this analysis, indicating that the Shields diagram breaks down for steep channels (Fig. 1). The success of the Shields diagram elsewhere probably follows from its past restriction to flows with gentle slopes where variations in the Shields dimensionless shear stress are small (Fig. 2). The Schoklitsch approach explains the major variations in both field and flume data satisfactorily but effects related to bed packing and other field conditions have still to be quantified (Fig. 3). A critical discharge equation is proposed for flows with slopes > 2%.

1 INTRODUCTION

For conventional open channel flows with gentle slopes, large ratios of depth to sediment size and fine bed materials (sand and fine gravels) the critical conditions at which sediment transport begins are often predicted using the Shields criterion (Shields 1936). However, several recent studies have shown that for flows with steep slopes (≳ 0.5%), with depths which are of the same order of magnitude as the bed material size and with coarse bed materials (gravel and boulders), this criterion does not apply. Here the reasons for this failure of applicability are considered and the suitability of the alternative Schoklitsch (1962) approach is investigated, using some preliminary results from a combined flume and field study of sediment transport in steep steams.

2 CRITERIA FOR SEDIMENT MOVEMENT

2.1 Basic criteria

The major forces which cause a sediment particle to move are drag, F_D, and lift, F_L, while those which counter the movement are submerged particle weight, W, and the sedimentary packing of the particles in the bed (Graf 1971: 84; Laronne & Carson 1976; Fortier & Scobey 1926, quoted by Graf 1971: 87-88). Critical conditions for the initiation of particle movement occur when these forces are in balance according to the equation

$$F_{Dc} + F_{Lc} \tan\phi$$
$$= W \cos\theta \tan\phi - W \sin\theta + X. \qquad (1)$$

Here ϕ = the angle of repose of the bed material; $\tan\theta = S$ = channel slope; X = bed packing factor; and the subscript c denotes critical conditions throughout this paper. Equation 1 allows for the downslope component of particle weight aiding particle movement.

2.2 Shields criterion

For this criterion the drag force on the particle is replaced by the function $\tau\ d_{50}^2$, where τ = shear stress acting on the bed and d_{50} = that size of particle median axis which, for a nonuniform particle size distribution, is larger than 50% of the particles in a sample. Usually lift force and bed packing effects are neglected and W is set proportional to $(\rho_s-\rho)\ d_{50}^3\ g$, so from Eq.1

$$\frac{\tau_c}{(\rho_s-\rho)\ g\ d_{50}\ (\cos\theta\ \tan\phi\ -\ \sin\theta)} = S_c \quad (2)$$

which is the traditional Shields formula with an allowance for the effects of channel slope and bed material angle of repose. In Eq.2 ρ = fluid density; ρ_s = particle density; and g = acceleration due to gravity. τ_c is usually defined as $\rho\ g\ D_cS$ where D_c = critical mean flow depth and S = channel slope. S_c is the critical Shields factor. Traditionally it is plotted against the shear Reynolds number $u_{*c}\ d_{50}/\nu$ and, based on experimental results, is held to have a constant value of about 0.04 - 0.06 for $u_{*c}\ d_{50}/\nu \gtrsim 500$. However, experimental results have not been obtained for $u_{*c}\ d_{50}/\nu \gtrsim 1000$. Here $u_{*c} = (\tau_c/\rho)^{1/2} = (g\ D_cS)^{1/2}$ and ν = kinematic viscosity.

There are two major objections to the use of this criterion, especially for steep flows and coarse bed materials.

1. For steep flows ($S \gtrsim 0.5\%$) and small ratios of depth to sediment size ($D_c/d_{50} \lesssim 10$), the Shields factor S_c is found to increase as D_c/d_{50} decreases (Ashida & Bayazit 1973; Mizuyama 1977; Bathurst et al 1979). S_c is also affected by the armouring or paving effects of the surface layer of bed material where the bed has a nonuniform size distribution (Egiazaroff 1965; Gessler 1971; Andrews 1981; Parker & Klingeman 1982). Neither of these processes is adequately catered for in the Shields approach

2. With $\tau_c = \rho\ g\ D_cS$, the Shields factor S_c varies essentially as $D_c\ S/d_{50}$ for most natural flows. Therefore plotting S_c against shear Reynolds number, which varies essentially as $(D_cS)^{1/2}\ d_{50}$, or against any flow parameter based on depth, does not provide a useful guide to the critical conditions for initiation of motion (Gessler 1971; Simons & Şentürk 1977: 412).

2.3 Approach based on the Schoklitsch criterion

An alternative approach can be developed by considering the processes described in Eq.1. The drag and lift forces may be expressed (Graf 1971: 85) as

$$F_D = C_D\ k_1\ d_{50}^2\ \rho\ U_b^2 \quad (3)$$

$$F_L = C_L\ k_2\ d_{50}^2\ \rho\ U_b^2. \quad (4)$$

In these equations U_b is the flow velocity around the particles and may be given by the mean flow velocity U where D/d_{50} is small. C_D and C_L are the coefficients of drag and lift respectively and, for steep flows with small values of D/d_{50}, may be expected to vary with particle Reynolds number $Re = U\ d_{50}/\nu$, with flow Froude number $Fr = U/(gD)^{1/2}$ and with D/d_{50} (Flammer et al 1970). k_1 and k_2 are particle shape factors. Remembering also the effects of nonuniform particle size distribution, Eq.1 can therefore be rewritten as

$$\frac{U_c^2}{(\rho_s/\rho\ -\ 1)\ g\ d_{50}\ (\cos\theta\ \tan\phi\ -\ \sin\theta)} = \text{fn}\ (Re_c,\ Fr_c,\ \frac{D_c}{d_{50}},\ k_i,\ \frac{d_{84}}{d_{50}},\ X) \quad (5)$$

where d_{84}/d_{50} represents the effect of particle size distribution and k_i represents particle shape. For the above it is assumed that flow width is large compared with depth and sediment particle size and that particle density ρ_s is approximately constant with a value such that $\rho_s/\rho \approx 2.6$.

Equation 5 may be simplified by noting that, for a given flow discharge and bed material size, the relationship between velocity and depth (and therefore between Froude number and D/d_{50}) is an empirical function of channel slope. Thus, transferring Froude number and D/d_{50} to the left hand side of the equation, assuming that they can be replaced on the right hand side by slope and making suitable mathematical rearrangements, Eq.5 can be rewritten to give the dimensionless factor q_{c*} as

$$q_{c*} = \frac{U_cD_c}{d_{50}^{3/2}[g(\rho_s/\rho\ -\ 1)\ (\cos\theta\ \tan\phi\ -\ \sin\theta)]^{1/2}} = \text{fn}\ (Re_c,\ S,\ k_i,\ \frac{d_{84}}{d_{50}},\ X). \quad (6)$$

Since U D = q = flow discharge per unit width of channel, this function is similar to the Schoklitsch (1962: 174) equation for the critical unit flow discharge

$$q_c = 0.26\ (\rho_s/\rho - 1)^{5/3} d_{40}^{3/2} S^{-7/6} \quad (7)$$

which was developed for gravel-bed rivers. (Here S is in % ÷ 100.)

The above approach, although only semi-empirical, should be more successful than the Shields approach for steep, shallow flows since it considers the relevant processes and also involves flow velocity as well as depth.

3 DATA COLLECTION

In order to elucidate the foregoing considerations, data are being collected in a specially designed flume at the Laboratoire d'Hydraulique of the EPF - Lausanne (Switzerland) and in the field by the Institute of Hydrology, Wallingford (UK).

The flume and its operation are described in the companion paper by **Bathurst et al** (1982). Briefly a gravel layer 100 mm thick was laid in a flume 0.6 m wide, the statistics of the gravel being d_{16} = 18 mm, d_{50} = 22 mm, d_{84} = 29 mm, $\tan\phi$ = 0.9 and ρ_s/ρ = 2.57. Measurements of flow conditions and sediment discharge (which was entirely bedload) were made over a range of water discharges 0.01 - 0.1 m^3/s at flume slopes of 3%, 5%, 7% and 9%. At the lowest flows at each slope there was no sediment transport. The critical unit water discharge, q_c, for initiation of sediment transport at each slope was obtained by back extrapolating the graphical plots of water discharge against sediment discharge. The values so obtained are estimated to be accurate to within 5%. Corresponding values of depth (and thence velocity) were obtained from the measured q-D relationship for each slope (Table 1).

Table 1. Critical flow conditions for the initiation of sediment transport in the EPFL flume (width = 0.6 m).

Flume slope %	Unit discharge m^2/s	Mean depth m	Mean velocity m/s	S_c (eq.2)	q_{c*} (eq.6)
3	0.092	0.080	1.16	0.076	7.60
5	0.056	0.058	0.97	0.096	4.68
7	0.033	0.045	0.73	0.108	2.79
9	0.024	0.034	0.71	0.108	2.06

Field data are currently available for the Tanllwyth, a small stream in the Institute of Hydrology's experimental catchments at Plynlimon, Wales (e.g. see Newson & Harrison (1978)), where bedload measurements have been carried out with a Helley-Smith sampler. Relevant details of the site and estimated critical values (obtained as for the flume data) are: S = 4%; d_{50} = 60 mm for the channel bed; d_{50} = 4-30 mm for the trapped bedload; channel width = 2.9 m; critical flow discharge = 1 - 1.5 m^3/s; q_{c*} = 6.1 - 9.1 based on d_{50} for the channel bed. Derivation of critical values for field sites is not as simple as for flumes because of sediment supply effects which can mean that a water discharge produces no sediment transport even though sediment movement can be observed at a lower water discharge in a different event. The critical values for the field data are therefore approximate. The basic data themselves were supplied by M. Newson and G. Leeks (Institute of Hydrology,UK).

Finally, a search through the literature for studies of flows with slopes S $\gtrsim$ 1% or with $D/d_{50} \lesssim 10$ revealed the additional flume data of Gilbert (1914), Meyer-Peter and Müller (1948), Ashida and Bayazit (1973) and Bathurst et al (1979) and the field data of Simpson (1978) and Haddock (1978). These were analysed using the same procedures as described above.

4 ANALYSIS OF DATA

4.1 Shields approach

The flume data are plotted on the traditional Shields diagram (using S_c from Eq.2) in Fig. 1. Also shown is the traditional curve of critical conditions (separating the zone with sediment movement from the zone without sediment movement) which is often plotted on the Shields diagram. As the curve has not previously been extended to Reynolds numbers greater than 1000 that limitation is also applied here. From previous studies it would be expected that the points in Fig. 1 (all of which represent critical flow conditions) should fall around, and exhibit the same trend as, the traditional curve. However, this is clearly not the case. In fact the pattern of separate but roughly parallel lines for each bed material complies mathematically with the earlier observation that the ordinate varies essentially as $D_c\ S/d_{50}$ while the abscissa varies essentially as $(D_c\ S)^{1/2}\ d_{50}$.

Some explanation for the difference between the observed patterns and the traditional curve (which is apparently supported by the results of many experimental studies) may be gained from Fig. 2 where S_c is plotted against D_c/d_{50}. Following the study of Ashida and Bayazit

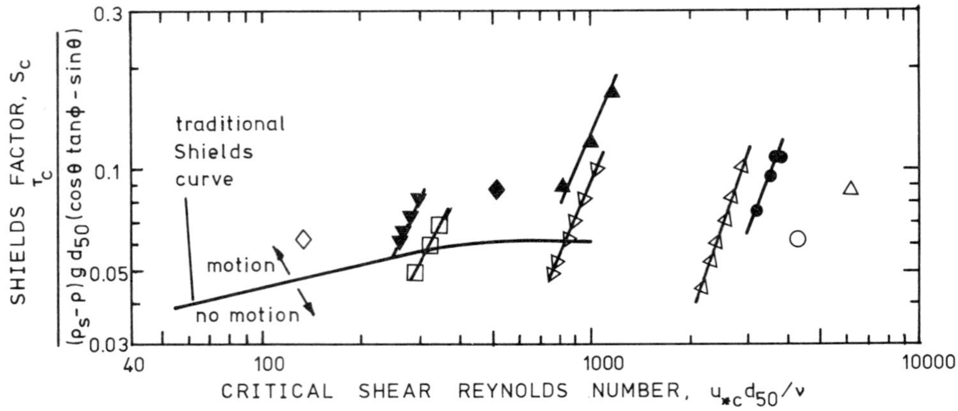

BED MATERIAL SIZE d_{50} (mm) AND DATA SOURCE

● 22	EPFL	◆ 7	Gilbert
▲ 8.8	Bathurst et al	○ 29	Meyer-Peter and Müller
△ 34	Bathurst et al	◁ 21	Ashida and Bayazit
◇ 3.2	Gilbert	▷ 10	Ashida and Bayazit
▼ 4.9	Gilbert	□ 5	Ashida and Bayazit

Fig. 1. Shields diagram of Shields factor or dimensionless critical shear stress (modified to include downslope component of particle weight) against critical shear Reynolds number; traditional Shields curve is compared with data points for critical flows in flumes with gravel sediment and slopes $\geq$ 1%.

(1973) it would be expected that S_c should decrease as D_c/d_{50} increases. Fig. 2 shows that this may be the case for a given bed material (see the solid lines) but that there is no common pattern for different bed materials. However, a more satisfactory mathematical explanation for the pattern in Fig. 2 is obtained by noting that the abscissa varies essentially as D_cS/d_{50} while the ordinate varies as D_c/d_{50} and that points of equal channel slope therefore all lie on a single line. (In Fig. 2 these lines, shown as broken lines, are not linear and free from discontinuities, as might be expected, owing to the effect of bed packing and other effects which are not included in the approach, as explained below.) The lower the channel slope, the lower is the gradient of these lines and the less is the change in S_c for a given change in D_c/d_{50}. Thus, for flows with $S \lesssim 1\%$, the gradients of these lines are so small that S_c varies only very gradually and is restricted in value to about 0.04 - 0.06. The traditional Shields approach (which assumes a constant value of S_c of about 0.04-0.06 at high shear Reynolds numbers) is therefore apparently based on the coincidence that most of the earlier studies concerned with evaluating S_c have involved values of channel slope small enough and values of D_c/d_{50} large enough that the real variations of S_c with flow and channel parameters have been too small to be commented on. Support for this may be gained from close examination of the results of earlier studies (see the data of C.R. Neill in Fig. 4.7 of Yalin 1977:82) which show that a pattern similar to that of the flume data in Fig. 1 is indeed evident but that the range of variation is small. The Shields diagram does not therefore apply to steep flows for which the range of values of S_c is large.

4.2 Schoklitsch approach

Before testing the discharge-based approach the dependencies on the right hand side of Eq.6 need to be examined.

For the flume data, the critical particle Reynolds numbers Re_c lie in the range 10^3 - 2.5×10^4 in which Reynolds number has only a secondary effect on the drag coefficient C_d (e.g. see the data of Flammer et al 1970). Bed material size distributions are relatively uniform (d_{84}/d_{50} $\lesssim$ 1.3) so the effects of a nonuniform size distribution should also be minimal. The

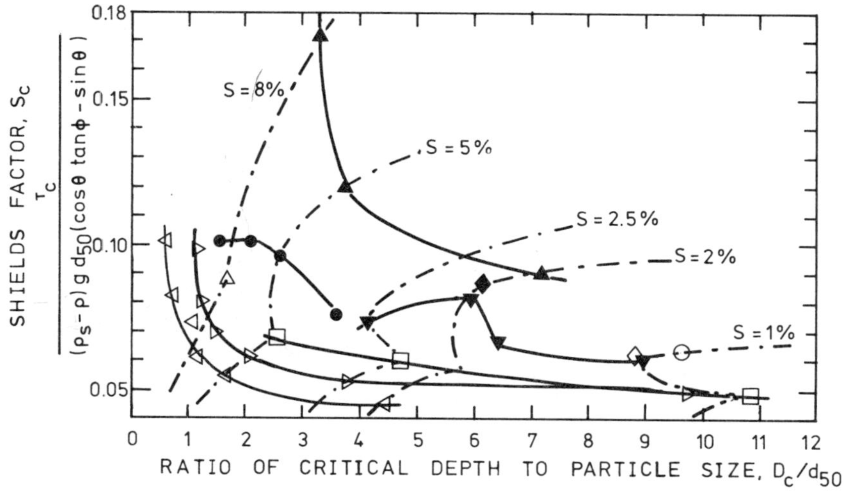

BED MATERIAL SIZE d_{50} (mm) AND DATA SOURCE

● 22	EPFL	◆ 7	Gilbert
▲ 8.8	Bathurst et al	○ 29	Meyer-Peter and Müller
△ 34	Bathurst et al	◁ 21	Ashida and Bayazit
◇ 3.2	Gilbert	▷ 10	Ashida and Bayazit
▼ 4.9	Gilbert	□ 5	Ashida and Bayazit

Fig. 2. Variation of Shields factor or dimensionless critical shear stress (modified to include downslope component of particle weight) with ratio of depth to sediment particle size for critical flows in flumes with gravel sediment and slopes $\geq$ 1%; solid lines join points for flows with same bed material in given experimental series; broken lines join points (measured or estimated) for flows with same flume slope S.

same may be said for bed material density since specific gravity $\rho s/\rho$ lies in the range 2.5 - 2.68. Bed material shape should also have only a secondary effect since the bed materials were all cohesionless gravels, mainly natural but consisting of crushed gravel chips in the case of Bathurst et al (1979). Thus, of the sedimentary factors, only the bed packing factor, X, is likely to have had a significantly varying effect on the critical flow conditions. For most of the studies the bed layers were several particle diameters thick and particles would therefore have been able to orient themselves into natural packing patterns. For Ashida and Bayazit's study, though, the bed layer was only two particles thick. This and other experimental arrangements may be expected to have produced a different bed packing pattern and therefore form the most likely explanation for the anomalous plots of the associated data in Fig. 2. In conclusion, the major determinants of the critical conditions for initiation of sediment transport for the flume flows (apart from those of Ashida and Bayazit's experiments) are, according to the approach in question, Froude number and the ratio of depth to sediment particle size (given on the left hand side of Eq.6) and channel slope (the only remaining dependency on the right hand side of Eq.6).

Fig. 3 shows the variation of q_{c*} with S using all the flume data except those of Ashida and Bayazit. The plot clearly exerts a unifying influence on the data which supports the Schoklitsch type of equation involving flow velocity or discharge as well as depth. For data points with S $\gtrsim$ 2% the relationship, determined by eye, is approximately

$$q_{c*} = 0.114\ S^{-1.2} \qquad (8)$$

which can be shown to compare favourably with Eq. 7. (In Eqs. 7 and 8, S is in % ÷ 100.) Scatter about this line depends on second order effects such as those mentioned earlier. For flows with S $\lesssim$ 2%,

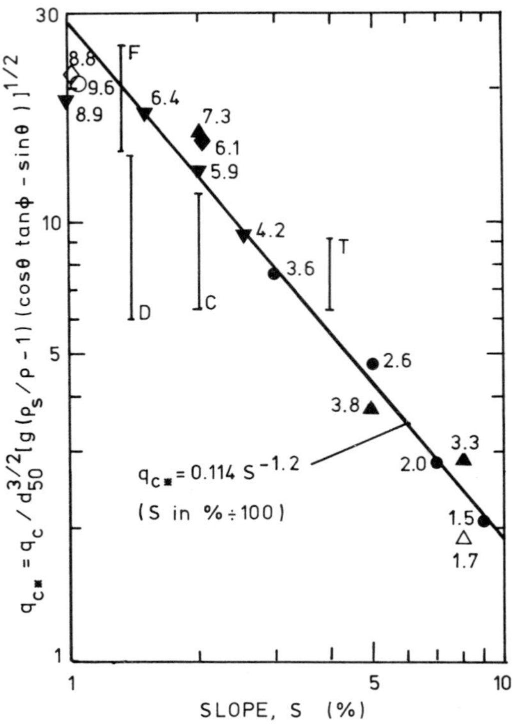

BED MATERIAL SIZE d_{50} (mm) AND DATA SOURCE

●	22	EPFL
▲	8.8	Bathurst et al
△	34	Bathurst et al
◇	3.2	Gilbert
▼	4.9	Gilbert
◆	7	Gilbert
○	29	Meyer-Peter and Müller
T	∿ 60	Newson and Leeks (Tanllwyth)
C	∿ 33	Haddock/Simpson
D	∿ 42-60	Haddock/Simpson
F	∿ 38	Haddock/Simpson

Fig. 3. Variation of dimensionless critical water discharge per unit width, q_{c*}, with channel slope for critical flows in flumes and rivers with gravel sediment and slopes ≥ 1%; line fitted by eye; error bars for rivers indicate range expected to contain true value of q_{c*}; D_c/d_{50} is noted by each point for flume flows.

the rate of change of q_{c*} with S appears to be less, suggesting that this slope (which coincides with values of $D_c/d_{50} \approx 5 - 6$) marks the boundary of the effects of steep flow with large-scale roughness which are described in Eq.5.

The field results are also plotted in Fig. 3, where the error bars (based on the observed variations in critical discharge referred to in Section 3) delineate the likely range in the value of q_{c*} for each field site. The greater scatter of the results is caused partly by the difficulty in specifying the critical conditions in the field and partly by effects related to the flow, sediment and bed packing characteristics not allowed for in Eq.8. Nevertheless, the results fall around the line of Eq. 8 and support the Schoklitsch or discharge-based approach as far as it goes.

5 CONCLUSIONS

Based on the limited available flume and field data, this study of the critical conditions for sediment transport in steep streams with small ratios of depth to sediment size indicates that:

1. Both hydraulic and bed packing effects are important.
2. The traditional Shields approach does not apply, partly because it does not include flow velocity.
3. The Schoklitsch approach, based on discharge, forms a satisfactory basis for predicting critical conditions.
4. The Schoklitsch approach requires further development to allow for the full range of natural field effects.
5. As Eq. 8 is process based, it should not be applied to flows with slopes S < 2% and $D_c/d_{50} \gtrsim 6$ in which the assumptions concerning sediment transport described above do not apply.

ACKNOWLEDGEMENTS

The authors thank Drs. W.R. White and R. Bettess (Hydraulics Research Station Ltd., Wallingford, UK) and Mr. R. T. Clarke and Dr. P.E. O'Connell (Institute of Hydrology, Wallingford, UK) for helpful discussions.

REFERENCES

Andrews, E.D. 1981. Entrainment of gravel and cobbles from a non-uniform streambed. EOS, Trans.Am.Geophys.Union 62(45):858.

Ashida, K. & M.Bayazit 1973. Initiation of motion and roughness of flows in steep channels. Intl.Assoc.Hydraul.Res., Proc. 15th Congress, Istanbul, Turkey 1:475-484.

Bathurst, J.C., R.M.Li & D.B.Simons 1979. Hydraulics of mountain rivers. Rep.CER78-79JCB-RML-DBS55, Dept.Civ.Eng., Colorado State Univ., Fort Collins, Colo., USA.

Bathurst, J.C., W.H.Graf & H.H.Cao 1982. Bedforms and flow resistance in steep gravel-bed channels. In B.M.Sumer & A.Müller (eds.) Euromech 156: The mechanics of sediment transport, Rotterdam, Balkema.
Egiazaroff, J.V. 1965. Calculation of non-uniform sediment concentrations. Proc. Am.Soc.Civ.Engrs., J.Hydraul.Div. 91(HY4).
Flammer, G.H., J.P.Tullis & E.S.Mason 1970. Free surface, velocity gradient flow past hemisphere. Proc.Am.Soc.Civ.Engrs., J. Hydraul.Div. 96(HY7):1485-1502.
Fortier, S. & F.C.Scobey 1926. Permissible canal velocities. Trans.Am.Soc.Civ.Engrs. 89.
Gessler, J. 1971. Beginning and ceasing of sediment motion. In H.W.Shen (ed.), River mechanics vol.1. Fort Collins, Colo.:H.W.Shen, P.O.Box 606.
Gilbert, G.K. 1914. The transportation of débris by running water. Prof.Pap.86, U.S.Geol.Surv., Washington DC, USA.
Graf, W.H. 1971. Hydraulics of sediment transport. New York:McGraw-Hill Book Co.
Haddock, D.R. 1978. Modeling bedload transport in mountain streams of the Colorado Front Range. M.Sc.thesis, Colorado State Univ., Fort Collins, Colo., USA.
Laronne, J.B. & M.A.Carson 1976. Interrelationships between bed morphology and bed-material transport for a small, gravel-bed channel. Sedimentology 23(1): 67-85.
Meyer-Peter, E. & R.Müller 1948. Formulas for bed-load transport. Intl.Assoc.Hydraul. Structures Res., Proc. 2nd Meeting, Stockholm, Sweden Appendix 2:39-64.
Mizuyama, T. 1977. Bedload transport in steep channels. Ph.D thesis, Kyoto Univ., Kyoto, Japan.
Newson, M. & G.Harrison 1978. Channel studies in the Plynlimon experimental catchments. Rep. 47, Institute of Hydrology, Wallingford, UK.
Parker, G. & P.C.Klingeman 1982. On why gravel streams are paved. Proc.Am.Soc. Civ.Engrs., J.Hydraul.Div. 108.
Schoklitsch, A. 1962. Handbuch des Wasserbaues. Vienna:Springer-Verlag.
Shields, A. 1936. Anwendung der Ähnlichkeitsmechanik und Turbulenzforschung auf die Geschiebebewegung. Mitteil.Preuss. Versuchsanst. Wasser, Erd, Schiffsbau, Berlin, no.26.
Simons, D.B. & F.Şentürk 1977. Sediment transport technology. Fort Collins, Colo.:Water Resources Publications.
Simpson, D.H. 1978. Modeling of bedload transport and hydraulic roughness in mountain streams. M.Sc.thesis, Colorado State Univ., Fort Collins, Colo., USA.
Yalin, M.S. 1977. Mechanics of sediment transport. 2nd ed. Oxford:Pergamon Press.

Bedforms and flow resistance in steep gravel-bed channels

J.C.BATHURST
Institute of Hydrology, Wallingford, Oxon, UK

W.H.GRAF & H.H.CAO
Laboratoire d'Hydraulique, Ecole Polytechnique Fédérale de Lausanne, Switzerland

ABSTRACT: For a bed material of median size 22 mm and a range of water discharges 0.01 - 0.10 m^3/s and at slopes of 3%, 5%, 7% and 9%, bedforms in a laboratory flume (0.6 m wide) are found to consist of antidunes and alternate bars while flow resistance is shown to depend on sediment movement and type of bedform. Following initiation of sediment movement (generally at Froude numbers >1 and consisting entirely of bedload) the bedforms develop immediately and prevent the retention of a plane bed. Antidunes appear at almost all flows but, at the lower discharges at the two higher slopes, they are replaced as the dominant but not exclusive form by alternate bars (Fig. 1). Because of the limited flume width, though, it is not possible to define the entire region in which bar development is possible. Wavelength of the bars varies inversely with channel slope (Fig.4). For the plane beds with little or no sediment transport the flow resistance varies smoothly and inversely with the ratio of depth to sediment size (Fig. 5). Once sediment transport and bedforms are generally established, the flow resistance is increased by the distortion of flow about the bedforms. At higher discharges the flow resistance varies according to the type of bedform and the measured sediment discharge (Figs. 5 and 6).

1 INTRODUCTION

By comparison with lowland rivers, which usually have gentle slopes, sand beds and large depths relative to sediment size, there has been relatively little research into the hydraulics of upland rivers with steep slopes ($\gtrsim$ 0.5%), coarse bed material (gravel and boulders) and flow depths of the same order of magnitude as the bed material size. In particular there is a poor understanding of : a) the bedforms characteristic of such rivers; and b) the relationships between water flow, sediment transport, bedforms and flow resistance. Here these problems are addressed using some preliminary results from a combined flume and field study of sediment transport in steep channels.

2 DATA COLLECTION

Data for the study are being collected in a specially designed flume at the Laboratoire d'Hydraulique of the EPF-Lausanne (Switzerland) and in the field by the Institute of Hydrology, Wallingford, (UK).

The flume, capable of tilting to slopes as high as 9.5 %, is 0.6 m wide, 16.8 m long and has glass walls 0.8 m deep. Water (and sediment up to 3mm in diameter) is recirculated and the water discharge is measured in the return pipe by an electromagnetic flowmeter. For the work reported here the flume bed was covered with a sediment layer about 100 mm thick, consisting of lake gravel with a specific gravity of 2.57 and a size distribution d_{16} = 18 mm, d_{50} = 22 mm, d_{84} = 29 mm according to a Wolman-style (1954) sample of the surface layer. (d_n = the size of median axis than which n% of a sample of bed material is finer.) Measurements were then made over a range of water discharges at each of four flume slopes, 3%, 5%, 7% and 9%. At each discharge either the cross-sectional area of the flow was measured using a point gauge or mean flow velocity was obtained by timing the movement of a salt solution between two pairs of electrodes placed either 3 m or 5 m apart along the flume. All other hydraulic quantities were calculated from these terms and the measured discharge. Volumetric sediment discharge (entirely bedload) was obtained from the rate at which sediment was trapped in a metal basket at the end of the flume. The sediment particles were too large to be recirculated with the water so, to maintain the bed thickness sediment was supplied to the flow by hand at the upstream end of the flume at a rate approximately equal to that at which it left the downstream end. Usually all the flume measurements were repeated at least once to ensure representative values.

The programme of field data collection was less advanced at the time of writing and the data presented here therefore relate only to bed morphology in the River Trannon, Wales. This is a small gravel-bed stream which has been straightened for flood control purposes and has subsequently developed a series of alternate bars, so that the lowest part of the stream bed winds between bars which occur at regular intervals on alternate sides of the stream. The data were collected by M. Newson and G. Leeks (Institute of Hydrology, UK).

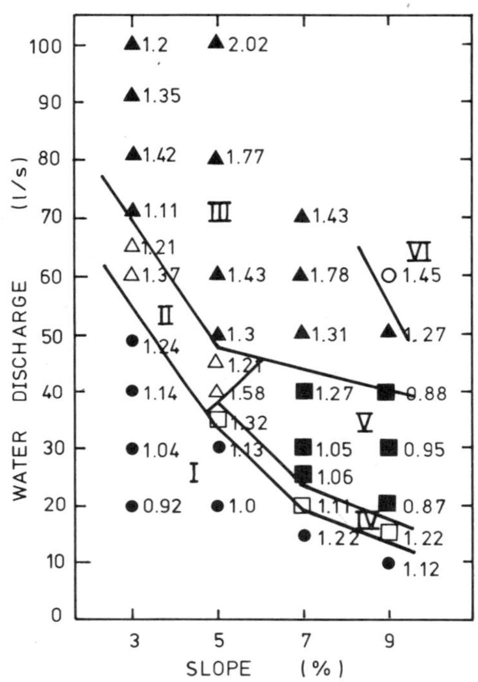

BEDFORMS

I ● Plane bed with no movement
II △ Irregular antidunes
III ▲ Antidune trains
IV □ Irregular alternate bars
V ■ Alternate bar series
VI ○ Plane bed with movement

Fig. 1. Presentation of flume data for gravel of size d_{50} = 22 mm; points mark discharges at which measurements were made at each slope; Froude number $U/(gD)^{1/2}$ is noted by each point; point symbol indicates observed bedform and numerals indicate regions of common bedform.

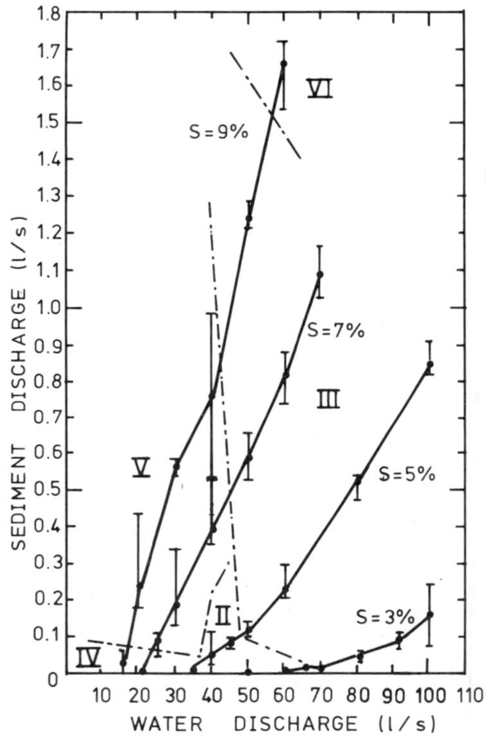

BEDFORMS

II Irregular antidunes
III Antidune trains
IV Irregular alternate bars
V Alternate bar series
VI Plane bed with movement

Fig. 2. Variation of sediment discharge with water discharge for the flume; lines join points of mean sediment discharge for each flume slope S; error bars indicate range of measured sediment discharges at each water discharge; regions of common bedform are indicated by numerals using same system as in Fig. 1.

3 DATA PRESENTATION

The basic flume data are summarized diagrammatically in Fig. 1, where data points mark the water discharge at which measurements were made at each slope. Superimposed on this distribution, regions having a common bedform are delineated using observations made during the experiments. Finally the calculated flow Froude number $U/(g\ D)^{1/2}$ is noted by each point in order to give an idea of flow conditions. (U = mean flow velocity; D = mean flow depth defined by A/W; A = flow cross-sectional area; W = flume width; g = acceleration due to gravity.)

For the range of observed flows, Froude numbers usually exceed 1, the major exceptions being at the 9% slope. On the whole the Froude number increases as discharge increases at each slope but sharp variations occur at the boundaries between different bedform regimes.

4 BEDLOAD TRANSPORT

Volumetric sediment discharge, Q_S, (entirely bedload) is plotted against water discharge, Q, in Fig. 2. The lines, plotted separately for each slope, join the mean values of sediment discharge at each discharge while the error bars encompass the range of measured sediment discharges. The large error bars apparent in the region typified by alternate bars reflect the abrupt (although probably temporary) increase in sediment discharge as pools were scoured during the formation of the bar series. Otherwise Fig. 2 clearly shows that sediment discharge increases with water discharge whatever the bedform but that the rate of change depends on flume slope.

5 DESCRIPTION OF BEDFORMS

Four major types of bedform were observed (Fig. 3):

1. Plane bed without sediment movement. As long as the water discharge was too weak to move sediment, the bed remained plane, as it had been prepared.

2. Antidunes. These appeared at nearly all flows at which sediment movement was possible, forming a train of bed waves migrating upstream with the water surface developing similar waves in phase.

3. Alternate bars. At the lower discharges at the 7% and 9% slopes these eventually became the dominant feature, replacing the antidunes which had formed initially.

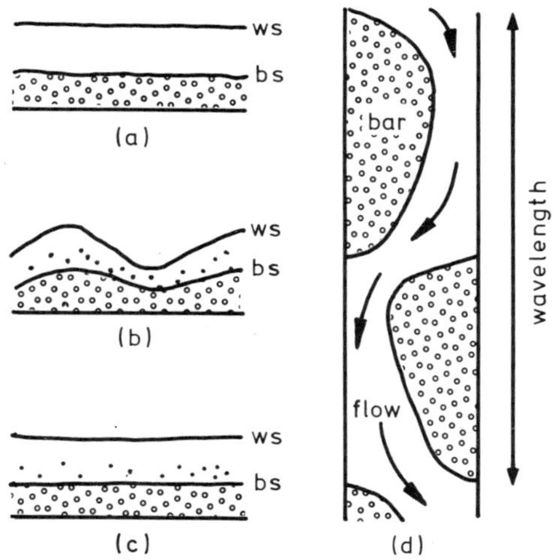

Fig. 3. Diagram of bedforms observed in the flume: (a) side view of plane bed with no sediment movement; (b) side view of antidune train; (c) side view of plane bed with sediment movement; (d) plan view of alternate bar series: ws = water surface and bs = bed surface.

4. Plane bed with sediment movement. This characterized the highest discharge at the 9% slope.

At all slopes the bedforms varied in strength, being irregular and often separated by reaches of plane bed at the lowest discharges, strongest once bedload movement was generally established and becoming weaker at still higher discharges. At the lowest discharges there was also a tendency for the flow to create its own channel with a width less than that of the flume. Similar variations in bedform have been noted in flume experiments by Mizuyama (1977).

Antidunes and alternate bars are not bedforms of the same scale and are not mutually exclusive. Thus antidunes not only preceded bar development but were sometimes noticed in the pools of the bar series. Antidunes (and plane beds with sediment movement) belong to that class of bedform which represents an interaction between the flow and the bed and which has an amplitude less than or equal to the depth of flow (e.g. Simons & Şentürk 1977 : Chap. 5). Although often observed in sand-bed channels, their appearance has been more rarely documented for gravel beds. Nevertheless their appearance here at Froude numbers of about 1 and higher conforms with the few other experimental observations (Fahnestock 1963; Cooper et al 1973; Shaw & Kellerhals 1977; Whittaker & Jaeggi 1982). Further, since sediment movement in flows with steep slopes and low ratios of depth to sediment size appears unlikely to occur until Froude numbers of about 1 are attained (see the data of Cooper et al 1973), these bedforms are likely to be the dominant features of their scale in steep channels.

Alternate bars are a larger-scale formation which represent an interaction between the flow and the channel planform and which have longstream wavelengths of a few channel widths (e.g. Simons and Şentürk 1977 : Chap. 5). They are a three-dimensional phenomenon and therefore depend in part on width-related characteristics of the channel (Sukegawa 1973; Mizuyama 1977). Consequently, as flume width was constant in this study, the region of bar formation delineated in Fig. 1 is peculiar to this study only and should not be thought of as generally definitive.

Wavelengths of the bar series (equal to two bar units (Fig. 3d)) are plotted against channel width in Fig.4, along with Lewin's (1976) river measurements, the Institute of Hydrology's measurements on the River Trannon and measurements made by the first author on a steep stream with alternate bars in mid-Wales. Channel slope is noted

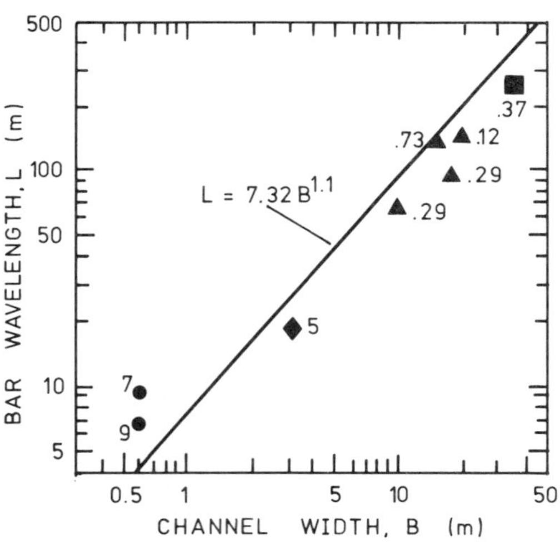

Fig. 4. Variation of alternate bar wavelength (2 x bar unit) with channel width; data from flume ● , River Trannon ▲ , Lewin (1976) ■ and Bathurst ◆ are compared with experimental formula of Leopold and Wolman (1957); channel slope S in per cent is noted by each point.

by each point and the equation of Leopold and Wolman (1957) (converted to metric units) which relates meander wavelength to channel width and is based on a wide range of data is represented by the solid line. The river data follow the general trend of the equation but for the flume, which has a constant width, slope is the important determinant, with bar wavelength increasing as slope decreases.

6 FLOW RESISTANCE

The effects of the large-scale bed roughness, steep channel slopes, bedforms and sediment discharge on the flow hydraulics can all be considered in terms of the flow resistance. This is quantified using the dimensionless Darcy-Weisbach resistance coefficient, f, defined by

$$\left(\frac{8}{f}\right)^{1/2} = \frac{U}{u_*} = \frac{U}{(g\,R\,S)^{1/2}}$$

where u_* = mean shear velocity; R = hydraulic radius; and S = slope.

Consideration of the processes of flow indicates that, for large-scale roughness, flow resistance is a function of the combined form drag of the sediment particles, their disposition in the channel, channel shape, bedform drag and sediment movement effects (Judd & Peterson 1969; Flammer et al 1970; Thompson & Campbell 1979; Bathurst et al 1979; 1981). As the ratio of depth to sediment size D/d_{50} plays an important role in all these processes, the function $(8/f)^{1/2}$ is plotted against D/d_{50} for the flume data in Fig. 5. For each flume slope the points are joined in order of increasing discharge. The values of $(8/f)^{1/2}$ (in the range 3 - 9) and D/d_{50} (in the range 1 - 6) compare well with other data for flows with large-scale roughness (e.g. Judd & Peterson 1969; Bathurst et al 1981).

The uneven variations are unlike the smoother relationships between $(8/f)^{1/2}$ and D/d_{50} often observed for flows with no sediment movement and with larger values of D/d_{50} (e.g. see the data of Hey 1979). They are not considered to be the result of experimental error since : a) similar variations have been reported elsewhere (Bathurst et al 1979); b) the same pattern is noted for each of the four flume slopes, suggesting that it is not caused by random measuring errors; and c) each experimental measurement was repeated to ensure that unrepresentative values were being avoided. It may be concluded therefore that the variations are the result of complex interactions between sediment movement and

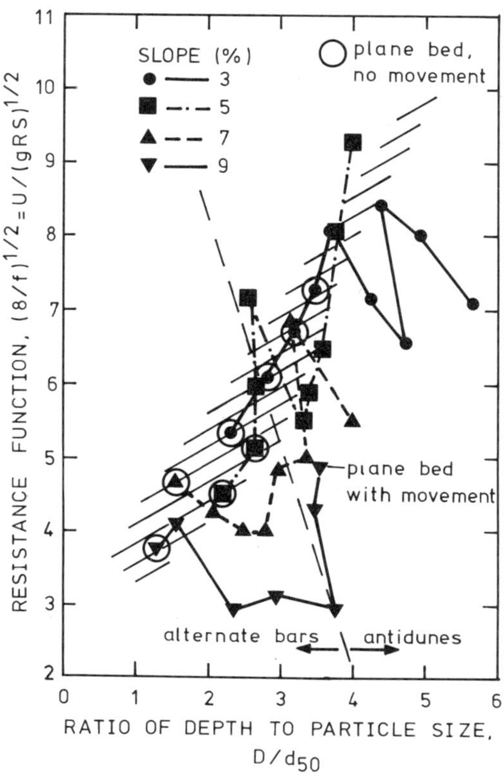

Fig. 5. Variation of resistance function $(8/f)^{1/2}$ with ratio of depth to sediment size D/d_{50} for the flume; for each flume slope points are joined in order of increasing water discharge from left to right; shaded band contains all points with no sediment movement but is not a region where sediment movement cannot occur; regions of common bedform as indicated.

large-scale roughness effects.

In view of this it would be meaningless to follow the approach used elsewhere (e.g. Griffiths 1981) and try to represent the data in Fig. 5 by a single line since no allowance would be made for the various processes of resistance. Instead each flume slope is considered separately and, further, distinction is drawn between flows with and without sediment movement.

6.1 Flow without sediment movement

Assuming that the bed is plane, flow resistance is determined largely by the boundary roughness and $(8/f)^{1/2}$ increases (ie resistance decreases) as D/d_{50} increases. The variation should be smooth but, since the form drag of the sediment particles varies with Froude number and sediment Reynolds number Ud_{50}/ν (where ν = kinematic viscosity) (Flammer et al 1970), there should be scatter about the trend depending on flow conditions (Bathurst et al 1981). In Fig. 5 the circled points representing flows without sediment movement, which are enclosed in the shaded band, conform to this pattern. The band is therefore used as a basis for comparing the effects of sediment movement on flow resistance.

6.2 Flows with sediment movement

Sediment movement increases flow resistance via the creation of bedforms (which increase energy losses due to boundary form drag, waviness of the water surface and internal flow distortion) (e.g. Simons & Richardson 1966) and via transfer of momentum from the flow to particles bouncing into the flow. Numerous methods have been devised to calculate the increased resistance (see Graf 1971 : Chap. 11 for a review) but these are mainly for sand-bed channels with large ratios of depth to sediment size and, from test applications, cannot explain the patterns of Fig. 5. The following explanation is therefore qualitative, considering the sequential changes which occur at each slope as discharge is increased.

At weak sediment discharges, flow resistance either varies as for the plane beds without movement or decreases slightly (e.g. the 5% slope). This could be because: a) flow resistance is barely affected by weak transport; or b) the weak transport is able to create a bed surface smoother than that obtained in the initial preparation of the bed.

With increased water and sediment discharges there is a sharp increase in flow resistance (i.e. a decrease in $(8/f)^{1/2}$). This region, coincident with the appearance of irregular bedforms, probably represents a transition from flows where sediment movement can be ignored to flows where the effects of sediment movement must be accounted for.

Following the transition to general bed-load movement and fully developed bedforms, the decrease in $(8/f)^{1/2}$ is halted. The region of bar formation at the two steeper slopes is generally characterized by high flow resistance relative to the plane bed without movement, presumably caused by the successive backwater and large-scale roughness features imposed by the bar series. However, at all slopes, once antidune trains become dominant, flow resistance decreases sharply, in some cases falling to levels comparable with that for a plane bed without movement. This agrees with observations made in sand-bed channels (e.g. Simons & Richardson 1966).

Further increases in flow resistance produce plane beds with sediment movement or weak antidunes and are characterized by flow resistances both lower and higher than that of a plane bed without movement. In the case of the 3% slope the observed increase in flow resistance may possibly be an effect of the low width/depth ratios (less than 6) at that slope. For the other slopes, though, as the bedforms were generally weak, it is more likely that the differences are the result of different rates of momentum transfer from flow to bedload.

The effect of sediment load is shown in Fig. 6 where $(8/f)^{1/2}$ is plotted against sediment discharge for four flows (one for each flume slope). As the flow discharge, velocity and depth are practically the same in each case and as the bedforms are generally weaker at the higher slopes, neither boundary roughness nor bedform resistance can be responsible for the observed increase in resistance with slope. It may therefore be concluded that flow resistance varies directly with sediment load. However, given the variations in Fig. 5, the relative effect of sediment load obviously depends on other flow conditions as well.

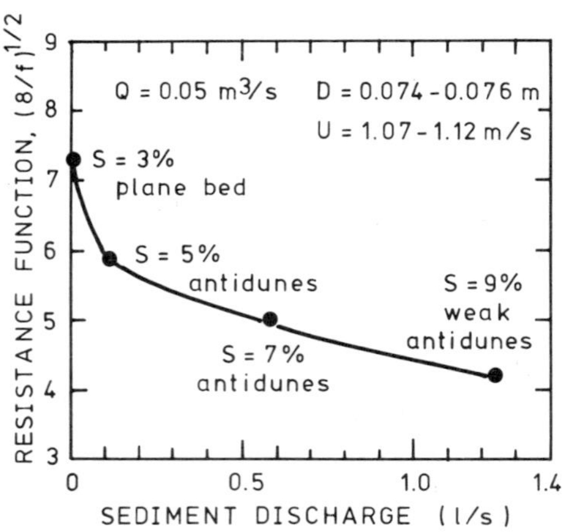

Fig. 6. Variation of resistance function $(8/f)^{1/2}$ with sediment discharge for a constant water discharge, Q, depth, D, and velocity, U, at the four flume slopes, S; bedforms as indicated.

7 CONCLUSIONS

Since results for only one flume and one sediment size are presented, general conclusions should be made with care. However, the study indicates that for steep channels with large-scale roughness:

1. The preferred small-scale bedforms in the presence of sediment movement are antidunes and plane beds;
2. Alternate bars can appear as a larger-scale but not mutually exclusive bedform;
3. Where channel width is constant, the wavelength of the alternate bars depends on channel slope;
4. Flow resistance of a plane bed without sediment movement is caused by boundary roughness and decreases smoothly as D/d_{50} increases;
5. Flow resistance of a mobile bed varies wildly as a function of bedform, boundary roughness and bedload transport rate.

ACKNOWLEDGEMENTS

The authors thank Drs. W.R.White and R. Bettess (Hydraulics Research Station Ltd., Wallingford, UK) and Mr. R.T.Clarke and Dr. P.E.O'Connell (Institute of Hydrology, Wallingford, UK) for helpful discussions.

REFERENCES

Bathurst, J.C., R.-M.Li & D.B.Simons 1979. Hydraulics of mountain rivers. Rep. CER78-79JCB-RML-DBS55, Dept. Civ. Eng., Colorado State Univ., Fort Collins, Colo., USA.

Bathurst, J.C., R.-M.Li & D.B.Simons 1981. Resistance equation for large-scale roughness. Proc.Am.Soc.Civ.Engrs., J.Hydraul.Div. 107(HY12) : 1593-1613.

Cooper, R.H., A.W.Peterson & T.Blench 1973. Closure to Critical review of sediment transport experiments by R.H.Cooper, A.W.Peterson & T.Blench. Proc.Am.Soc. Civ.Engrs., J.Hydraul.Div. 99(HY7) : 1161-1164.

Fahnestock, R.K. 1963. Morphology and hydrology of a glacial stream - White River , Mount Rainier, Washington. Prof. Pap. 422-A, U.S.Geol.Surv., Washington DC, USA.

Flammer, G.H., J.P.Tullis & E.S.Mason 1970. Free surface, velocity gradient flow past hemisphere. Proc.Am.Soc.Civ. Engrs., J.Hydraul.Div. 96(HY7) : 1485-1502.

Graf, W.H. 1971. Hydraulics of sediment transport. New York : McGraw-Hill Book Co.

Griffiths, G.A. 1981. Flow resistance in coarse gravel bed rivers. Proc.Am.Soc. Civ.Engrs., J.Hydraul.Div. 107(HY7) : 899-918.
Hey, R.D. 1979. Flow resistance in gravel-bed rivers. Proc.Am.Soc.Civ.Engrs., J. Hydraul.Div. 105(HY4) : 365-379.
Judd, H.E. & D.F.Peterson 1969. Hydraulics of large bed element channels. Rep.PRWG 17-6, Utah Water Research Lab., Utah State Univ., Logan, Utah, USA.
Leopold, L.B. & M.G.Wolman 1957. River channel patterns : braided, meandering and straight. Prof.Pap.282-B, U.S.Geol. Surv., Washington DC, USA.
Lewin, J. 1976. Initiation of bedforms and meanders in coarse-grained sediment. Geol.Soc.Am.Bull. 87 : 281-285.
Mizuyama, T. 1977. Bedload transport in steep channels. Ph.D.thesis, Kyoto Univ., Kyoto, Japan.
Shaw, J. & R.Kellerhals 1977. Paleohydraulic interpretation of antidune bedforms with applications to antidunes in gravel. J.Sedim.Petrol. 47(1) : 257-266.
Simons, D.B. & E.V.Richardson 1966. Resistance to flow in alluvial channels. Prof.Pap.422-J, U.S.Geol.Surv., Washington DC, USA.
Simons, D.B. & F.Şentürk 1977. Sediment transport technology. Fort Collins, Colo.: Water Resources Publications.
Sukegawa, N. 1973. Condition for the formation of alternate bars in straight alluvial channels. Intl.Assoc.Hydraul. Res.,Proc.Intl. Symposium on River Mechanics, Bangkok, Thailand 1 : 679-689.
Thompson, S.M. & P.L.Campbell 1979. Hydraulics of a large channel paved with boulders. Intl.Assoc.Hydraul.Res., J.Hydraul.Res. 17(4) : 341-354.
Whittaker, J.G. & M.N.R.Jaeggi 1982. Origin of step-pool systems in mountain streams. Proc.Am.Soc.Civ.Engrs., J.Hydraul.Div. 108(HY6) : 758-773.
Wolman, M.G. 1954. A method of sampling coarse river-bed material. Trans.Am. Geophys.Union 35(6-1) : 951-956.

Euromech 156: Mechanics of Sediment Transport / Istanbul / 12-14 July 1982

First experiences measuring coarse material bedload transport with a magnetic device

PETER ERGENZINGER
Free University of Berlin, Germany
STEFAN CUSTER
Montana State University, USA

ABSTRACT: Using a magnetic device for monitoring bedload transport a first test with artifical magnetic tracers in cobbles was undertaken in 1980 at Fiumara Buonamico in Calabria/Southern Italy. In a second test naturally magnetic andesitic pebbles and cobbles were used as tracers at Squaw Creek, headwaters of Missouri, Gallatin County in Montana/USA. The experiences are promising.

1 INTRODUCTION

The only reliable data on coarse bedload transport (particles larger than 32 mm) are geodetic measurements of delta aggradation. The steeper the channel gradient, the coarser the bed material, the less data are published on bedload transport. Since coarse material transport is an important part of many fluvial research problems such as reservoir filling rate, bridge erosion, and trout spawning habitat, new measuring techniques are urgently required.

Several samplers were tested (US Geological Survey 1978) without striking effect. All calibrated samplers (e.g. Helley-Smith sampler) are restricted to coarse sand and finer materials (Emmett 1980). This restriction is the result of several specific problems in coarse-bed streams. Adequate contact with the stream bed during sampling is especially difficult to achieve. Large samplers are necessary for coarse material since representative samples are heavy and hard to handle. Finally, large samplers disturb flow. A new technique is needed.

2 PRINCIPLES OF THE MAGNETIC TRACER TECHNIQUE

The new Magnetic Tracer Technique is based on Faraday's laws of induction: whenever a magnet is moved with a certain velocity along a coil crossing the field lines a voltage will be induced. It makes no difference whether the coil is moved or the magnet. A close relationship exists between inductivity and magnetic field force. The greater the magnetic field force, the greater the inductivity. Inductivity is also increased by bigger coils, higher velocity of the moving magnet, and a short distance between coil and magnet.

Assuming that the rolling magnetic cobbles have the same average distance from the coils, an average velocity and a known inductivity (measured in Gauss), it is possible to obtain coarse material transport rates using the Magnetic Tracer Technique.

3 RESULTS FROM BUONAMICO IN CALABRIA/ITALY

In Calabria (Fiumara Buonamico) artificial magnetic tracers were used (Ergenzinger, Conrady 1982). The task was to determine the starting conditions of cobble transport. Since there is no natural magnetic material in the cobbles, holes were drilled through granitic cobbles with an average width of 55 mm, bar magnets (diameter 6 mm, length 25 mm) were inserted and fixed by an adhesive close to the gravity center

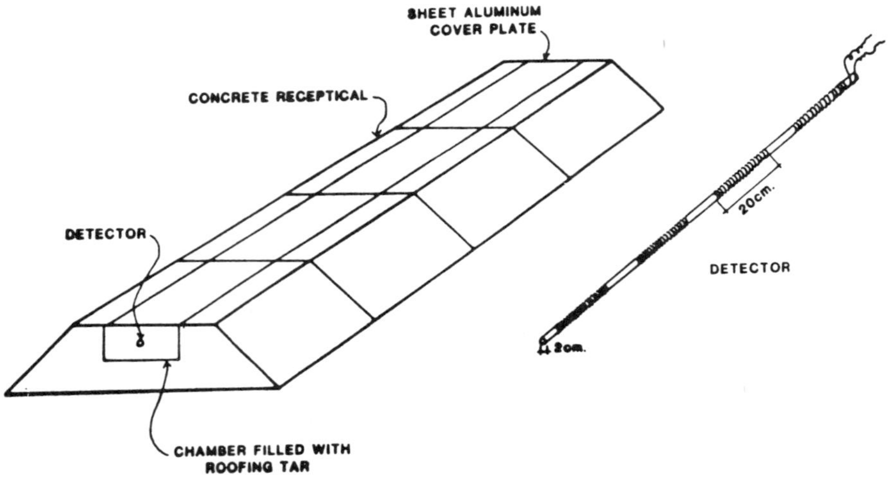

Fig. 1: Concrete block with magnetic detector

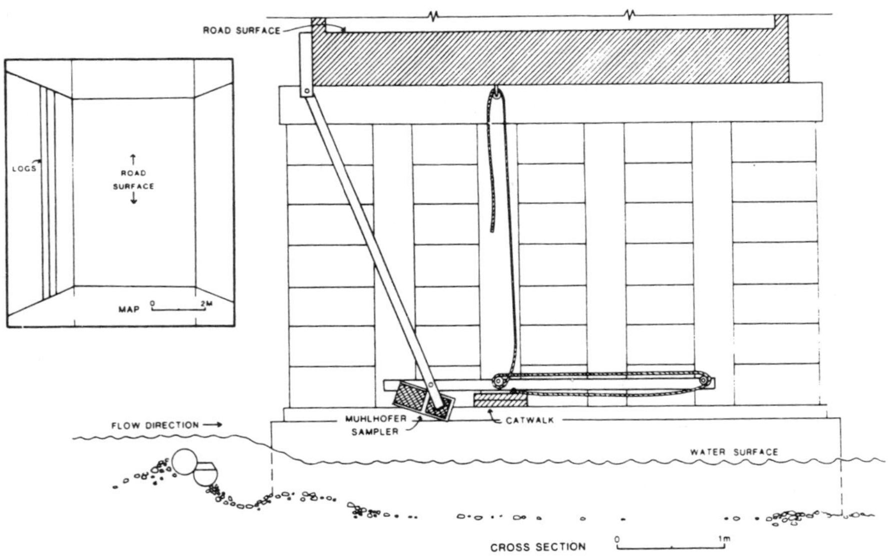

Fig. 2: Measuring systems at Squaw Creek (Bible Camp Bridge)

of the stone. By this procedure the median weight of the cobbles of 246 g was increased by about 1.3 %.

The circuit of the detector is comparable to a normal iron detector. The stable magnetic field was achieved by a coil with diameter 250 mm holding 5000 m of copper wire (Cu 0.3). By this arrangement it was possible to induce a voltage of close to 1 mV whenever a magnetic cobble was in transit over a distance up to 0.7 m. Therefore the coiling system was mounted on aluminium girders above the river channel. The signals created by the rolling magnetic cobbles were amplified and registered by recorders.

With 100 magnetic-tracered cobbles we succeeded in determining the starting conditions of cobble transport in Buonamico (Ergenzinger, Conrady 1982, 79-80). Unfortunately the determination of the transportation rate of cobbles with this technique is quite expensive and time-consuming, since there should be a continuous input of magnetic cobbles during the floods.

4 RESULTS FROM SQUAW CREEK, HEADWATERS OF MISSOURI IN MONTANA/USA

In contrast to Calabria there is a high quantity of naturally magnetic cobbles in many streams in the northern Rocky Mountains in North America. To test the feasibility of detecting the movement of naturally magnetic cobbles and pebbles on a stream bottom, a reconnaisance survey was carried out on several streams in the region near Bozeman, Montana. Whenever the coarse material contains particles sufficiently magnetic to attract a free swinging small bar magnet the magnetic tracer technique is promising.

For testing purposes we chose Squaw Creek, draining the andesitic volcanic terrace of the Gallatin Range, Gallatin County, Montana. The stream has approximately 20 years of crest stage records and close to the confluence with Gallatin River there is electricity and a shelter for field research. A detector installed in Squaw Creek successfully sensed coarse-particle movement and was used to estimate bedload transport during summer 1981.

The detector consists of four wire coils wrapped around an iron bar (1.1 m long, diameter 0.02 m). Each coil consists of 9000 windings of 0.2 mm copper wire. The distance between the coils is 140 mm. The coils are connected in series and protected with several coats of silicone. The iron bar with the coils is bolted in a slotted concrete block (1.25 m long, 0.2 m wide, 0.15 m high). The slot which contains the iron bar is filled with roofing tar and is covered with an aluminium sheet to protect the detector from cobble impact and water (compare Fig. 1).

The detectors were installed close to the Bible Camp bridge (Fig. 2) below a log across the entire stream on the upstream side of the bridge. Bed material accumulated upstream of the log, consequently there is no obstruction by the concrete block installed just below the log. The signals of the detector are amplified, filtered and recorded in the nearby shelter.

For calibration a Mühlhofer-style basket (Mühlhofer 1933) with a 1 cm^2 mesh was modified and installed under the bridge in such a manner that it could be swung forward to touch the upper edge of the log.

During the spring flood unfortunately the men were unable to force the basket into measuring position against the flow of the stream. The calibration of the magnetic signals by the basket failed in 1981.

Although the mechanical samplers failed to measure bedload transport a crude relationship between magnetic signals and transport rate was established, assuming that grain size, particle mass and inductivity are statistically connected (compare table 1). Grab samples were taken from an unvegetated gravel bar just upstream of the detector and, under the hypothesis that this material is representative for bedload material of Squaw Creek, the physical and magnetic properties were determined:

- The detector is at present incapable of detecting any inductivity when pebbles smaller than 16 mm swing 1 cm above the detector.

- Signals larger than 1 mV are produced by particles with more than 800 gamma inductivity, signals between 0.2 and 1 mV are produced by particles with 240 to 800 gamma inductivity.

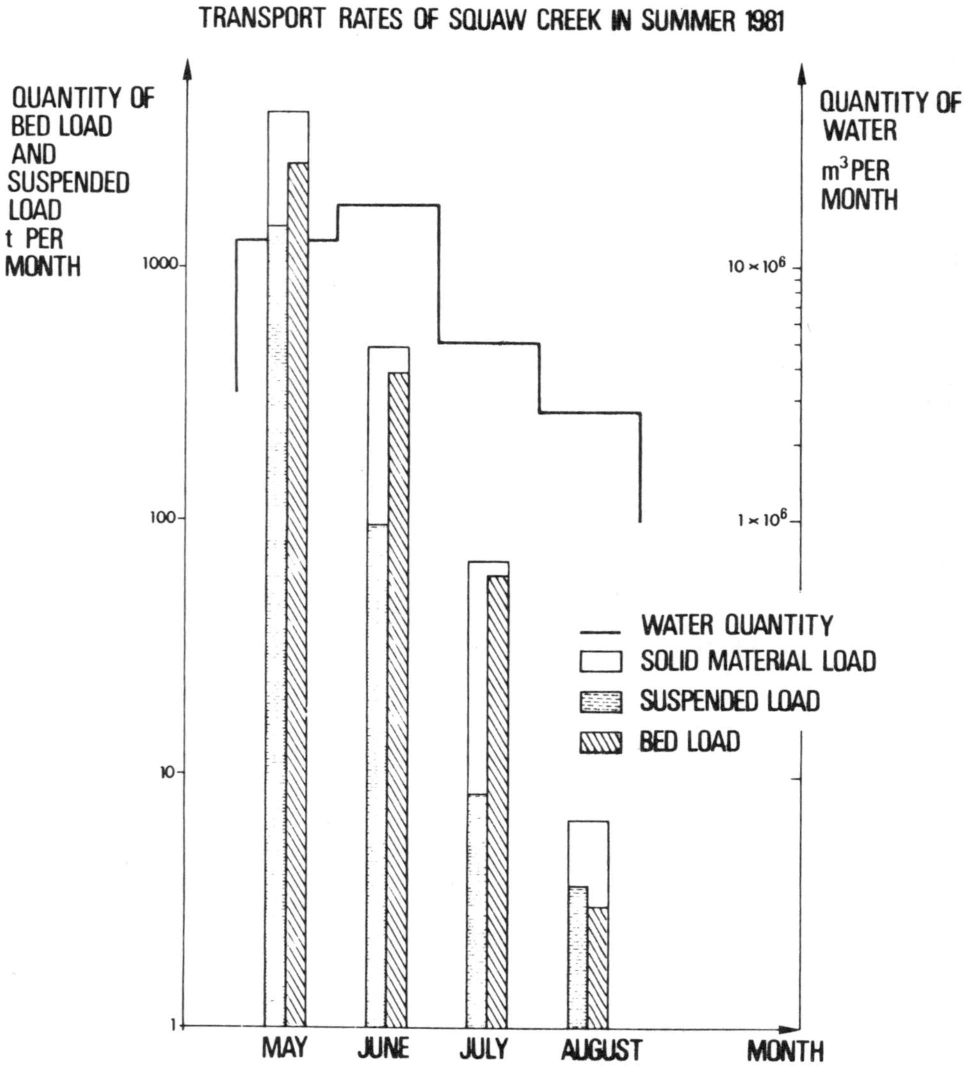

Fig. 3: Transport rates of Squaw Creek during summer 1981

TABLE 1. Physical and magnetic properties of coarse material from Squaw Creek

Grain size (mm)	Number	Mean weight (kg)	Stand. dev. (kg)	Max. Magnetic Inductivity at a distance of 10 mm (gamma) 240 %	240-800 %	800 %
64	100	0.989	0.580	80.0	16.0	4.0
32-64	145	0.164	0.100	89.6	9.0	1.4
16-32	139	0.032	0.017	100.0	-	-

- The signals above 1 mV are related with 74 % probability to cobbles with an average weight of 990 g and with 26 % probability to pebbles with an average weight of 164 g. Therefore every signal above 1 mV represents a particle with 775 g weight.

- The signals between 0.2 and 1 mV are related with 64 % probability to cobbles and with 36 % probability to pebbles. Every signal represents a particle with 692 g weight.

- The content of magnetic material in particles larger than 32 mm is 15 % and the coarse material comprises only 33 % of the bar material. Therefore the signals have to be multiplied by 6.7 x 3 = 20.1.

- The width of bedload transport is 7 m, the measuring width 1.25 m, the signal is again multiplied by 2.8.

- Every signal larger than 1 mV was therefore multiplied by (0.775 x 20.1 x 2.8) = 44 and the small signals by (0.692 x 20.1 x 2.8) = 39. The resulting figure is the bedload transport in kg.

While there are several assumptions in this deduction (e.g. the composition of the gravel bar is representative throughout bedload transport, the signals were only slightly influenced by the increase of mean velocity of the particles during flood ...) the resulting transport rates for bedload in Squaw Creek (Fig. 3) are only first estimates and need further adjustments by better trapping systems. Nevertheless the great importance of bedload transport in comparison with suspended load is remarkable.
During the flood of May 22 the maximum rate of bedload was close to 200 metric tons per hour. During the flood (20-23 May) about 2200 metric tons were transported as bedload or approximately 66 % of the total solid material load (3400 metric tons). During the summer of 1981 (May 1 to August 30) 33.5 x $10^6 m^3$ water, 3200 t bedload and 1500 t suspended load were exported by the Squaw Creek basin (106 km^2).

5 CONCLUSION

The magnetic tracer technique is a promising measuring technique for coarse-bedload transport. The most effective measuring sites are streams with a certain amount of naturally magnetic particles. The technique should be applicable in all basins containing basaltic or andesitic material.

The measuring site must have a protected underground. Erosion should not occur during floodtime. The detector must further be mounted in such a manner that there is no movement of the coils in the magnetic earth field. The sensitivity of the detector may be improved in several ways (more windings, vertical arrangement of the axis of the coils). Better calibration systems with big samplers are required. With more natural measurements of coarse material transport rates a better understanding and better models will result.

ACKNOWLEDGEMENTS

Funds for this research were provided by the Deutsche Forschungsgemeinschaft, Bonn, and the Montana State University, Department of Earth Sciences in Bozeman. J. Conrady and S. Saradeth helped us with field work.

REFERENCES

Emmett, W.W. 1980. Bedload Sampling in Rivers. Proceedings of the International Symposium on River Sedimentation. Beijing, China, Vol. 2: 991-1017.

Ergenzinger, P. & J. Conrady 1982. A new tracer technique for measuring bedload in natural channels. Catena 9:77-80, Braunschweig.

Ergenzinger, P. 1982. Über den Einsatz von Magnettracern zur Messung des Grobgeschiebetransportes. Beiträge zur Geologie der Schweiz - Hydrologie, Bd. 28II:483-91, Bern.

Mühlhofer, L. 1933. Untersuchungen über Schwebstoff und Geschiebeführung des Inns nächst Kirchbiehl. Wasserwirtschaft Nr. 1-6.

U.S. Geological Survey 1978. National Handbook of Recommended Methods for water data acquisition. 3.29 - 3-79. Washington D.C.

6 Other sediment-transport problems

Euromech 156: Mechanics of Sediment Transport / Istanbul

Longitudinal sorting of grain sizes in alluvial rivers

ROLF DEIGAARD
Danish Hydraulic Institute

ABSTRACT: A sorting model based on a sediment transport model is formulated. The model describes the development of grain sizes of the bed material and the longitudinal profile of a river. Two time scales exist in the model: one for the grain sorting and one for the change in the profile, the latter being by far the largest. This implies that for a given longitudinal profile of a river the variation of the mean grain size along the river is uniquely determined. The analysis is in agreement with the common observation that the grain size decreases in the down stream direction of a river. Results from the model are compared with observations from three rivers and with laboratory experiments. The model also describes the grain size distribution curves. The grain size distributions given by the model have properties which are frequently observed in natural sediments e.g. break points in log-normal plots.

INTRODUCTION

It is a common observation that the grain size of bed sediments decreases in the downstream direction of rivers. The phenomenon has been explained by abrasion of the sediment grains as they move down through the rivers. But all experiments indicate that this effect is far too weak to account for the size reduction observed in nature - especially in case of less coarse sediments : sand and gravel.

Another explanation of the decrease in particle size is that it is caused by some kind of hydraulic sorting process, related to spatial variations of the stream's capability to carry sediments of different sizes. These variations have been attributed to longitudinal changes in the slope and the specific water discharge. A majority of the rivers in the world have their bed slopes to decrease more or less gradually from source to mouth.

This paper considers the grain sorting in rivers by looking at the development with time of both the longitudinal profile and the grain size. Both processes are described by use of a sediment transport formula. The time scale for changes in the profile is found to be much larger than the time scale for current sorting of the grains. This implies that the distribution of the mean grain diameter along a river always will correspond to the longitudinal profile.

The current sorting model does not cover local sorting (sorting due to dunes and meanders) and the variation in grain size can only be found as a trend over a distance of approximately one meander length.

THE SEDIMENT TRANSPORT MODEL

In the current sorting model the sediment transport is described by the sediment transport model developed by Engelund and Fredsøe (1976). The model treats the bed load and suspended load separately, which is important in this case, because the material moving in suspension is finer than the bed load. By assuming that the bed material consists of a number of different size fractions, the model has been modified slightly in order to describe the composition of the sediment load. The procedure for calculation of the transport of graded material is described in detail by Deigaard and Fredsøe (1978), and Deigaard (1980).

The bed of the alluvial river is assumed to be covered by dunes, and the hydraulic resistance is calculated by the method proposed by Engelund (1966).

ᴴE GRAIN SORTING MODEL

The grain sorting is found by considering the development with time of an idealized river, Fig. 1.

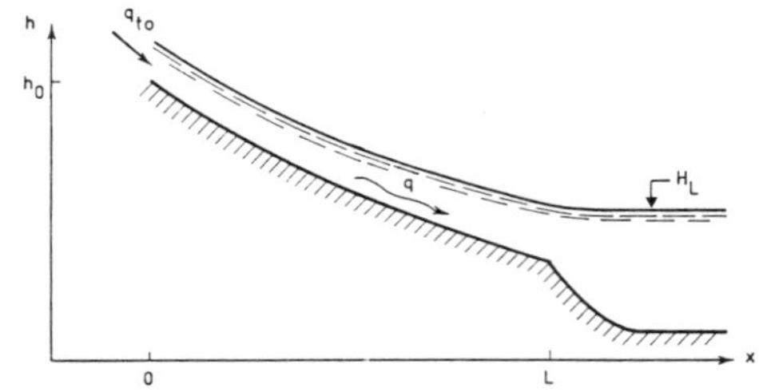

Fig. 1 : The idealized river treated in the sorting model.

At time equal zero the bed level, h, of the river is described by the equation : $h = h_o \exp(-\alpha x)$, x is a coordinate in the downstream direction, α is a constant and h_o is the bed level at the upstream cross section (x = o). The bed material is the same in the whole length of the river with a log-normal grain size distribution characterized by the median diameter, d_{mo} and geometric standard deviation, σ_{go} The width and water discharge, q, is kept constant along the river.

At x equal zero the sediment load is maintained constant with time (amount and composition), which means that the bed material and the slope are not changing either. At the downstream end of the river (x = L) the water level is assumed to be constant.

The development of the longitudinal profile and of the composition of the bed material is found by use of the sediment transport model and continuity equations.

The continuity equation for changes of the bed level is :

$$\frac{\partial h}{\partial t} = - \frac{1}{(1-n)} \frac{\partial q_s}{\partial x}$$

n is the porosity of the bed material, q_s is the sediment transport.

The changes in the composition of the bed material are calculated with the continuity equation for each size fraction.

The continuity equations are made into finite differences by considering sections of the length = ΔL. The sediment coming into a section is assumed instantly to be mixed with the material in an active layer in this section with a thickness equal to the dune height.

For each time step, ΔT, the continuity equations are applied on all the sections (with the length ΔL), and thus the hydraulic sorting and longitudinal profile are computed.

CHANGES IN THE GRAIN DIAMETER WITH TIME

Some of the results of a computation of the development with time of the system described above are illustrated in Fig. 2.

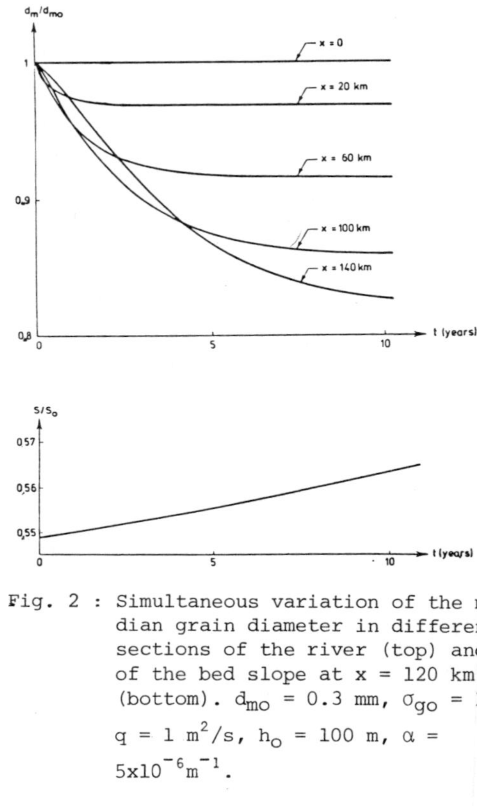

Fig. 2 : Simultaneous variation of the median grain diameter in different sections of the river (top) and of the bed slope at x = 120 km (bottom). d_{mo} = 0.3 mm, σ_{go} = 1.3, $q = 1\ m^2/s$, h_o = 100 m, α = $5 \times 10^{-6} m^{-1}$.

It can be noted that the grain diameter everywhere becomes nearly constant, faster in the upstream than in the downstream parts of the river. It will also be noted that the values which the diameters approach are decreasing in the downstream direction. The downstream variation in the grain size is closely related to the longitudinal variation of the bed shear stress and therefore also to the shape of the longitudinal profile.

The simultaneous changes in the slope at x = 120 km are shown in the lower graph in Fig. 2. These changes are rather insignificant compared to what has happened to the median diameter. This means that the current sorting can be treated as a quasi-stationary process: the river will continue to be in current sorting equilibrium (once it has been obtained) while the longitudinal profile is developing. This quasi-stationarity has been found in all the cases that have been treated. In all cases the size of the river was comparable with natural streams.

In the present idealized case, where the width and the discharge are assumed to be constant, the ultimate equilibrium profile will be a straight line with the slope, $I_o = h_o \cdot \alpha$. This is the slope that initially was at the upstream section. In the same way the median diameter, d_m, will everywhere approach the original value, d_{mo}. This extremely slow process is, however, not of interest for the current sorting. The primary results are the (quasi-stationary) variation of the median diameter, when the current sorting equilibrium corresponding to the actual longitudinal profile has been obtained.

VARIATION OF THE GRAIN DIAMETER ALONG THE RIVER

By the quasi-stationary current sorting equilibrium predicted by the model, the distribution of the median diameter along a river corresponding to its present longitudinal profile can be found. This is the result from the model that can most readily be compared with measurements from rivers.

The field data required are rather extensive, and only in three cases has it been possible to obtain sufficient data in order to make the comparison. The three rivers are: The Niger, The Mississippi and The Rio Grande.

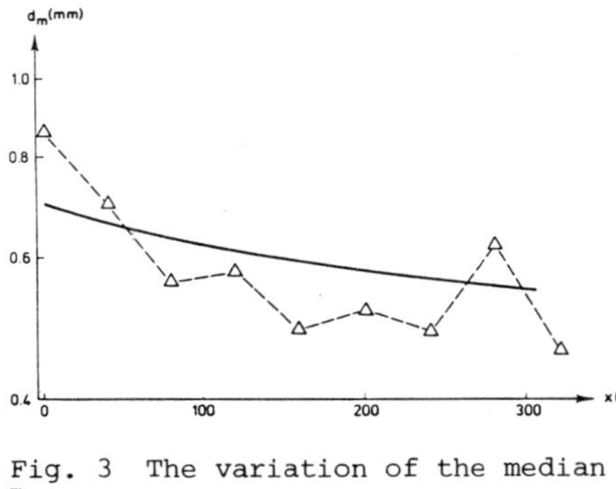

Fig. 3 The variation of the median diameter along the lower river Niger: $q = 10.6\ m^2/s$, $I_o = 1.07 \times 10^{-4}$ $\alpha = 1.38 \times 10^{-6} m^{-1}$.

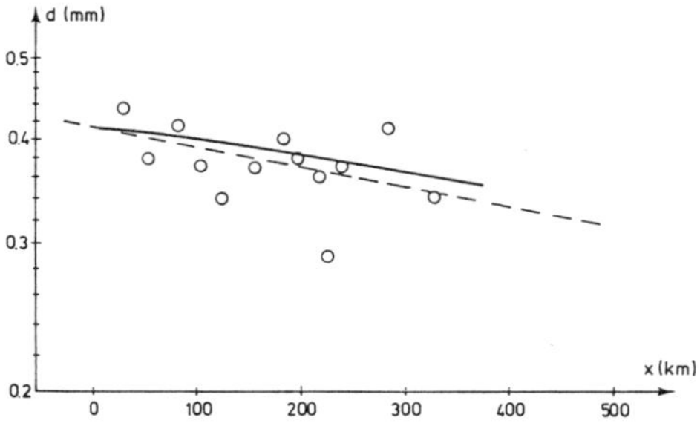

Fig. 4 Variation of sand size 1966, Mississippi river, Vickburg district. Full drawn line: Exp. fit by Anding (1970) : $q = 18.5\ m^2/s$ $I_o = 6.3 \times 10^{-5}$, $\alpha = 10^{-6}\ m^{-1}$.

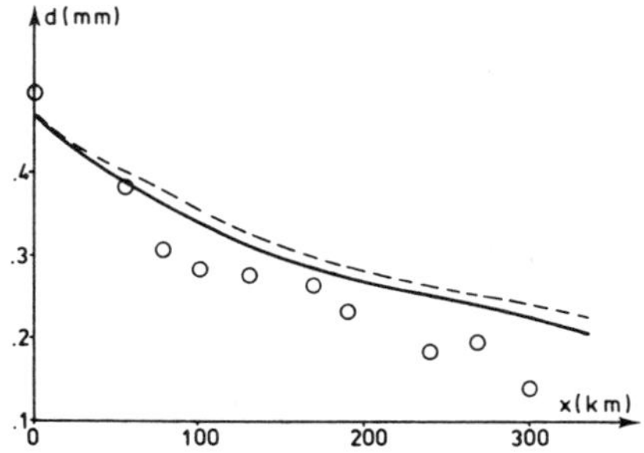

Fig. 5 Variation of grain sizes along the Rio Grande: $q = 3.8\ m^2/s$, $I_o = 2.2 \times 10^{-3}$, $\alpha = 5.7 \times 10^{-6} m^{-1}$.

The measured and calculated variation of the median diameter along the rivers is shown in Fig. 3, 4 and 5. The main parameters used in the model are given in the caption below each figure: The dominant discharge, q, the energy slope at the upstream cross section, I_o, and the parameter, α, in the longitudinal profile of the river. In all cases the sections of the rivers, which are described by the model, fulfil the requirement of a constant specific discharge, q. The data has been obtained from: The Nedeco Report (1959), Anding (1970) and Rafay (1964).

It appears that there is a considerable scatter in the measurements, probably due to local sorting. But it can be seen that the agreement between the predicted downstream decrease in grain size and the general trend in the measurements is rather good.

EXPERIMENTS ON GRAIN SORTING

The grain sorting has been studied in a laboratory flume. The variation in bed shear stress has been obtained by making the width of the 10 m long test section vary from 25 cm in the upstream section to 50 cm in the downstream, Fig. 6. The sediment transport was kept rather low in order to avoid any suspended load transport. The sand which covered the bed of the flume had a median diameter of 0.17 mm and a geometric standard deviation of 2.2.

Two experiments were made. In experiment I no sediment was added to the flume, in experiment II the sediment that had been eroded in the upstream part of the flume was replaced.

Samples of the sand in the surface layer were taken at intervals of about one hour. The samples were taken in the upstream part of the flume (station 1) the middle part (station 2) and the downstream part (station 3).

The development with time of the median diameter through the flume has been calculated by use of a modified version of the sorting model. The changes in the model has been described by Deigaard et al. (1978).

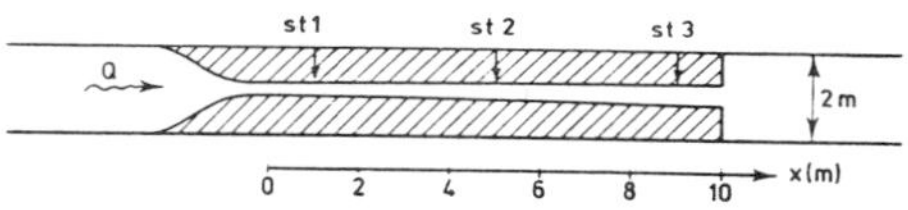

Fig. 6 The geometry of the diverging flume.

The model gives the same results for the two experiments, because it does not distinguish whether the sediment removed from the upstream part of the flume has been fed into it or has its origin from erosion of the bed.

The variation of the median diameter of the bed material in station 1, 2 and 3 is shown in Fig. 7. The discrepancy between the measurements and the theoretical development in station 3 occurs, because the bed shear stress is very close to the critical, and the model is then very sensitive to small changes in the median diameter.

The variation of the median diameter along the flume at the end of the experiments is shown in Fig. 8.

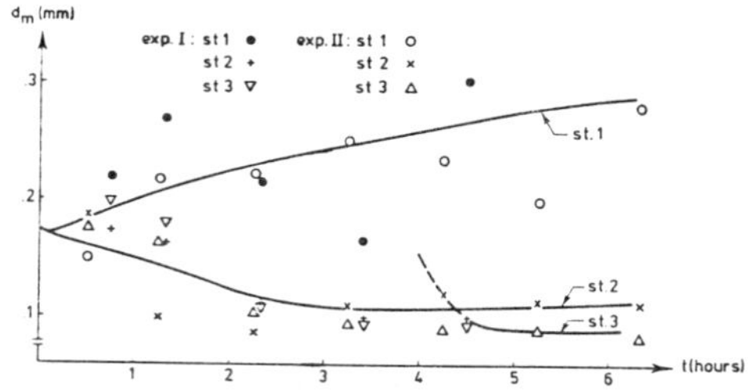

Fig. 7 Development with time of the median diameter.

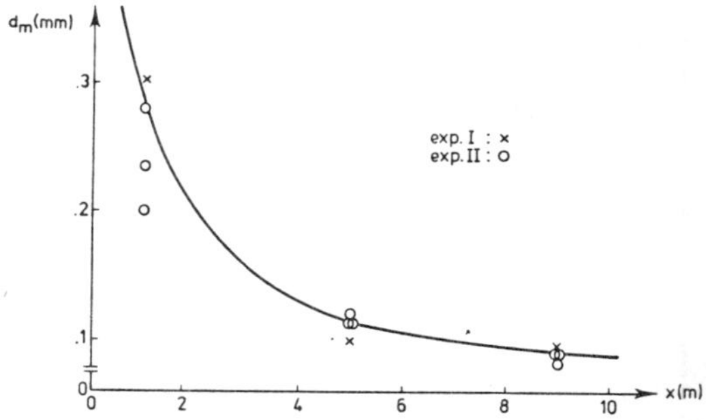

Fig. 8 Longitudinal distribution of the median diameter, curve: Theory at time = 6 h, points: measurements for time > 4 h.

THE GRAIN SIZE DISTRIBUTION

The bed material in the upstream section of the river is in the model always assumed to have a log-normal grain size distribution. But in other parts of the river deviations from this distribution are predicted by the model.

A feature often met in natural sand deposits is, that when the grain size distribution is plotted on log-probability paper, it will appear as several straight line segments. The line segments are often assumed to represent grain populations which have been transported by the same mechanism (saltation, bed load etc.), Visher (1969).

Some grain size distributions from The Rio Grande obtained by the model are shown in Fig. 9. The two curves (x = 150 and 330 km) show a break point and appear as line segments. For comparison, measured grain size distributions from x = 150 km are also shown. The measurements have been presented by Nordin (1964). It is seen that hydraulic sorting of the sediment can be responsible for the occurrence of break points on the log-normal plots of grain size distributions.

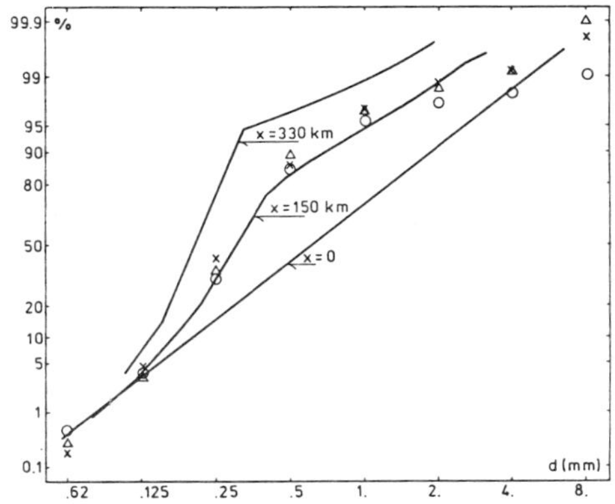

Fig. 9 Grain size distributions from the Rio Grande. Curves: results from sorting model. Points: measurements near Bernalillo (x = 150 km).

Another deviation from the log-normal distribution for natural sands has been observed by Bagnold (1971). If the diameter and the density function are plotted on double logarithmic paper the log-normal distribution will become a parabola, while the measured distribution often will be a hyperbola. These logarithmic-hyperbolic distributions have been presented by Barndorf-Nielsen (1976).

If the downstream decrease in grain size is not too large, the present model will often yield results analogous with Bagnold's observations.

An example is given in Fig. 10 where results from a grain sorting experiment made by Bagnold in a wind tunnel is shown together with results from the present grain sorting model. Direct comparison is not possible because the mechanism in aeolian sand transport differs from sediment transport by water.

CONCLUSION

A model for current sorting has been formulated. The grain size of the bed material of a river is found to be closely related to the longitudinal profile. The predicted downstream decrease in mean grain size compares well to measurements, although the measured mean grain sizes show a considerable scatter.

The grain size distributions obtained from the model exhibit characteristics familiar from natural sediments.

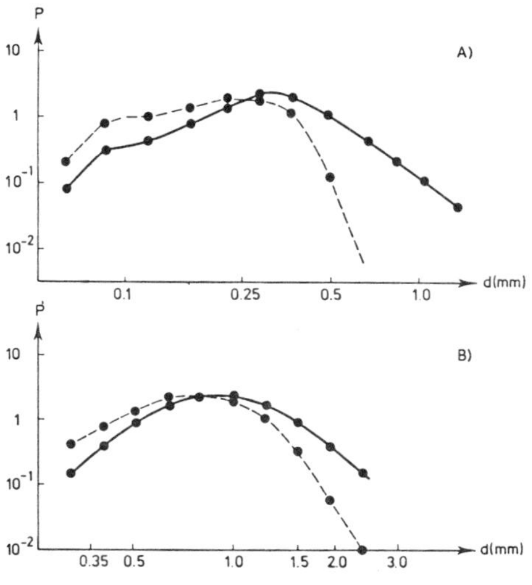

Fig. 10 A: Double - logarithmic plot of sand size distributions from Bagnold's experiment.
B: Double - logarithmic plot of sand size distributions from the sorting model at x = 0 km and x = 600 km; d_{mo} = 1 mm, σ_g = 1.5, q = 4.7 m^2/s, $I_o = 6.5 \times 10^{-4}$, $\alpha = 1 \times 10^{-6} m^{-1}$

ACKNOWLEDGMENTS

The present work was carried out during a PhD study at the Institute for Hydrodynamics and Hydraulic Engineering, Tech. Univ. Denmark.

REFERENCES

Anding, M.G. 1970. Hydraulic characteristics of Mississipi River channels. U.S. Army Engineer district, Potamology investigations, Report 19-3. Vicksburg. Miss.

Bagnold, R.A., 1971. The physics of blown sand and desert dunes. Chapman and Hall, London.

Barndorff-Nielsen, O., 1976. Exponentially decreasing distributions for the logarithm of particle size. Proc. R. Soc. , London (A) 353.

Deigaard, R. and Fredsøe, J., 1978. Longitudinal grain sorting by current in alluvial streams. Nordic Hydrology, Vol. 9 No. 5.

Deigaard, R., Fredsøe J., Møller, J.S. and Østerbye, B., 1978. Experiments on grain sorting in a diverging flume. Prog.Rep.45,

Inst. Hydrodyn. & Hydr. Engrg., Tech. Univ. Denmark.

Deigaard, R., 1980. Longitudinal and transverse sorting of grain sizes in alluvial rivers. Series Paper 26, Inst. Hydrodyn. & Hydr. Engrg., Tech. Univ. Denmark.

Engelund, F., 1966. Hydraulic resistance of alluvial streams. J. Hyd. Div., ASCE, Vol. 92, Hy 2.

Engelund, F. and Fredsøe, J., 1976. A sediment transport model for straight alluvial channels. Nordic Hydrology, Vol. 7, No. 5.

NEDECO, 1959. River studies and recommendation on improvement of Niger and Nenue. North-Holland Publishing Company, Amsterdam.

Nordin, C.F., 1964. Aspects of flow resistance and sediment transport. Rio Grande near Bernalillo, New Mexico. U.S. Geological Survey, Water supply Paper 1498-H.

Rafray, T., 1964. Analysis of change in size of bed material along alluvial channels. M.S. Thesis, Colorado State University.

Visher, G.S., 1969. Grain size distributions and depositional processes. Journal of sedimentary Petrology, Vol. 39. No. 3.

Euromech 156: Mechanics of Sediment Transport / Istanbul / 12-14 July 1982

Degradation of river beds and associated changes in the composition of the sediments

R.BETTESS
Hydraulics Research Station Ltd., Wallingford, UK

ABSTRACT: A series of laboratory experiments on the degradation of a non-uniform gravel bed are described. The observed changes in the composition of the bed are compared with the results of a numerical simulation of the experiments.

INTRODUCTION

Many laboratory investigations of sediment transport have used sediments of narrow grading so that the sediment is approximately uniform. This has been done so that the effect of sediment size can be correctly identified. Such work has greatly contributed to our understanding of the mechanisms of sediment transport. Most naturally occurring sediments, however, exhibit a much wider grading and so recently attention has turned to the behaviour of more widely graded sediments in an attempt to identify the similarities and differences in behaviour between sediments of narrow and wide grading.

In alluvial streams, where the composition of the bed sediments is rarely uniform, differential sediment transport rates of different sediment sizes can lead to changes in the composition of the material on the surface of the bed. This is often referred to as sediment sorting or, in particular circumstances, as armouring. The most familiar example arises during degradation downstream of a recently constructed dam. The reservoir will trap a proportion of the sediment load of the river and as a consequence degradation may take place downstream of the dam. The fine material in the bed downstream of the dam will be transported more readily than the coarse and, as a result, the sediment material in the surface layers becomes progressively coarser. This phenomenon of armouring can have a significant effect on the river system by limiting the amount of degradation that takes place.

Although the process of sediment sorting and its importance has long been known, until recently it has not been possible to make quantitative predictions about the changes in sediment composition. With the development of numerical models to predict the changes in bed level in alluvial rivers it has become more important to be able to make quantitative statements about sediment sorting. Recently a number of methods of calculating sediment sorting have been proposed (Ashida and Michiue, 1971; Thomas, 1980; Bettess and White, 1981). Before any confidence, however, can be placed in the use of such methods to make predictions it is necessary to test their validity. As an attempt to gain greater insight into the phenomenon of sediment sorting and to enable comparisons to be made between numerical calculations and experiments a series of experiments were performed at HRS involving degradation and armouring. In this paper the experiments are described and the results are briefly compared with those obtained from a numerical model.

EXPERIMENTAL PROCEDURE

The experiments were performed at HRS using a 2.46 m wide, recirculating, tilting flume. The working length of the flume was 18 m. At the downstream end of the sediment bed were hoppers into which sediment settled. This sediment was then pumped continuously underneath the flume channel to re-enter the main channel through a set of eight nozzles located downstream of the main discharge pump entrances and just at the beginning of the sediment bed. The discharge was measured using a flat-V weir located at the downstream end of the flume,

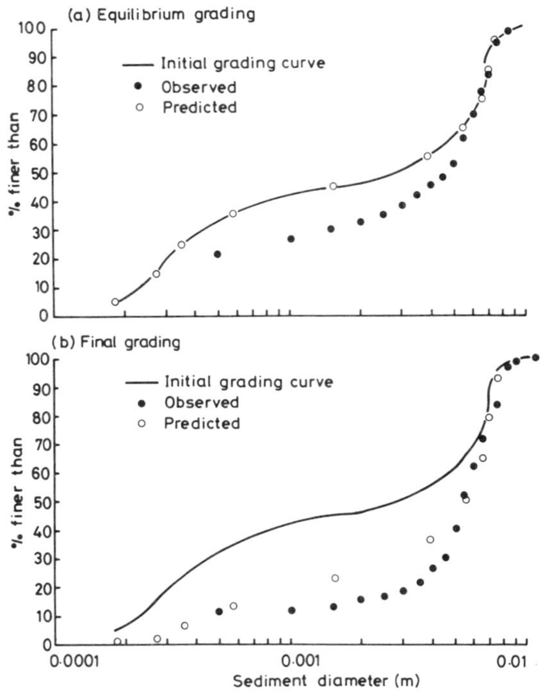

Fig 1

Test 4: Sediment grading curves, equilibrium and final.

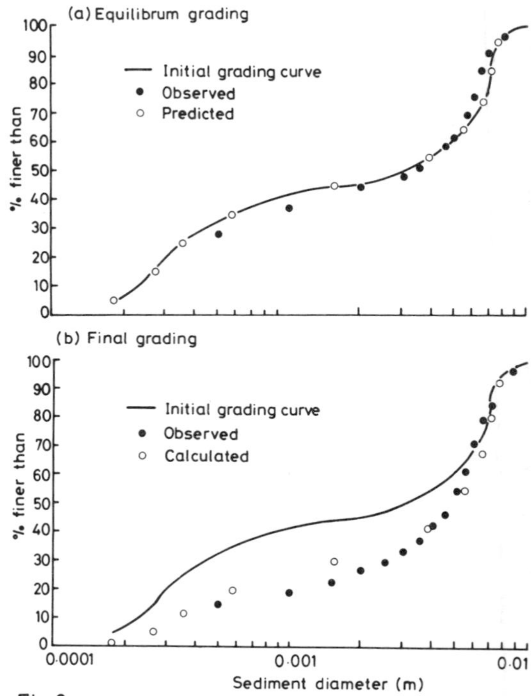

Fig 2

Test 5: Sediment grading curves, equilibrium and final.

two metres beyond the sediment hoppers. For both tests the discharge was kept constant at approximately 0.2 m^3/s. The main pump discharge system entered the flume about 2m upstream of the sediment return nozzles. Water surface slope was measured using five stilling pots fitted with point gauges. The inlets to the stilling pots were located at 3.25m intervals along one side of the flume. Measurements of water depth and bed level changes could be made using metre rules attached to the flume at 1m intervals along both sides of the 10m long glass walls of the channel.

Initially a sediment bed was laid in the flume and smoothed out by template. The grading curve of a bulk sample analysed by weight is shown in Fig 1. The bed was then wetted and the flume filled with water. The discharge was then adjusted to the required value. The sediment load was measured at 2 hour intervals as it returned to the upstream end of the flume. The sampling period was 3 minutes. At intervals the flow was stopped by raising the tailgate so that the water depths were increased to such a level that sediment transport stopped and then the discharge was turned off. The bed was then photographed so that the composition of the surface layer could be determined. The flow was then restarted and the tailgate lowered to obtain the previous flow conditions. After equilibrium conditions had been achieved the sediment in the sediment return system was prevented from returning to the flume. Sediment samples were taken at hourly intervals for the first 8 hours and then at successively longer and longer intervals. The sampling period was increased as the sediment transport rate dropped. At intervals the bed was photographed so that the composition of the surface layer could be determined. After the end of the experiment the bed was reinstated and the experiment repeated with the flume set at a steeper slope.

SEDIMENT BED COMPOSITION

Throughout the experiment photographs were taken of the surface of the sediment bed. Each photograph included a square 10cms by 10cms. For analysis purposes the photographs were enlarged to twice actual size. A regular grid of 340 points at 10mms spacing was super-imposed on the photographs and the size of the stone under each point was measured. Particles small than 0.2mm in diameter could not be measured so that any such particles were put in a single category. Thus the grading curve could not be distinguished for sizes less than 0.2mm. The sizes were then analysed by number. This should give the same grading curve as a bulk sample analysed by weight (Hey, 1981) This was confirmed by analysing the initial bed composition using both methods.

NUMERICAL MODEL

A one-dimensional, numerical, mobile-bed river model has been developed at HRS (Bettess and White, 1981). The fluid flow is calculated using a simple backwater calculation. The roughness of the channel is related to the flow and the sediment transport rate using the White, Paris and Bettess theory (1980). The sediment transport rate at each section is calculated using the Ackers and White sediment transport theory (1973). A sediment continuity equation is then used to determine the changes in bed level at each section. Such a model can be used to make useful predictions about the long term changes in bed level that may occur if the sediment regime of a river is altered in any way (HRS, 1982; Peter 1981). In many potential applications, however, it is apparent that effects due to the wide grading of the sediment material, such as armouring, will be important and that to make meaningful predictions a model must be capable of predicting the variation in sediment grading caused by the flow.

To model sediment sorting one must be able to calculate the amount of each sediment size that is moving. The Ackers and White sediment transport theory only provides an expression for the total load. White and Day (1980), in their study on the movement of graded sediments, suggested a procedure for using such a formula for the prediction of the movement of mixtures. They proposed that the transport rate of a particular size fraction could be calculated by assuming that the whole of the bed consisted of material of that size and then multiplying the resulting transport rate by the proportion of that size fraction that was present in the bed. On the basis of their experimental results they postulated that in mixtures the movement of one sediment size was affected by the presence of the other size ranges. They included this effect into the Ackers and White sediment transport theory by modifying the parameter on the mobility A to make it dependent upon the sediment grading curve. If we denote the modified parameter for the sediment size D by A' then the two different equations that have been proposed are

$$\frac{A'}{A} = 0.4 \left[\frac{D}{D_{50} \; 1.62 \left(\frac{D_{84}}{D_{16}} \right)^{-0.275}} \right]^{-0.5} + 0.6$$

$$\text{and} \quad \frac{A'}{A} = \left(\frac{D}{D_{50}} \right)^{-0.2}$$

Using both the standard Ackers and White equations and the same equations including eq (1) as a method for determining transport rates a model for sediment sorting has been developed. There are two main assumptions in the model: that at any given time only sediment to a particular depth in the bed, called the active depth, is affected by the water flow, and that within this active layer the sediment grading curve does not vary with depth. The active depth is related to the height of bed features present, which in turn is assumed to be proportional to the effective roughness height. These assumptions together with equations of continuity for each sediment size provide enough information to determine how the composition of the bed varies. The model keeps an inventory of the sediments at different elevations within the bed and uses the appropriate sediment sizes as erosion or accretion takes place. For more precise details of the model see Bettess and White (1981).

The numerical model was first used to simulate the equilibrium phase of the experiments. The initial sediment grading curve, water discharge, bed slope and downstream depth were specified. The sediment transport rate at the upstream end of the flume was set equal to that at the downstream end, to simulate the sediment return system. The model was then run until equilibrium was achieved. To simulate the degradation, the equilibrium conditions were used as initial conditions and the model was run again but with the upstream sediment inflow set to zero.

RESULTS

Results for two tests are given with a comparison with the corresponding numerical simulations. The tests were at slopes of 0.0025 and 0.0015 but the discharge, downstream depth and sediment were the same in both cases. Figures 1 and 2 show the grading curves for the mid-point of the flume after equilibrium had been achieved for Tests 4 and 5 respectively. There is a major qualitative difference between the two tests. In Test 5 the grading curve for the equilibrium state is close to that of the original parent material whereas in Test 4 the equilibrium grading is very much

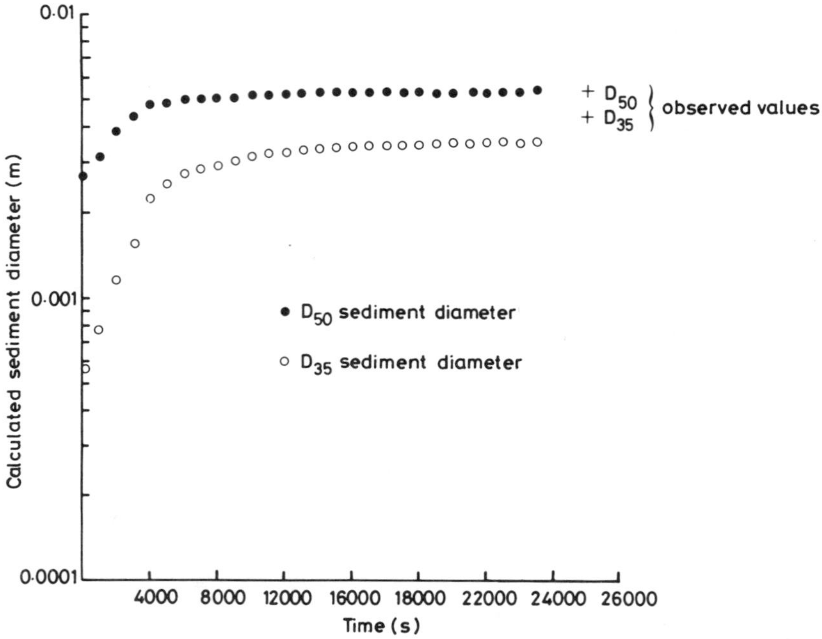

Test 4: Change in calculated sediment diameter with time. Fig 3

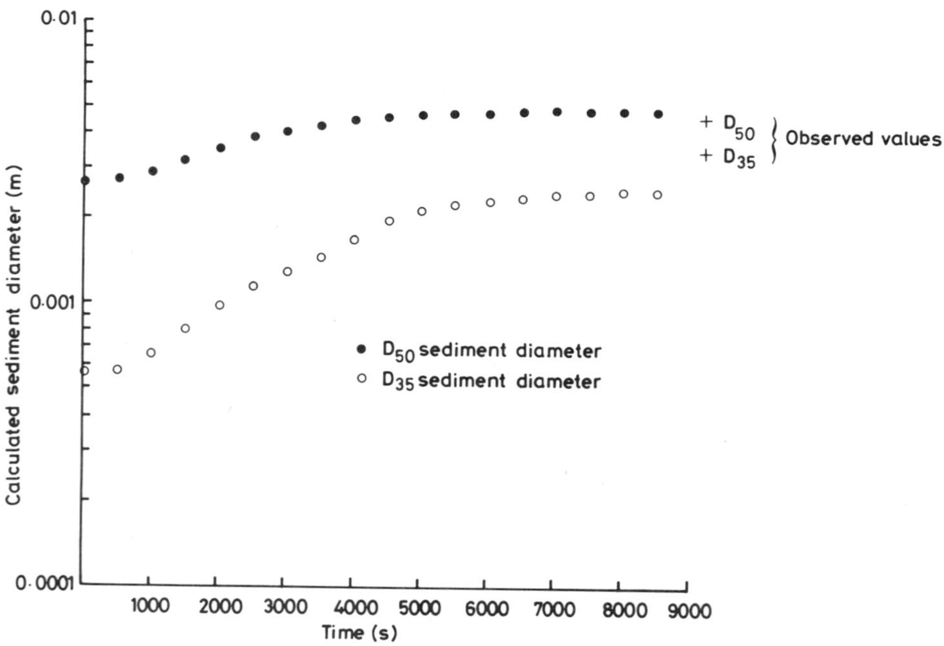

Test 5: Change in calculated sediment diameter with time. Fig 4

(White and Day, 1980) (1)

(Ackers and White 1980) (2)

coarser. For example in Test 4 the D_{35} of the parent material was 0.57mm while at equilibrium it was 2.5mm. At the moment I can offer no explanation for this difference between the two tests. Figures 1 and 2 also show that in both tests there is a significant change in grading following the stopping of the sediment return system though the armouring is less marked in Test 4 as the bed had partly armoured in achieving the equilibrium state. In both cases there is little change in the grading curves for sizes larger than approximately 6mm. The first photographs of the bed were taken about 7 hours after the sediment return was prevented and by this time the bed had virtually achieved its final armoured state. The flume results thus indicate that the armouring was achieved quite rapidly. At the downstream end of the flume the transport rate was observed to gradually fall after the sediment return had ceased.

The numerical simulation provided grading curves which are shown in Figures 1 and 2. These were obtained using the standard Ackers and White equations. The simulation including the modified mobility threshold A' lead to very reduced transport rates for the finest size fractions. As a result of this the calculated proportions of these sediment sizes increased as more of the slightly coarser material was eroded. These results suggest that the work of White and Day (1980) may only apply over a limited range. Figs 1 and 2 show a sediment grading curve for the equilibrium conditions close to that of the original parent material. This agrees well with the observed results from Test 5 but not with those of Test 4. The same figures show reasonable agreement between the predicted and observed grading curves for the final states. In Test 4 the predicted D_{35} and D_{50} sizes were 3.6mms and 5.4mms respectively while the observed values were 4.7mms and 5.5mms. The corresponding results for Test 5 are predicted 2.7mms and 4.9mms; observed 3.2mms and 4.6mms. To indicate the variation of the sediment sizes against time Figures 3 and 4 show the D_{35} and D_{50} sizes each as a fraction of the time after the sediment return ceased. The initial change is quite rapid and the final value is approached quickly corresponding with the observation that such a change took place in less than

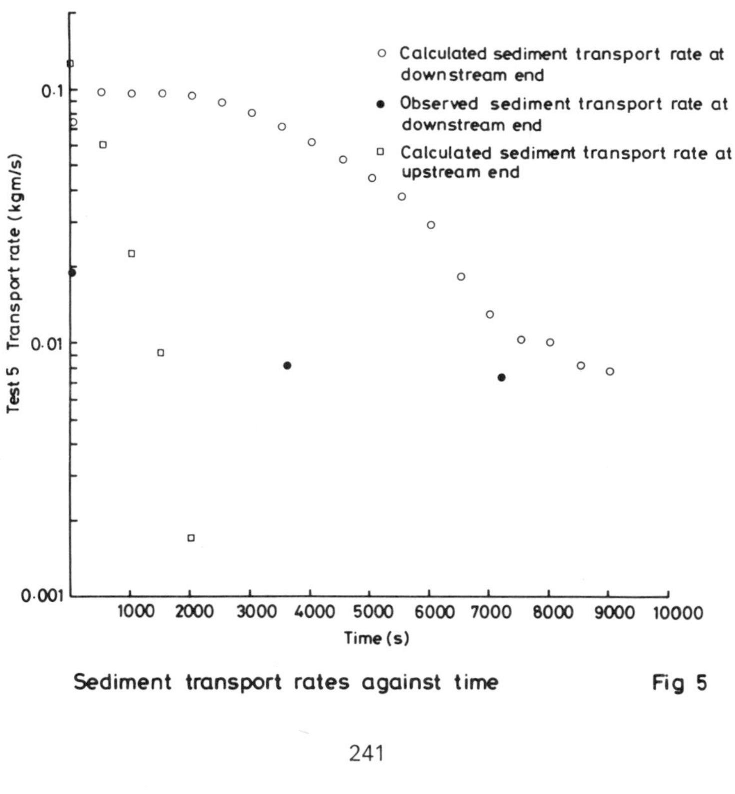

Sediment transport rates against time Fig 5

7 hours. Figure 5 shows how the various sediment transport rates varied with time. It shows the observed transport rate at the downstream end of the flume and the calculated transport rate at both the upstream and the downstream ends. The calculated transport rate at the upstream end drops quite rapidly while the downstream rate shows a distinct plateau lasting for approximately 30 min. This corresponds to the length of time required for information that the sediment return to the upstream end has ceased to propagate down the flume to the downstream end. The observed transport rate reduces in a similar manner to the calculated transport role but is noticeably less than the calculated value.

The numerical computations show that the observed lack of change at the coarse end of the grading curve is because the larger sediment sizes were not transported by the flow.

CONCLUSIONS

A careful set of laboratory experiments on degradation of a non-uniform gravel material have been described. The changes in bed sediment composition have been successfully simulated using a numerical model indicating that numerical techniques can be used to predict such changes due to differential sediment transport rates. The problems experienced when using the White and Day proposal for the modified mobility threshold in the Ackers and White equation suggest, however, that more work is required to elucidate the transport mechanisms in graded mixtures.

ACKNOWLEDGEMENTS

This work was carried out at the Hydraulics Research Station in Dr. W.R. White's section of Mr. A.J.M. Harrison's Rivers and Drainage Division. The author would like to thank Mrs. P.M. Brown and Mr. P.G. Hollinrake who were responsible for all the experimental work and Dr. E. Paris of the University of Florence who assisted with the numerical computations.

REFERENCES

Ackers, P. & W.R. White 1973, Sediment transport : new approach and analysis ASCE, JHN,99,HY11:2014-2060.

Ackers, P. & W.R. White 1980, Bed material transport : a theory for total load and its verification, Int.Symp. on River Sedimentation, Beijing, China.

Ashida, K. & M. Michiue 1971, An investigation of river bed degradation downstream of a dam. IAHR, Proc.14th Congress, Paris, Pape C-30.

Bettess, R. & W.R. White 1981, Mathematical simulation of sediment movement in streams, Proc.ICE,Part 2,71:879-892.

Hey, R.D. 1981, Sampling and analysis of gravel-bed material.

Hydraulics Research Station, 1982, Degradation of the River Nile, Report EX1040.

Peter, W. 1981, Numerical modelling of the Alpine Rhine, Hydraulics Research Station Report IT220.

Thomas, W.A. 1980, Calculation of sediment movement in gravel bed rivers, Proc. Workshop on Engineering problems in the management of gravel bed rivers, Newtown, Wales, UK.

White, W.R., E. Paris & R. Bettess, 1980, The frictional characteristics of alluvial streams : a new approach, Proc ICE, Part 2,69,:737-750.

White, W.R. & T.J. Day, 1980, Sediment transport in gravel bed rivers, Proc. Workshop on Engineering problems on the management of gravel bed rivers, Newtown, Wales.

Laboratory and insitu bed shear stress measurements

HORST U.OEBIUS
Institute for Hydraulic Research & Naval Architecture, Berlin, Germany

ABSTRACT: Devices to measure the shear stress have been developed by Preston (1954), Vennard (1961), Strauss (1964) and Blau (1972). These instruments are limited to smooth and rigid walls and to cases where the flow direction is known. They work well under model condition. The differences, however, between measurements of sediment transport and numerical forecasts made it worthwhile to develop a shear stress meter to be deployed in rivers and coastal waters, capable to work during sediment transport and under wave action. According to experiences with the Strauss-stress-meter, the new instrument should behave like a single sediment grain or a group of particles, should be capable to measure the horizontal shear force on a sediment surface of given grain diameter as well as the vertical one, due to percolation from wave action, should be able to tolerate sediment transportation and should be easily deployable. These criteria had to be met with the technical construction and mounting possibilities, as well as reliability and accuracy demands. According to these conditions a shear stress meter similar to the Strauss-proposal was mounted in a casing of 350 mm diameter and 64 mm height, sensor plate and casing being built from sand pasted with polyester to permit percolation. All interspaces in the instrument, not covered by mechanical or electronic transducers, are also filled with porous material, to become as sedimentlike as possible.

1 INTRODUCTION

One of the scientifically and from the navigational and engineering viewpoint most interesting but least accessible phenomena in inland and coastal waters is the current induced sediment transport. The valleys, banks and plains along rivers, lakes and estuaries, generated by erosion and sedimentation processes during eones of time have been favourable settling areas for mankind ever since, and it has been one of the most important tasks of men to protect these areas against the endeavor of nature, to change these places by floods, erosion and sedimentation. Therefore both, the scientist and the engineer, have been closely working together to define mathematical approaches to forecast these attacks and therefrom to develop combat strategies. Despite all efforts and the development of more and more sophisticated measuring instruments, they have not succeeded so far, however, to make the sediment transport processes and their physical correlations universally more transparent. It was one of the objectives therefore, to develop a measuring device to enable the research engineer to measure real locally optional forces upon the autochthonous sediment in order to provide him with the desired loadings for his constructions, and the scientist, respectively, to compare this numerical approaches with nature and, if necessary, to improve his understanding of the physical correlations.

2 STATE OF THE ART

Non cohesive, cohesive and bedrock sediments can only be moved from their stable gravity-dependent positions by an additional force. In natural waters this force is equivalent to the friction-related resistance of the phase sediment against the phase water. Since du Boys (1879) the force to overcome this resistance is called "shift force" if an individual grain is concerned, and "shearing stress"* if this force is acting on a distinct area. Today's approaches to calculate sediment transportation essentially vary in the manner how this shift force or shearing stress, respectively, are introduced. Agreements exist only concerning the moment of incipient sediment transport, although even here no universally valid equations could be deduced, due to the heterogeneity of the soil mechanical parameters to be considered. This handicap is related to all direct and/or indirect measuring methods, too* (see Annex).

Despite of all resentments the shear stress seems to be an appropriate reference value to which all known procedures can be reduced. It therefore appears explicitly or implicitly in all sediment transport formulae. It is calculated from the measured velocity profile of any existing current by means of empirically developed transfer functions. Although in some special cases empiricism and nature seem to harmonize, these functions often fail if applied to natural rivers or to coastal areas. Therefore analytical prognoses about transported sediment rates can be afflicted with errors of unknown magnitude. Efforts to measure the shearing stress under laboratory conditions have been undertaken by Preston (1954), by Vennard (1961), by Strauss (1964), by Blau (1972) and by Wagner/Bürger (1980). Their shear meters are limited to rigid and smooth walls mainly and to cases where the flow direction is known. They fail in cases where non cohesive, moveable walls are concerned. To close this gap, it seemed worthwhile to try to develop a shear meter capable to work on moveable beds and to measure the horizontal as well as the vertical wave induced shearing forces before and during sediment transportation.

3 DESIGN OF A SHEAR METER

The design criteria for a shear meter are dominated primarily by three specifications:

1. it has to be as analogous to nature as possible, i.e. the instrument should be able to be integrated into the natural environment and should behave like it in order not to cause extra uncontrollable disturbances,
2. it must be able to measure forces from arbitrary directions; and
3. there must be a distinct relationship between the hydrodynamic procedures, causing the forces, and the shearing forces measured by the shear meter, i.e. the instrument or its peripheric parts must be capable to measure current velocities and wave heights, which can be unambigously collated to the measured shearing forces.

3.1 Measuring system

Analogous to the instruments designed by Vennard, Strauss, Blau and Wagner/Bürger a similar sensor plate was chosen also for the new shear stress meter. The condition of natural similarity required the application of sediment as construction material, with grain size and porosity as main parameters. To provide the same conditions under influence from any direction the casing and the sensor plate of the instrument were designed circular. The maximum diameter resulted from the requirement to be able to measure locally restricted hydrodynamic processes, such as secondary currents etc. From laboratory tests with wave lengths of L = 2 m a maximum sensor diameter of D_o = 0.5 m was estimated, which should be suitable for the measurement of wake and other rotational currents, too.

According to pilot tests with a Strauss shear stress meter, cantilever beams with strain gauges proved to be a convenient measuring device. As the construction height of the new shear stress meter had to be minimized, vertical cantilever beams were not applicable. Such systems would also lead to disturbing tilting motions of the sensor plate. Instead, four horizontal cantilever beams, mounted crosslike, were chosen. They consisted of laminated springs, clamped in a center block and connected to the sensor plate at its rim by a specially developed power transmission system. The actual lengths of the cantilever beams resulted to L_B = 120 mm from the required minimum dimensions of the strain gauges, the technical constructional restrictions, and the accuracy of measurement desired. The height of the laminated springs was determined to H_B = 10 mm. From these dimensions a diameter of the sensor plate of D_o = 246 mm was determined. The overall diameter including the sealing device and the outer casing of the instrument was designed to D = 350 mm. The maximum permissible bending, i.e. the maximum displacement of the sensor plate was determined to 0.2 mm. Assuming a maximum horizontal current velocity of u_∞ = 2 m/s the shearing force resulting therefrom requires a minimum width of the springs of B_B = 4 mm. B_B = 5 mm was chosen. These dimensions correspond to a maximum admittable shear force at the cantilever beam ends of

$$P_{adm} = 6 \text{ N} \qquad (1)$$

and a permissible shearing stress at the sensor plate of

$$\tau_{adm} = 1.24 \cdot 10^{-4} \text{ N/cm}^2 . \qquad (2)$$

As the strain gauges have a measuring range of 10^{-4} μm/m, shearing stresses down to

$$\tau_{min} = 1.2 \cdot 10^{-8} \text{ N/cm}^2 \qquad (3)$$

can be measured, no losses in the electronic equipment assumed. For the vertical shear force measuring system an analogously dimensioned cross of cantilever beams was selected. The overall height of the instrument then resulted from the individual heights of the sensor plate, the horizontal and vertical measuring device and the casing to H = 64 mm. A schematic impression of the measuring system is shown in Figure 1.

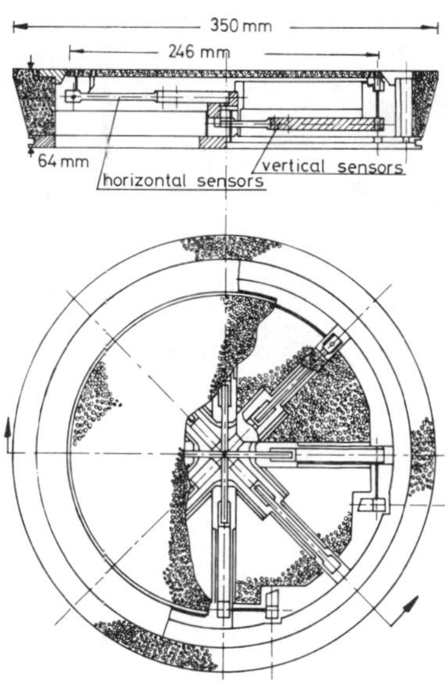

Fig. 1. Measuring system.

3.2 Casing

The presupposition to optimally integrate the shear stress meter into the natural environment at the measuring site involves a similar construction of the instrument casing as this environment. This refers to the grain size and the porosity as well as other soil mechanical parameters. Unfortunately those parameters vary locally and are obtainable only in special cases. It seems to be opportune therefore, to select a material representative for as wide a spectrum of grain fractions as possible. For the tests at the German coast in the North Sea a quartz-sand material of a mean diameter of D_{50} = 0.80 mm was chosen. To control the influence of grain size and porosity a second casing made from material of a grain size of D_{50} = 3 mm in diameter was constructed.

The form of the casing resulted from that of the mechanical sensor and power transmission system and the electronic equipment to be integrated into the casing. Figure 2 shows these parts.

Fig. 2. Mechanical measuring device.

Fig. 3. Inner filling pieces, casing and sensor plate.

Figure 3 gives an impression of the inner filling pieces, the outer ring of the casing and the sensor plate built from sand pasted together by synthetic resin.

3.3 Field instrument

The shear stress meter designed for the field tests consists of a spider built from stainless steel (Figure 2). The center of this spider carries the clamp block for the cantilever beams, which are connected by thin hollow needles to the stainless steel ring, housing the sensor plate. The gap between this ring and the casing is sealed by a specially developed seam ribbon made from silicon caoutchouc. This seal prohibits sediment to penetrate into the casing. It does not transmit horizontal or vertical forces.

The strain gauges of two opposite cantilever beams of the horizontal measuring system are electronicly connected totally compensated by bridge circuits to exclude other influences than those resulting from shear forces. The same applies to the vertical measuring system, only that here all four strain gauges are connected totally compensated. This methods allows only vectors to be measured. The cables from the strain gauges are bundled in the center block and led to a groove at the bottom of the instrument. In this groove the bundle runs to a connector receptacle outside the instrument. The cables are fixed and sealed on their way by plexiglass grouting.

If the instrument can be deployed in transient dry areas of rivers, lakes and the coastline or by divers, the abovementioned shear meter meets all requirements. If it has to be deployed from the sea surface, however, an instrument carrier becomes essential (Figure 4).

As an imbedding of the instrument in the moveable seafloor is impossible this way, a transition structure from the sensor plate to the seafloor in form of a lenslike body was designed and built from the same porous material as used for the casing of the shear meter. Due to stability problems this lens had to be placed on a carrier structure built from steel. The overall height of the shear meter plus lens then amounts to H = 75 mm. This lens also shelters the connector receptacle, which houses two accelleration pickups with a re-solution of 0.5° inclination for the determination of the position of the sensor plate re-

Fig. 4. Instrument carrier.

lated to the horizontal plane, and a pressure meter with a measuring range from 0 to 50 PSI ≙ 0 to 35 m hydrostatic pressure for wave height measurements (Figure 5).

Fig. 5. Lens, shear meter and connector receptacle.

All measured data are transferred by underwater cable from the lens to an amplifier set-up. If the measuring site is not far from a landbased station, the energy supply and the data acquisition, amplification and recording can be accomplished on land. This method is limited to approximately 200 m distance

Fig. 6. Telemetric buoy.

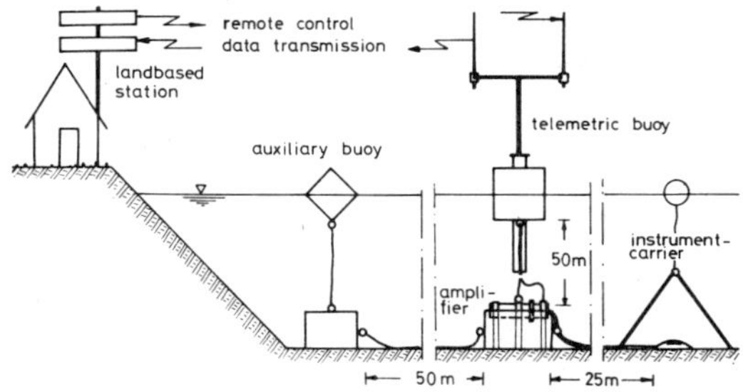

Fig. 7. Schematic presentation of the instrument chain.

from the coast. If the distance is more than that, either a data acquisition under water or a telemetric transfer of the amplified data to a landbased station and recording there have to be regarded. For the field measurements off the isle of Norderney a telemetric unit was chosen due to power supply problems in case of continuous measurements. The telemetric unit was equipped with a remote control which enables the user to optimally utilize the available battery power by switching on the instruments only at hydrodynamically interesting times. The telemetric unit was mounted in a specially designed buoy (Figure 6). The data assembly and transmission set-up is shown schematically in Figure 7.

The parameters to be measured during the field test were the horizontal and vertical shift force, the inclination of the sensor plate and the wave heights. Originally, these data were regarded sufficient to give a clear picture of the shear force conditions at the seafloor. During the field tests it proved however, that the current velocity near the instrument and the current direction relative to the main axes of the sensor plate as well as the actual wave heights above the shear stress meter are unalterable process variables too, because they cannot be derived from the afore mentioned data. Appropriate instruments will have to be found.

4 FUNCTIONAL LABORATORY TESTS

Two instruments were used for the laboratory tests with rigid and porous casing walls respectively, built from different grain sizes to investigate into the influences of different constructional characteristics upon the measured data. Such characteristics were the roughness of the sensor plate, the influence of tight and porous casing parts on the measured variables in twodimensional, horizontal currents or in threedimensional, wave induced currents and when percolation occurs. Other investigations concerned the influence of the transition lens and its porosity on the horizontal and vertical shear forces. All tests were carried out in different porous media under sediment transport conditions varying from zero to fully developed sediment movement.

4.1 Horizontal shear stresses with buried instrument

The functional tests concerning the influence of horizontal currents were accomplished in the 16 m long Shields flume of the VWS. This tilting flume has glasswalls along both sides of the measuring section of 0.4 m width and 0.6 m height. It was favoured because the tests carried out by Shields in this flume

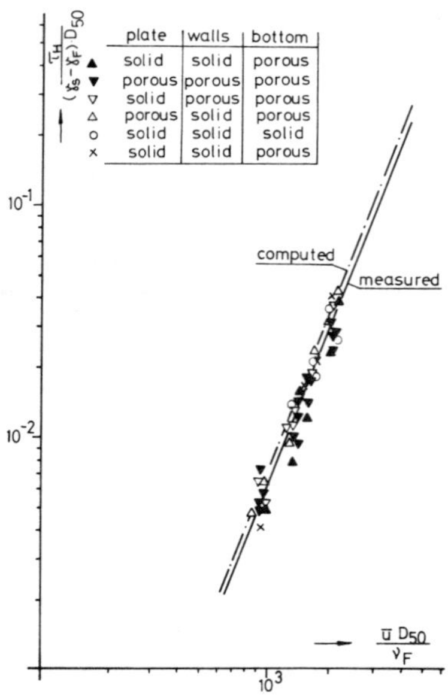

Fig. 8. Horizontal shear stress τ_H.

offered an ideal comparison and calibration possibility.

The following variations were investigated and combined with each other, bottom porous and tight, casing side walls porous and tight and sensor plate porous and tight. For the graphs the dimensionless Shields form of presentation of the functional correlation between the Froude number and the Reynolds number, both based on the sediment grain roughness

$$\frac{\tau_o}{(\gamma_S - \gamma_F)} \cdot D_{50}^{-1} = f\left(\frac{v_* \cdot D_{50}}{\nu_F}\right) \qquad (4)$$

was chosen. The result of the tests is shown in Figure 8. It reveals a fairly good arrangement of computed and measured data in one curve. This means, that there is no percolation due to horizontal currents. The shear stress is merely dependent on the roughness of the sensor plate and its surrounding. This was stressed by the fact that no vertical forces were measured at lower current velocities. During sediment motion, however, a scour in front of the casing was found which caused wake flows. This rotational current induced a low pressure field above the instrument, which expressed itself by upward directed vertical forces upon the sensor plate. As the scour formation was stochastic the appearance of vertical forces was stochastic, too. The threshold for a clearly recognizable occurance of vertical force components was at approximately $u_\infty = Q/F = 0.5$ m/s. The fairly good agreement between measured and computed data demonstates the reliability of the shear meter as far as horizontal forces are concerned. It moreover proves the reliability of the Shields approach to incipient sediment movements.

4.2 Vertical shear forces with buried instrument

It is well known from geologic and soil mechanic investigations that long period waves like tides induce pressure differences in the ocean floor. These pressure gradients induce currents, the velocity of which is mostly unknown. Tests in the VWS have revealed that also short period waves like normal swell can induce currents in the upper sediment layer of the sea-floor, which is equivalent to percolation due to wave action. The tests concerning vertical shear components were carried out in the small wave tank of the VWS. For the reproduction or results the functional correlation between the Shields parameter

$$F_* = \frac{\tau_V}{(\gamma_S - \gamma_F) \cdot D_{50}}$$

and the Ursell-parameter

$$U_r = \frac{H \cdot L^2}{h^3}$$

with H = wave height, L = wave length and h = water depth were chosen. The tests were accomplished on a 0.12 m high sediment layer from quartz sand of a mean diameter of $D_{50} = 0.8$ mm. The same material was selected for the sensor plate of the laboratory instrument. The casing walls were built from sediment of the diameter $D_{50} = 0.8$ mm and $D_{50} = 3$ mm respectively.

The results of the measurements with wave lengths of L = 2.10 m to 3.20 m and wave heights of H = 0.05 m to 0.20 m are given in Figure 9. As the test data can be represented by one curve there

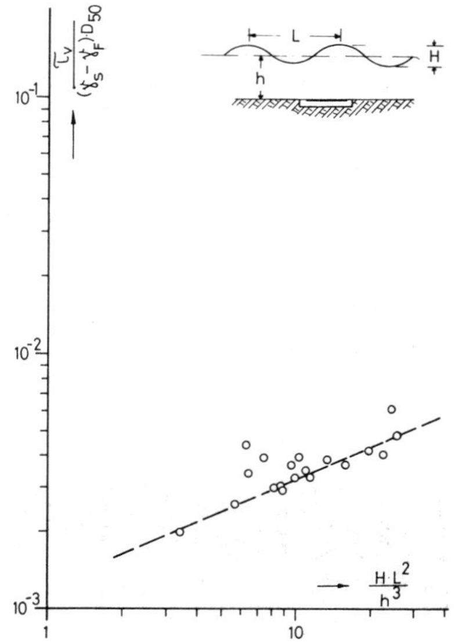

Fig. 9. Vertical shear force τ_V.

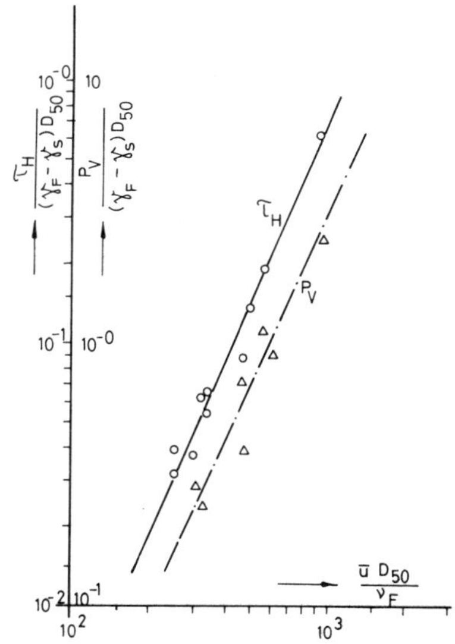

Fig. 10. Horizontal shear stress: shear meter with lens.

has to exist percolation but no influence from the casing walls or at least no measurable ones.

4.3 Horizontal shear stresses: shear meter with lens

The investigation concerning the influence of the lens body around the shear meter took place in the Shields flume too. As the width of this flume is much smaller than the lens diameter, parts of the lens had to be cut off. The remaining segment was then mounted totally buried into and placed on the sediment layer in the flume. The tests were carried out analogously to those without lens. The results are shown in Figure 10. It can be seen that the test data also seem to follow one curve. An influence of the different grain diameters of the surrounding sand layer could not be realized but lift forces as soon as the lens protruded from the sand layer. In this case the lens represents an airfoil, which reacts with a lift force on the current deflection. These lift forces differ completely from those due to percolation and should not be mixed up with them. They are approximately ten times higher than the horizontal shear forces. P_V has the same dimension as τ_V.

4.4 Vertical shear forces: shear meter with lens

These tests were again accomplished in the small wave tank of the VWS, with the field instrument and two lens bodies built from different sediments of D_{50} = 0.8 mm and D_{50} = 3 mm grain diameter. The wave lengths of L = 6.25 m, L = 4.00 m, L = 2.45 m and L = 1.75 m remained constant during one test series while the wave heights varied between H = 0.05 m and H = 0.22 m. The results as total force, i.e. the difference between maximum and minimum force corresponding from wave crest to wave trough are plotted in Figure 11, in fact the wave induced horizontal and vertical components. As expected the test data follow distinct curves, which indicate that the horizontal and the vertical shear forces (including lift force from horizontal currents) are of same order.

4.5 Sediment transport

As the instrument was tested during fully developed sediment transport, also special modes of sediment movement in form of dunes occurred, which migrated slowly across the sensor plate. Figure 12 shows a foto together with a recording of such event. The evaluation of this figure

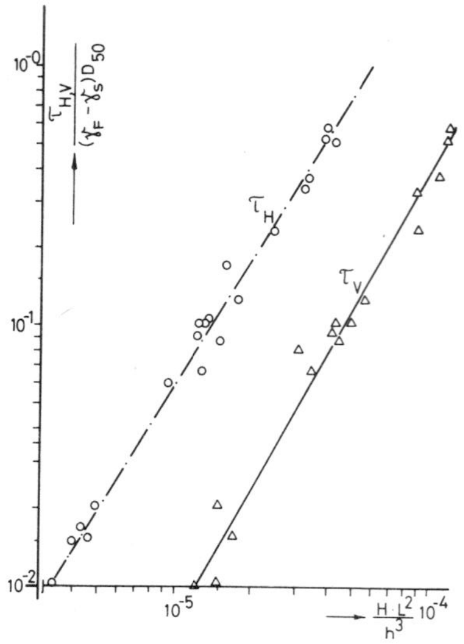

Fig. 11. Vertical shear force: shear meter with lens.

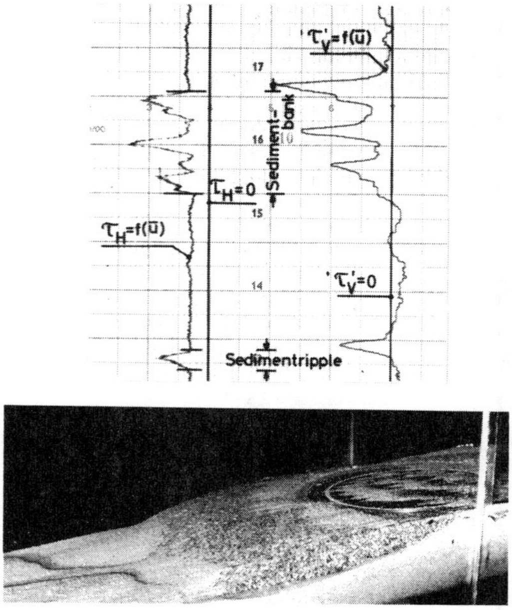

Fig. 12. Sediment dune migration across the instrument.

leads to two peculiarities. The recording of the normal shear force is accompanied by high frequency overtones, which could be corresponding to macro-turbulent oscillations. More intensive investigations into their character, their interpretations and the possibilities of application of such measurements have not been accomplished today. The recording moreover shows extreme forces horizontally as well as vertically. These correspond to sediment dunes travelling across the sensor plate. The measured forces are induced by the weight of the dune body resting on the plate and the vertical component of the resistance force of the dune against the current in vertical direction and the horizontal component of the resistance in horizontal direction. The differentiation of the two vertical forces is not possible yet. These recordings prove however, that sediment transport does not affect the reliability of the shear meter and that it is possible to identify the moment of coverage by ripples and dunes.

5 FIELD EXPERIMENT

Before starting the field experiment all parts of the instrument chain, the data transmission and the recording set were coupled, functionally tested in the wave tank and the amplifier adjusted to the expected measuring ranges. For transportation purposes special disconnecting points were foreseen, which represented individual check points simultaneously.

The offshore experiment required a vessel to transport the instruments to the selected location, to deploy them and to rescue them after test completion. This vessel and all services connected with the experiment were provided by the Forschungsstelle Norderney. Together with the director of this institute a suitable location for the measurements was determined at the position 53°44'40"N 7°7'20"E, approximately 4 km off the west coast of Norderney. This position was selected because of its waterdepth (ca. 10 m below chard data) and tidal currents, waves and the sediment movement to be expected.

The field tests started on June 1st, 1982, when the research vessel, loaded with all components belonging to the instrument chain offshore, anchored at the selected position. After deploying final functional tests proved the accurate condition of the chain.

The long-term tests started the following day by measuring the inclination of the instrument, the wave height H, the

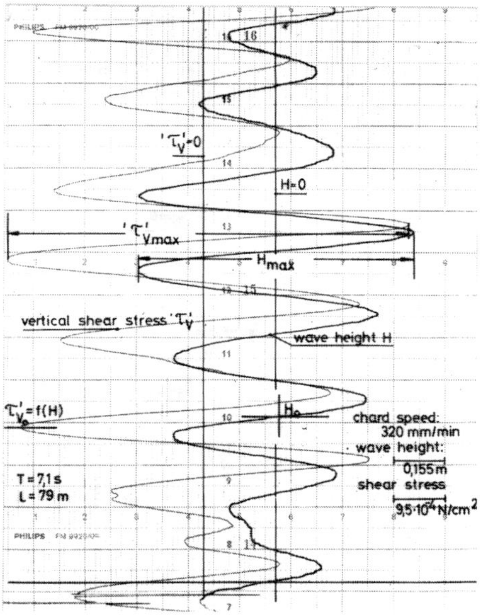

Fig. 13. Field experiment, H and $"\tau"_{V}$.

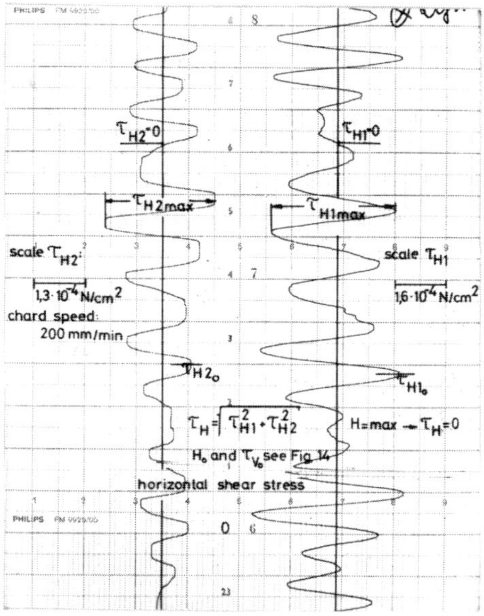

Fig. 14. Field experiment, τ_{H1} and τ_{H2}.

vertical shear force $"\tau"_{V}$ and the horizontal shear stress

$$\tau_H = \sqrt{\tau_{H1}^2 + \tau_{H2}^2}$$

with τ_{H1} = north - south coordinate of the instrument and τ_{H2} = east - west coordinate. Simultaneously the main wave direction was determined to compare the instrument north pole with the geographic one in order to define the current direction at the sea-floor. Functional criteria to be investigated except those of the shear meter and the peripheric instruments were those of the data transmission, as well as the interaction and the handling of the instrument chain.

During the following time until the end of the test series on June 8th, 1982 the chain was started by the remote control unit when there occurred interesting tidal or wave phases. The clock inside the control unit automatically disengaged the electronic equipment after seven minutes. This time was absolutely sufficient to get a clear picture of the momentary hydrodynamic situation.

The results of these measurements are given in Figures 13 and 14 for a relatively calm sea with wave heights of about H = 0.08 m, approximately 0.5 hours before turn of the ebb-tide, in Figures 15 and 16 for wave heights up to H = 0.50 m, at the turn of the ebb-tide and in Figures 17 and 18 for heights up to H = 0.60 m, approximately 3.5 hours after the turn of the ebb-tide. They demonstrate the close interaction between wave formation, vertical shear force and horizontal shear stress, as well as the reliability of the shear stress meter concerning its operation. A more detailed evaluation of the recorded data suffers from the restrictions defined from the laboratory tests and from peculiarities in the recording not identified yet. The figures, however, underline the important role that waves play concerning sediment transportation at the coast. Although tidal currents could not be defined clearly during this test and can only be derived from earlier current measurements at the same place, their influence seems to be much less on sediment movement than that of waves.

During the tests the instrument was buried under a transported sand dune. This could be derived from the fact that all components except the wave sensor transiently recorded maximum deflections. After about ten minutes the recordings returned to normal values.

There occurred no problems whatsoever with the telemetric transmission of data and the remote control of the instrument chain.

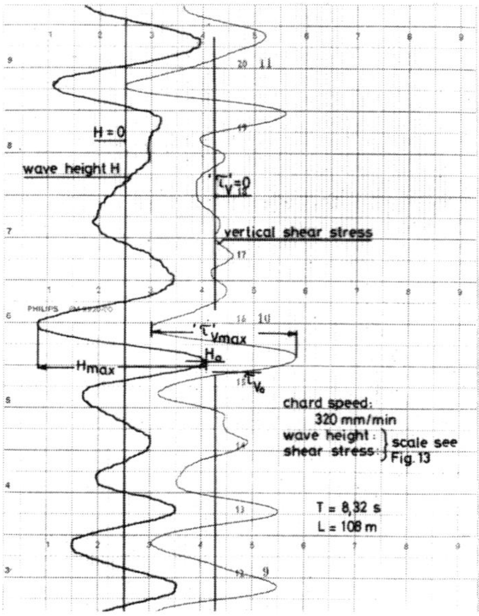

Fig. 15. Field experiment, H and "τ"$_V$.

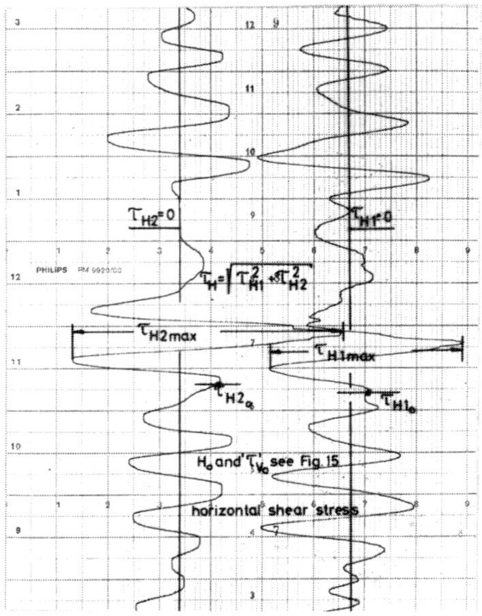

Fig. 16. Field experiment, τ_{H1} and τ_{H2}.

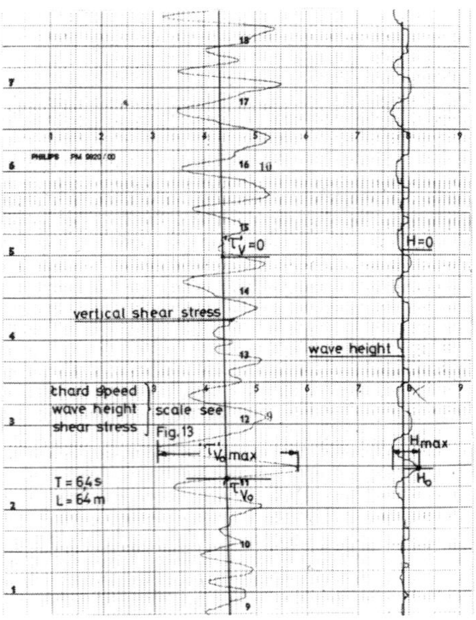

Fig. 17. Field experiment, H and "τ"$_V$.

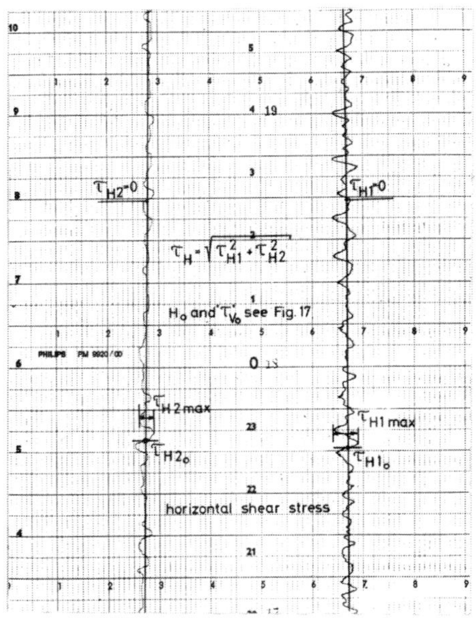

Fig. 18. Field experiment, τ_{H1} and τ_{H2}.

6 CONCLUSION AND OUTLOOK

The designed shear stress meter passed the first tests satisfactorily. It proved the ability to measure the inclination of the instrument, wave heights and horizontal and vertical shear forces. The recorded data represent actual loadings on the instrument.

Problems arise with the interpretation of these data. It is known that pressure meters have to be adjusted to the water depth and the density of the water as transport medium for the pressure wave. From a reference wave measurement at the sea surface, no remarkable difference between the two wave heights could be detected, however.

The recording of the tidal current or its identifivation from the horizontal shear stress measurements failed totally. This underlines the essential necessity of an extra current speed and direction recorder and a compass combined with the instrument.

Although the vertical shear force measurements must be influenced by the horizontal currents (capt. 4.3), the recording of oscillating vertical shear forces with nearly equal upward and downward components indicates, however, that this influence can apparently be neglected here. More detailed investigations will have to clarify this problem.

The laboratory and insitu tests of the shear stress meter revealed many new possibilities of approach to the sediment transport phenomenon, which will be part of another research programme. We hope that other institutes will participate in this research and that the dialogue concerning the results will help to make the sediment transport problem more transparent.

7 ACKNOWLEDGEMENT

This research was supported by a grant of the German Research Foundation. The support of Dr. G. Luck and Dipl.-Ing. H.D. Niemeyer of the Forschungsstelle Norderney are gratefully acknowledged. The author is grateful to the technical staff - J. Hohnberg und H. Strube - whose assistance made this work possible.

*ANNEX:

The development of a shear stress meter raised the question, in how far the expression "shear stress" in combination with this instrument is suitable and permissible.

In hydromechanics, too, shear stresses are tangential force-originated tensions in a continuum. They depend very much on the viscosity of the real fluid and are defined at a plane wall by the equation

$$\tau_0 = \varrho_F \cdot \nu_F \left(\frac{du}{dy}\right)_0$$

with $\varrho_F \cdot \nu_F$ = friction-proportionality factor or viscosity. Exept those tangential stresses there also occur normal forces or viscous pressure forces due to static friction at the wall. These forces normally are hardly to compute due to the mostly turbulent state of real fluids.

In case of a sediment layer overflown by water analogies to plate friction can easily be established. The afore mentioned equation then has to be extended by Prandtl's mixing law

$$l^2 \left(\frac{du}{dy}\right)$$

and by the friction factor λ. Analogously but not identically to Hooke's law three states can be defined for the sediment layer under the influence of a horizontal current, too:

1. The "elastic" phase, where individual sediment grains do not change place and position,
2. "yield-point" = moment of incipient motion, where individual grains start to change place and/or position sporadically and in small steps, and
3. "fracture point" = fully developed sediment transportation with its own peculiarities.

The instrument senses these tensions and reacts with displacements of the sensor plate. In this case the expression "shear stress meter" is permissible.

In cases of vertical percolation of the sediment layer, however, the viscous forces exceed the tangential stresses, in fact relative to the spatial concentration of a unit volume. The greater the ratio of the sediment covered area to the pore area is, the greater is also the ratio of dynamic pressure to shearing stress. It reaches a maximum value when the percolation area is totally closed by sediment grains. In this case the instrument should rather be called "shear force meter".

To avoid different expressions for the same instrument, in the report the expression "shear stress meter" is used, bearing in mind, that this instrument measures horizontal shearing stresses but vertical shear forces.

REFERENCES

Blau, E. 1972. Die Verteilung der Wandschubspannung in offenen Gerinnen. Wasserwirtschaft-Wassertechnik 22(7): 236

du Boys, 1879. Le Rhône et les rivières à lit affouillable. Annales des Ponts et Chaussées, No. II.

Oebius, H. & S. Schuster, 1976. Analytische und experimentelle Untersuchungen der Auswirkung von Flachwasserwellen auf die Reststromkomponente am Meeresboden. Versuchsanstalt für Wasserbau und Schiffbau, VWS-Bericht Nr. 764/76.

Preston, J.H., 1954. The determination of turbulent skin friction by means of pitot tubes. Journ. Royal Aeron. Society, 58.

Shields, A., 1936. Anwendung der Ähnlichkeitsmechanik und der Turbulenzforschung auf die Geschiebebewegung. Mitt. der Preußischen Versuchsanstalt für Wasserbau und Schiffbau, Nr. 36.

Strauss, V., 1964. Möglichkeiten der direkten Messung von Schubspannungen im hydraulischen Versuchswesen. Scient. Hydr. Conf: Bukarest.

Vennard, J.K., 1961. Elementary fluid mechanics. New York: Wiley.

Wagner, H. & W. Bürger, 1980. Entwicklung und Einsatz eines neuen Schubspannungsmeßgerätes, 10:357.

Transition in oscillatory boundary layers

M.SERT
TUBITAK, Marmara Research Institute, Turkey

ABSTRACT: In studies of coastal sediment transport by wave action, the nature of the flow near the bed becomes significant. In the past, considerable amount of research effort has been directed towards understanding the mechanism of transition in oscillatory boundary layers. However, the results obtained display discrepencies and the existing delineations of the flow regimes remain insufficient. In this study, an effort is made to provide a general account of the laminar-turbulent and smooth-rough transition in oscillatory boundary layers on the basis of some fundamental criteria and the available experimental evidence. The laminar-turbulent and smooth-rough transitions have been considered separately over a wide flow range depending on two basic flow parameters. Thus, the plane formed by these two parameters are divided into two in two different ways by the laminar -turbulent and smooth-rough transition curves. Hence the domains for various flow regimes (smooth laminar, rough laminar, smooth turbulent, rough turbulent) are determined by the intersections of these two curves.

1 INTRODUCTION

In studies of coastal sediment transport by wave action and for other related coastal phenomena, the nature of the flow near the bed becomes significant. Therefore it is necessary to define dividing lines between various flow regimes. In the past, considerable amount of research effort has been directed towards undestanding the mechanism of transition in oscillatory boundry layers. However, the results obtained display discrepancies and the existing delineations of the flow regimes remain insufficient.

As it has been pointed out for the first time by Reynolds (1883), laminar flow becomes unstable when certain flow parameters are increased beyond certain limiting values. Several theoretical studies have been carried out on the stability of the fluid flow subjected to small disturbances, aimed at finding out the critical point at which a small instability would be increased in amplitude causing the scale of mixing to be transformed from molecular size to larger lumps of fluid. However, stability analysis does not always seem to be reliable. For oscillatory boundary layer flow, Collins (1963) carried out an approximate analysis using the method put forward by Lin (1955) for flat plates in steady flow, and later Kerczek and Davis (1974) attempted a more rigorous stability analysis, although no definite conclusions could be drawn from these.

Transition from a hydraulically smooth bed to a rough bed has to be considered separately, since, despite the formation of vortices around the roughness particles, the flow in the bed boundary layer may remain essentially laminar. The mechanism of vortex formation, deterioration of the vortex structure into random eddies, the way the eddy size is controlled by the flow parameters and the mechanism of mixing with the flow above are the intermediate questions that have to be answered before a full explanation can be given for the transition phenomena. At present, for realistic results of practical value, experimental evidence evaluated on the basis of some fundamental criteria appears to be essential and such an approach have been adopted in this study.

2 DESCRIPTION OF THE FLOW CHARACTERISTICS

When the surface profile for oscillatory, progressive, two dimensional waves are expressed in sinusoidal form (small amplitude waves), the velocity just outside the boundary layer on a plane, horizontal bed is given by

$$u_\infty = \bar{u}_\infty \cos(\sigma t - kx) = \frac{\sigma a}{\sinh kh} \cos(\sigma t - kx)$$

where a=wave amplitude, $\sigma = \frac{2\pi}{T}$ = radial frequency, T=wave period, $k = \frac{2\pi}{\lambda}$ = wave number, λ=wave length, h=water depth, t=time, x=horizontal coordinate along the bed, and, a bar over a symbol indicates amplitude. Hence, writing $\bar{u}_\infty = \sigma b$, the amplitude of the fluid motion just outside the bed boundary layer is $b = \frac{a}{\sinh kh}$. Solution of the linearized boundary layer equation

$$\frac{\partial u}{\partial t} = \frac{\partial u_\infty}{\partial t} + \nu \frac{\partial^2 u}{\partial y^2}$$

gives the velocity profile in the bed boundary layer as,

$$u = \bar{u}_\infty(\cos(\sigma t - kx) - e^{-\beta y}\cos(\sigma t - kx - \beta y))$$

where y=vertical coordinate measured upwards from the bed, $\beta = (\frac{\sigma}{2\nu})^{\frac{1}{2}}$ (β^{-1} being a measure of the boundary layer thickness) and ν=coefficient of kinematic viscosity. This expression is valid on the assumption that $(\frac{k^2\nu}{\sigma})^{\frac{1}{2}} << 1$ and corresponds to a first order solution in $\frac{a}{\lambda}$, (Basset (1888) Hough (1896)). The fluid motion near the bed is very similar to that near a smooth flat plate oscillating in its own plane in still fluid (except for pressure and phase variations along the bed) for which the solution was obtained by Stokes (1851). The linearized solution (Stokes solution) over a smooth bed remains valid almost as long as the flow remains laminar, since the non-linear terms in the boundary layer are inherently small. There is no general account of the oscillatory flow over rough beds and for turbulent flow, although some significant work on these have been done (Jonsson (1963,1966a) Kajiura (1968)). It can be seen from dimensional considerations that for a bed of roughness diameter d, two dimensionless parameters will control the flow and these may be obtained as $R = \frac{\bar{u}_\infty d}{\nu}$ and $\beta d = (\frac{\sigma}{2\nu})^{\frac{1}{2}} d$, (or alternatively $\frac{b}{d} = \frac{R}{2\beta^2 d^2}$ and $\beta b = \frac{R}{2\beta d}$). A third parameter, σt, will account for temporal variation within a flow cycle.

3 TRANSITION CRITERIA

3.1 Laminar-Turbulent Transition

Transition from laminar to turbulent flow will be defined depending on a critical value of a Reynolds number based on the boundary layer displacement thickness and the velocity at the edge of the bed boundary layer. The boundary layer displacement thickness δ_* will be defined as

$$\delta_* = \frac{1}{\bar{u}_\infty} \cdot \text{Amp.} \int_o^\delta (u_\infty - u)dy$$

where δ=boundary layer thickness. However, from the equation of motion, (noting also that $\tau = \mu \frac{\partial u}{\partial y}$ = shear stress, $\mu = \rho\nu$ = coefficient of dynamic viscosity, ρ=fluid density),

$$\frac{\partial}{\partial t}(u - u_\infty) = \frac{\partial}{\partial y}(\frac{\tau}{\rho})$$

and, by integrating across the boundary layer thickness δ this becomes,

$$\frac{\tau_o}{\rho} = \int_o^\delta \frac{\partial}{\partial t}(u - u_\infty)dy$$

where τ_o is the shear stress on the bed. Thus, defining friction velocity as $\bar{u}_* . u_* = \frac{\tau_o}{\rho}$ (and $\bar{u}_*^2 = \frac{\bar{\tau}_o}{\rho}$) it can be seen that

$$\frac{\tau_o}{\rho} = \bar{u}_* . u_* = \sigma \int_o^\delta (u_\infty - u)dy.$$

Hence, it follows that

$$\text{Amp.} \int_o^\delta (u_\infty - u)dy = \text{Amp.} (\frac{\bar{u}_* . u_*}{\sigma}) = \frac{\bar{u}_*^2}{\sigma}$$

and substituting this in the above equation for δ_* gives

$$\delta_* = \frac{1}{\bar{u}_\infty} \cdot \frac{\bar{u}_*^2}{\sigma} .$$

However, if the bed friction coefficient

C_f is defined as

$$C_f = \frac{\overline{\tau}_o}{\frac{1}{2}\rho \overline{u}_\infty^2}$$

it can be seen that

$$\overline{u}_*^2 = (\frac{1}{2} C_f)\, \overline{u}_\infty^2$$

hence, substituting for $\overline{u}_*^2$ in the expression for δ_* gives

$$\delta_* = \frac{1}{2} C_f . \frac{\overline{u}_\infty}{\sigma} = \frac{1}{2} C_f . b.$$

Now, the transition Reynolds number Re will be defined as

$$Re = \frac{\overline{u}_\infty\, \delta_*}{\upsilon} = \frac{\sigma b}{\upsilon} (\frac{1}{2} C_f . b) = C_f . \beta^2 b^2$$

and the transition from laminar to turbulent flow will be expected to occur at a certain critical value of Re, that is

$$Re_c = C_f .\ \beta^2 b^2.$$

3.2 Smooth-Rough Transition

For the smooth-rough transition, two basic criteria will be employed. These are,

1. The roughness particles in the bed boundary layer remain in the laminar sublayer,
2. No vortices are formed around the roughness particles when the flow is laminar.

For the first of the above criteria, the nominal laminar sublayer thickness δ' will be defined similar to the steady flow case, as $\frac{\overline{u}_* \delta'}{\upsilon} = 11.6$, and expressing the condition for the roughness particle to remain in the laminar sublayer as $\frac{d}{\delta'} \leq 1$, the condition for a hydraulically smooth bed becomes

$$R_* = \frac{\overline{u}_* d}{\upsilon} \leq 11.6.$$

However, since $\overline{u}_* = (\frac{1}{2} C_f)^{\frac{1}{2}} \overline{u}_\infty$, (hence $R_* = (\frac{1}{2} C_f)^{\frac{1}{2}} R)$, it follows that

$$(\frac{1}{2} C_f)^{\frac{1}{2}} \frac{\overline{u}_\infty d}{\upsilon} = (\frac{1}{2} C_f)^{\frac{1}{2}} R \leq 11.6 .$$

The second criterion involves the conditions for vortex formation around the roughness particles. It is known that in time dependent motion, vortex formation can not take place before the fluid particles travel a certain distance, whatever the value of R, (e.g. Goldstein (1938)). Therefore, for oscillatory flow, vortex formation is expected to be controlled not only by R as in steady flow, but a second perameter $\frac{b}{d}$ should be significant. Experiments carried out by Sert (1976) have shown that for spherical particles (of 6.1 mm dia, βd=5.5-12) in oscillatory flow near the bed of a wave tank, seperation on the sphere surface could be observed for about $\frac{b}{d}$=0.3 and vortex formation occurs (provided that R>25) at about $\frac{b}{d}$ =0.5. Therefore, even if the roughness particles remain outside the laminar sublayer, the bed may be considered smooth if there is no vortex formation and the condition for this may be written as

$$\frac{b}{d} = \frac{R}{2\beta^2 d^2} \leq 0.5 .$$

Thus, in conclusion, conditions for transition from a smooth to a rough bed in oscillatory flow will be expressed as

$$(\frac{1}{2} C_f)^{\frac{1}{2}} R \geq 11.6,\quad \frac{b}{d} = \frac{R}{2\beta^2 d^2} \geq 0.5,\quad R \gtrsim 25.$$

3.3 Bed Friction Coefficient

Since the expressions for the transition criteria defined above have been based on the bed friction coefficient, a brief account of it will be given.

The bed friction coefficient for oscillatory laminar boundary layer on a smooth bed can readily be evaluated from the velocity profile (given by the Stokes solution) of the form

$$C_f = \frac{\overline{\tau}_o}{\frac{1}{2} \rho \overline{u}_\infty^2} = \frac{\sqrt{2}}{\beta b} = 2\sqrt{2}\ \frac{\beta d}{R}$$

and it is shown by several investigators that this expression remains valid for a large flow range, up to about βb = 200, after which C_f takes gradually higher values with increasing values of βb, (Iwagaki, Tsuchiya, Chen (1967), Kajuira (1968)). It is expected that this expression will also be essentially valid for a bed with rough-

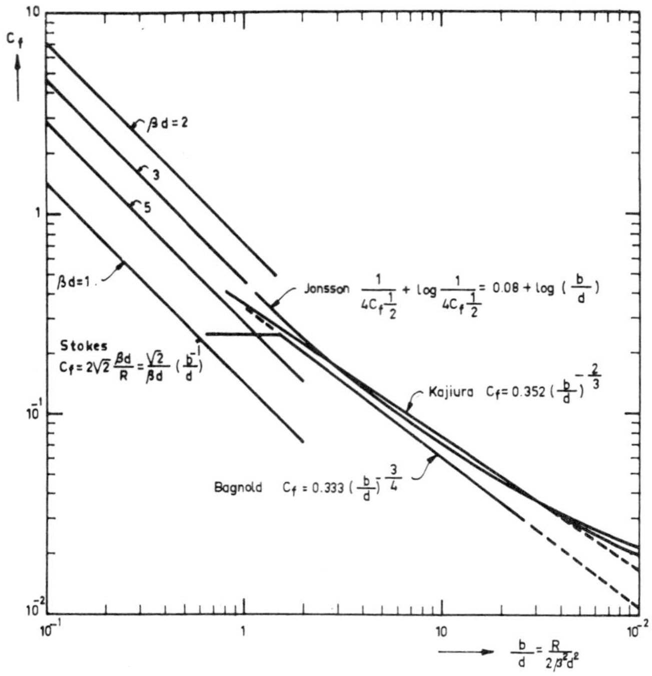

Figure (1) Friction coefficient on a bed in oscillatory flow.

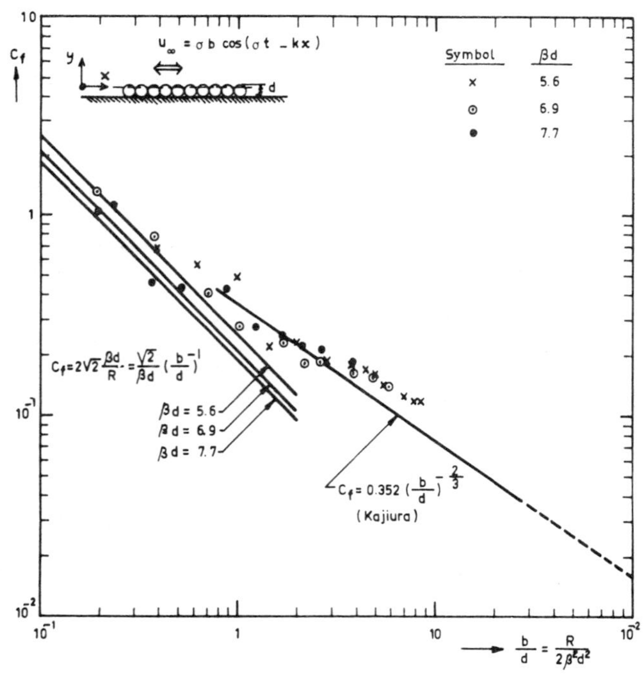

Figure (2) Friction coefficient evaluated from force measurements on roughness porticles.

ness particles before (and at very early stages of) vortex formation around these roughness particles.

For a rough bed, several investigators carried out experimental and semi-empirical theoretical studies, and amongst these are the studies of Bagnold (1946), Jonnson (1966a) and Kajiura (1968). These are shownin Figure (1), expressing C_f as a function of $\frac{b}{d}$, and appear to be in agreement in general terms.

Sert (1980), from earlier force and phase measurements (Sert (1976)) on spherical bed particles of 6.1 mm in diameter arranged on the bed of a wave tank in square packing (centers of adjecant spheres on the bed surface forming a square), has shown that the bed friction coefficient can be evaluated from the expression

$$C_f = \frac{\pi}{4} C_D \cos \phi$$

where C_D is the total drag coefficient of a spherical particle (including the acceleration component of the drag) and ϕ is the phase difference between the drag and the velocity just outside the bed boundary layer (where the phase variation within the boundary layer could be omitted). The values of C_f thus obtained are plotted in Figure (2). Although there is some experimental scatter, it is seen that for small values of $\frac{b}{d}$ (for $\frac{b}{d} < \sim 0.5$) C_f shows strong dependence on βd parameter and appear to be consistent with the laminar flow friction coefficient expression. However, for larger values of $\frac{b}{d}$, (for $\frac{b}{d} > \sim 1$) dependence on βd is weakened, and, for $\frac{b}{d} > \sim 5$, C_f appear to be dependent on $\frac{b}{d}$ only and rather close to C_f values obtained by Kajiura (1968) and other investigators for a rough bed. An interesting point however, (which is unlikely to be due to experimental scatter) is that for a short interval around $0.5 < \frac{b}{d} < 2$, C_f remains nearly constant at about $C_f = 0.2$. (A similar phenomenon had been observed by Bagnold (1946) within a close flow range in experiments with an oscillating rippled plate).

Thus, in the light of the brief explanation given above, C_f in oscillatory flow is defined as follows:

For a smooth bed:

$$C_f = \begin{cases} \frac{\sqrt{2}}{\beta b} & \text{for } \beta b \leq \sim 200 \\ C_f\,(\beta b) & \text{for } \beta b > \sim 200 \end{cases}$$

For a rough bed:

$$C_f = \begin{cases} \frac{\sqrt{2}}{\beta b} = 2\sqrt{2}\ \frac{\beta d}{R} & \text{for } \frac{b}{d} < \sim 0.5 \\ \sim 0.2\ (0.15\text{–}0.4) & \text{for } \sim 0.5 \leq \frac{b}{d} \leq \sim 2 \\ 0.352\ (\frac{b}{d})^{-\frac{2}{3}} & \text{for } 2 < \frac{b}{d} \leq 30 \\ 0.2\ (\frac{b}{d})^{-\frac{1}{2}} & \text{for } 30 < \frac{b}{d} \leq 100 \\ 0.06\ (\frac{b}{d})^{-\frac{1}{4}} & \text{for } 100 < \frac{b}{d} \leq 10^4 \end{cases}$$

4 EVALUATION AND DISCUSSION OF THE RESULTS

Experimental data for laminar-turbulent transition on a smooth bed in oscillatory flow appear to be much more consistent (compared with rough beds) and therefore this will be used in evaluation of the critical value of $Re = \frac{\bar{u}_\infty \delta_*}{\nu}$.

Experiments by Li (1954) and Manohar (1955) with oscillating smooth plates in still water show that laminar-turbulent transition occurs at about $\sqrt{2}\ \beta b = 400$ or $\beta b = 283$, (fully developed turbulence). In addition, experiments for oscillatory flow in smooth pipes by Sergeev (1966) gives $\beta b = 250$ and by Hino, Sawamoto, Takasu (1976) gives $\beta b = 280$ for transition from laminar flow to "fully developed" turbulence, and, Clarion and Pelissier (1975) found a transition range at about $\beta b = 130\text{–}335$. On the other hand, "bursts of turbulence" (followed by decay and relaminarization) in oscillatory flow has been observed by Merkli and Thomann (1975) at about $\beta b = 142$ and by Ramaprian and Mueller (1980) at $\beta b = 185$. It may be noted however that, Collins (1963), (from measurements of mass transport velocity at the edge of the bed boundary layer in a wave tank) proposed a much lower value ($\beta b = 80$) for transition, although the technique employed is somewhat questionable.

Therefore, for a smooth bed, it may be taken that $\beta b = 283$ at laminar-turbulent transition. Using the smooth bed friction coefficient expression $C_f = \frac{\sqrt{2}}{\beta b}$ gives $Re = C_f.\ \beta^2 b^2 = \sqrt{2}\ \beta b$, hence the critical value of Re at transition becomes

$Re_C = \sqrt{2} \times 283 = 400$. Thus, assuming that this critical value is invariant, it may be used in evaluating transition criterion from the expression $C_f.\ \beta^2 b^2 = Re_C = 400$ for the entire flow range.

Using the corresponding expressions for C_f in appropriate flow ranges, the transition criteria for laminar-turbulent and smooth-rough transitions have been evaluated in terms of the two basic flow para-

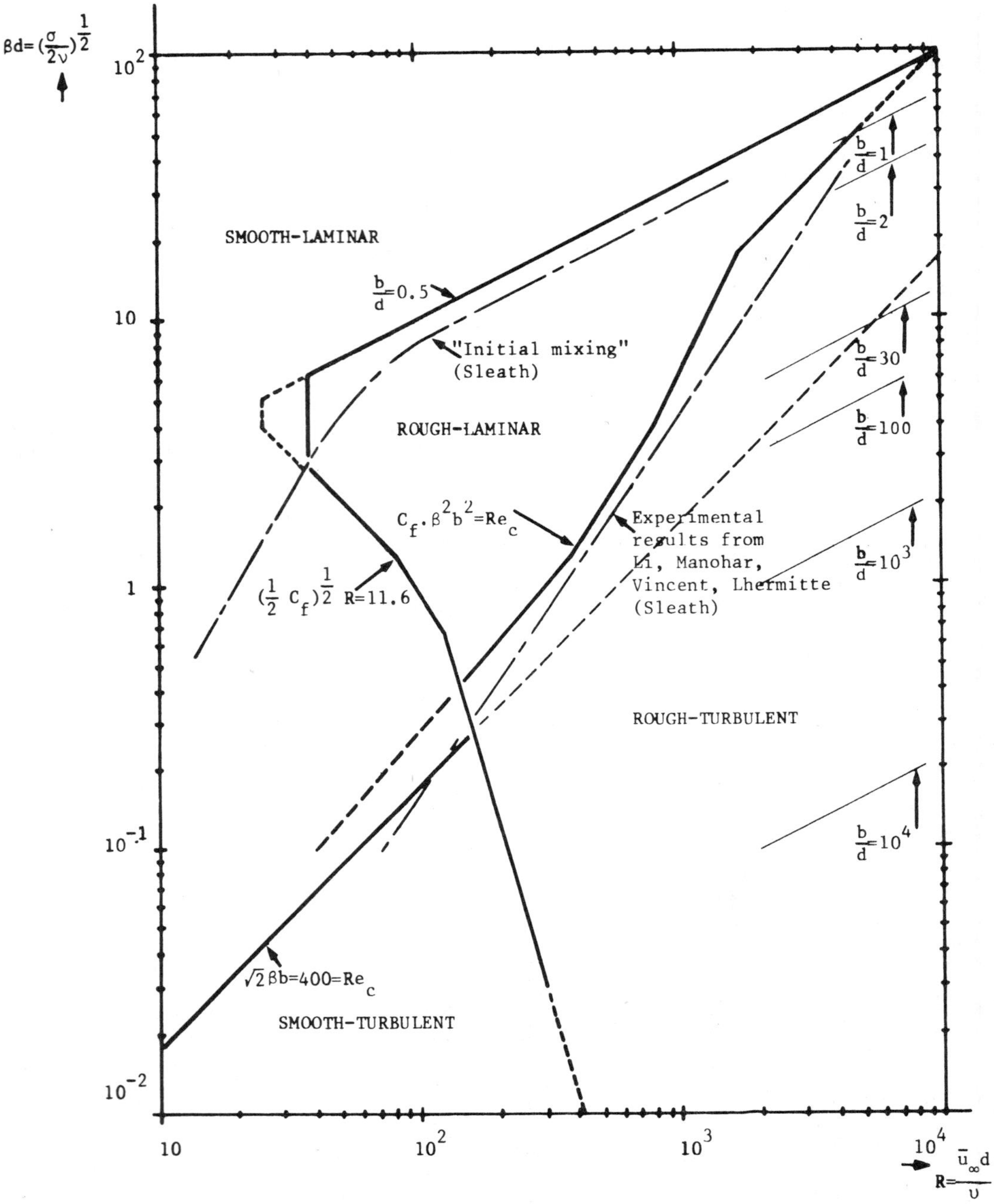

Figure (3) Transition in oscillatory boundary layers.

meters, R and βd, and plotted in Figure (3). Thus, smooth-laminar, rough-laminar, smooth -turbulent and rough-turbulent flow regimes have been delineated by intersections of these two transition curves.

Previous experimental studies on transition for rough beds have usually been evaluated on emprical basis and display large discrepencies at places. Li (1954), from oscillating plate experiments with two and three dimensional roughness, defines laminar-turbulent transition in a range $R = \frac{\bar{u}_\infty d}{\nu}$ = 100-1500, and smooth-rough transition in the range βd = 0.153-0.249. Manohar (1955) confirms Li for the smooth -rough transition (from experiments with the same technique) and puts forward an empirical expression for the laminar-turbulent transition of the form $\frac{\bar{u}_\infty d^{0.2}}{\nu} = 1.15 \times 10^3$ (CGS units, d = 0.235-1.981 mm.). However Kalkanis (1964), again using the same technique for experiments, proposed R = 100 for laminar-turbulent transition. Vincent (1957), carried out tests in a wave tank with a sand bed and proposed an emprical expression for laminar-turbulent transition of the form $\bar{u}_\infty = 1.2\, d^{-0.5}$ (CGS units, d = 0.1-1.17 mm). Lhermitte (1958) also made tests in a wave tank with a sand bed and his findings appear to be around the same range as Vincent. Sleath (1974) plotted all the experimental results of Li (1954), Manohar (1955), Vincent (1957) and Lhermitte (1958) for laminar-turbulent transition and passed an approximate empirical curve through these. This is shown in Figure (3) within the range it is assumed to be valid and it appears to be close to the transition criterion obtained in this study. Sleath (1974) also attempted to study analytically the mechanism of vortex formation and mixing on a bed of two dimensional roughness by numerical solution of the equations of motion, however, physical interpretation in relation to the existing experimental results could not be made. This curve for vortex formation and "initial mixing" obtained by Sleath is also shown in Figure (3), and it is seen that it corresponds approximately to the smooth-rough transition obtained in this study. It may also be noted that smooth-rough transition defined by Li (1954) and Manohar (1955) appears to be near the intersection of the smooth-rough and laminar-turbulent transition curves, (around $\beta d = \sim 0.3$).

In previous delineations of flow regimes by Jonsson (1966,b) and Kamphius (1975) (from empirical and experimental values with steady flow analogy) upper and lower limits had been specified for the transition criteria. These appear to be arbitrary and somewhat unnecessary, as the question arises about the nature of the flow between these limits. However, it has some significance in that, as observed in several experimental studies, "bursts of turbulence" occurs at certain stages of a flow cycle and usually dies out within the cycle before "fully developed turbulence" occurs at higher values of the flow parameters, and, upper and lower limits for transition may be taken to apply to these "initial" and "final" stages. On the other hand there appears to be no objective criterion for defining these limits. Thus, some discrepencies amongst the experimental results may be due to the different "degress of turbulence" observed by the experimenters. (Usually, experiments were based on the visual observation of the characteristic mixing of dye filaments near the boundary, although hot wire measurement and laser doppler anemometry were also used). In addition, it appears that in some cases, vortex formation around the particles (smooth-rough transition) may have been misinterpreted for turbulence. Thus, it is put forward that a new parameter must be defined as a measure of the total turbulence which takes place along the time scale within one flow period. For example, such a non-dimensional parameter may be of the form (in one dimension)

$$\frac{1}{T\delta} \int_0^T \int_0^\delta \frac{u'v'}{u^2}\, dy.dt$$

where u(y,t) is the mean oscillatory velocity in horizontal direction at a given distance from the bed and at a given instant within the flow period, and u'(y,t), v'(y,t) are the random velocity compenents, in horizontal and vertical direction due to turbulence (averaged over a number of cycles for every y,t). Although this would be a parameter rather difficult to measure, it could bring a definition to the "degree of turbulence" and clarify some discrepencies.

As a final remark, it will be noted that a laminar-turbulent transition Reynolds number based on the amplitude of the fluid displacement and velocity at the edge of the oscillatory boundary layer may be obtained as

$\frac{\bar{u}_\infty b}{\nu} = 2\beta^2 b^2 = 1.6 \times 10^5$, (for βb = 283).

This may be compared with the criterion used for the development of turbulence on a smooth flat plate in steady flow at about $\frac{Ux}{\nu} = 5 \times 10^5 - 2 \times 10^6$ (where U is the velocity at the edge of the boundary layer and x is the distance from the leading edge of the plate). It is seen that these values are comparable in magnitude, and in oscillatory flow, the negative acceleration field towards the end of the fluid displacement may be expected to facilitate transition. In addition, it is known that in steady flow, roughness elements on the flat plate facilitates transition depending on the roughness size compared with the boundary layer thickness (Dryden (1953)) and this also seems to be in agreement with results obtained in this study.

5 CONCLUSIONS

In this study, laminar-turbulent and smooth-rough transitions in oscillatory boundary layers have been examined on the basis of some fundamental criteria and compared with the available experimental evidence.

For the laminar-turbulent transition, the existing experimental results are examined within a unified framework. The transition criterion have been defined depending on a critical value of a Reynolds number based on the boundary layer displacement thickness and the velocity just outside the bed boundary layer. Then, the boundary layer displacement thickness, hence the transition criterion, have been evaluated depending on the bed friction coefficient. It is not suggested that this can be the only criterion, but the use of boundary layer displacement thickness as defined in this study has the advantage of being a time invariant quantity and the results appear to be in general agreement with the experimental findings. It is also put forward that a new parameter must be defined as a measure of the total turbulence which takes place along the time scale within one flow period, in order to reach a complete unification and explain discrepencies amongst various studies.

For the smooth-rough transition, two basic criteria have been employed. These are,

1. The roughness particles in the bed boundary layer remain in the laminar sub-layer,
2. No vortices are formed around the roughness particles when the flow is laminar.

The first of the above criteria can be expressed as a condition involving the bed friction coefficient and thus the smooth-rough transition can be evaluated in appropriate flow ranges depending on the friction coefficient. The second criterion involves the examination of vortex formation around roughness particles in oscillatory laminar flow, and this is controlled not only by the Reynolds number (based on particle diameter and a typical velocity), but a second parameter defined as the ratio of the flow amplitude to the particle diameter becomes significant. Both of the above criteria are employed in appropriate flow ranges so that they display continuity and produce a single delineation curve for the smooth-rough transition, which also appears to be consistent with the existing studies.

REFERENCES

Bagnold R.A. 1946. Motion of waves in shallow water: Interaction between waves and sand bottoms, Proc.Roy.Soc. London A, 187:1-18.

Basset A.B. 1888. A Treatise on Hydrodynamics, 2:313. Cambridge: Deighton Bell.

Clarion C., Pelissier R. 1975. A theorotical and experimental study of the velocity distribution and transition to turbulence in free oscillatory flow, J.Fluid Mech. 70,1:59-79

Collins J.I. 1963. Inception of turbulence at the bed under periodic gravity waves, J.Geoph.Res. 68,21:6007-6014.

Dryden H.L. 1953. Review of published data on the effect of roughness on transition from laminar to turbulent flow. J.Aeronau. Sc. July 1953:477-482.

Goldstein S. (ed.) 1938. Modern Developments in Fluid Dynamics, Oxford: Clarendon Press.

Hino M., Sawamoto M., Takasu S. 1976. Experiments on transition to turbulence in oscillatory flows, J.Fluid Mech. 75,2:193-207.

Hough S. 1896. On the influence of viscosity on waves and currents, Proc.London Math.Soc. 28:264.

Iwagaki Y., Tsuchiya Y., Chen H. 1967. On the mechanism of laminar damping of oscillatory waves due to bottom friction, Bull. Disast. Prevention Res.Inst. Kyoto Univ. 16,3(116):49

Jonsson I.G. 1963. Measurements in the turbulent wave boundary layer, Proc.Intern. Assoc.Hyd.Res. 10th Congress, 1:85.

Jonsson I.G. 1966 a. On the existance of universal velocity distributions in an oscillatory turbulent boundary layer, Basic Research Progress Rep.No.12 Coastal Eng.Lab., Tech.Univ.of Denmark.

Jonsson I.G. 1966 b. Wave boundary layers and friction factors, Proc.10th conf. on Coastal Eng. (Tokyo):127.

Kajiura K. 1968. A model of bottom boundary layer in water waves, Bull.of Earthquake Res.Inst., 46:75-123.

Kalkanis G. 1964. Transportation of bed material due to wave action, Coastal Eng. Res.Center, Tech.Memo No.2, Corps of Engineers, U.S.

Kamphius J.W. 1975. Friction factor under oscillatory waves, J.of the Waterways Harbours and Coastal Eng.Div.ASCE,101, WW2:135-144.
Kerczek von C., Davis S.H. 1974.Linear stability of oscillatory Stokes layers, J. Fluid Mech., 62,4:753-773.
Lhermitte P. 1958. Constribution à l'etude de la couche limite des houles progressives, comité central d'océanographie et d'etudes des cotes No.136, Ministére de la Défense Nationale, France.
Li H. 1954. Stability of oscillatory laminar flow along a wall, Beach Erosion Board, Tech.Memo No.47, Corps of Engineers, U.S.
Lin C.C. 1955. The Theory of Hydrodynamic Stability, Cambridge University Press.
Manohar M. 1955. Mechanics of bottom sediment movement due to wave action, Beach Erosion Board, Tech.Memo No.75, Corps of Engineers, U.S.
Merkli P., Thomann H. 1975. Transition to turbulence in oscillatory pipe flow, J. Fluid Mech., 68,3:567-575.
Ramaprian B.R., Mueller A. 1980. Transitional periodic boundary layer study, J.of the Hydraulics Div. ASCE,106,HY12:1959-1971.
Reynolds O. 1883. An experimental investigation of the circumstanceswhich determine whether the motion of water shall be direct or sinous, and the laws of resistance in parallel channels, Phil.Trans.Roy.Soc. London, 174:935
Sergeev S.I. 1966. Fluid oscillations in pipes at moderate Re-numbers, Fluid Dyn. 1:121.
Sert M. 1976. Investigation of the force on a particle near a bed acted upon by water waves, Ph.D. Thesis, University of Cambridge.
Sert M. 1980. Coefficient of friction on beds under wave action, TUBITAK VIIth Science Congress (Eng.Res.)- in Turkish- .
Sleath J.F.A. 1974. Stability of laminar flow at seabed, J.of the Waterways, Harbours and Coastal Eng. Div. ASCE,100, WW2:105-122.
Stokes G.G. 1851. On the effect of internal friction of fluids on the motion of pendulums, Trans.Camb.Phil.Soc., 9:8.
Vincent G.E. 1958. Contribution to the study of sediment transport on a horizontal bed due to wave action, Proc.6th Conf. on Coastal Eng. Florida :326-355.

Laboratory study of breaker type effect on longshore sand transport

E.ÖZHAN
Middle East Technical University, Ankara, Turkey

ABSTRACT: The type of wave breaking has an influence on various hydrodynamic phenomena in the breaker area. In this study, the effect of the breaker type on longshore sand transport was investigated through laboratory experiments using regular waves. Tests were run with plunging, collapsing and surging breakers on an initially plane shaped model beach with slope 1:10. The experimental data are analyzed in accordance with the Inman-Bagnold model for the longshore sand transport rate ($I_\ell = K_\ell P_{\ell,b}$). The variation of the proportionality parameter K_ℓ with the wave steepness and with the Irribaren number are discussed.

1 INTRODUCTION

During the last decade, the observations by a number of researchers have shown that the manner in which the wind waves break (the breaker type) is an important factor affecting various characteristics of the hydrodynamic phenomena in the breaker area. These include the nature and the level of turbulence induced by wave breaking (Sawaragi & Iwata 1974); the longshore current and the nearshore circulation cell structure (Huntley & Bowen 1975); the edge wave motion (Guza & Bowen 1976); the uprush-backwash velocities in the swash zone (Kemp 1975); and the sediment suspension (Kana 1979). In spite of these observations, the most widely used model today for the estimation of longshore sand transport rate:

$$I_\ell = K_\ell P_{\ell,b} \tag{1}$$

where; I_ℓ is the submerged weight rate of the longshore sand transport; $P_{\ell,b}$, the longshore component of the mean wave energy flux at breaking; and K_ℓ, a proportionality parameter; does not differentiate between the breaker types. Eq.1 will hereafter be referred to as the Inman-Bagnold model (Inman & Bagnold 1963).

Although the origin of the Inman-Bagnold model is purely empirical, it was later shown that the correlation of I_ℓ with $P_{\ell,b}$ rests on a theoretical framework (Bagnold 1963, Komar 1971, Dean 1973). However, these theoretical work which all follow the energetics approach of Bagnold (1963), do not end up with a constant value of the parameter K_ℓ as it has been empirically suggested. The most popular empirical value of K_ℓ is 0.77 at present (Komar & Inman 1970).

When the data obtained from the past laboratory experiments and field studies are plotted in accordance with Eq.1, a large scatter of the data points is readily observed (see Fig.8-2 of Komar 1976:207). For the laboratory points, the scatter is about 50 times whereas for the field points of much less number, it is about 6 times. A number of factors may contribute to this scatter. Among these are the inaccuracies of different measurement and data analysis techniques, the differences in the assumptions to compute the value of $P_{\ell,b}$, the differences in the sand transport rates of regular and irregular waves, and the different levels of scale effects in some laboratory experiments. Of course, the fact that the parameter K_ℓ may not be constant, is also a factor causing the scatter. As the breaker type affects various hydrodynamic phenomena in the breaker area, it is worthwhile to study its probable influence on the longshore sand transport rate as well.

2 BREAKER TYPES AND LONGSHORE SAND TRANSPORT MECHANISMS

Galvin(1968) identified four breaker types on a concrete laboratory beach. These are named as spilling, plunging, collapsing, and surging breakers. Battjes(1974), by

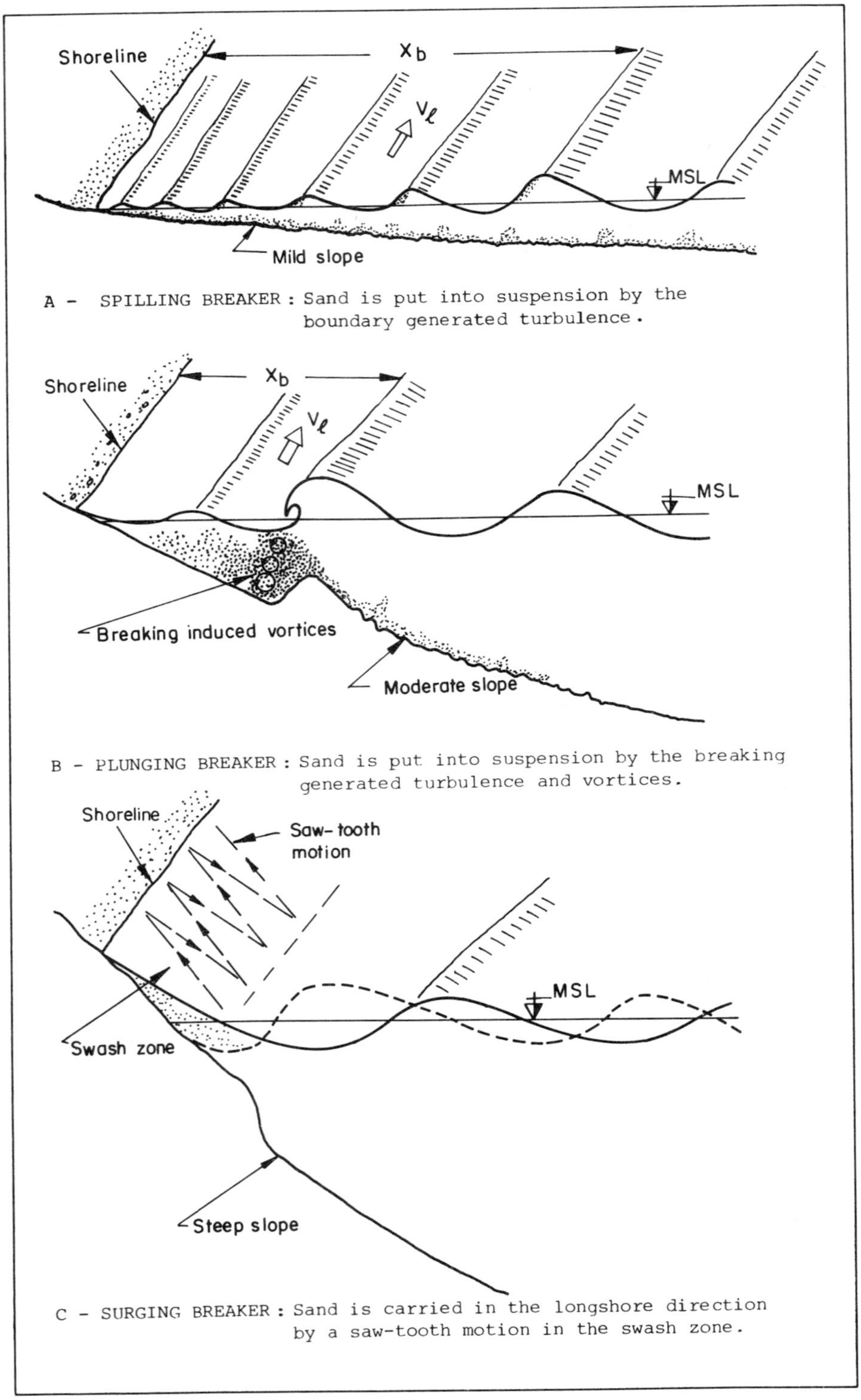

A - SPILLING BREAKER : Sand is put into suspension by the boundary generated turbulence.

B - PLUNGING BREAKER : Sand is put into suspension by the breaking generated turbulence and vortices.

C - SURGING BREAKER : Sand is carried in the longshore direction by a saw-tooth motion in the swash zone.

Fig.1 - Longshore sand transport mechanisms and the breaker types for which they dominate.

reanalyzing Galvin's data, suggested the Irribaren number:

$$\xi_o = \frac{m}{(H_o/L_o)^{\frac{1}{2}}} \qquad (2)$$

where: m is the beach slope; H_o, deep water wave height; and L_o, deep water wave length, to be used as the breaker type parameter. Battjes' classification is given in Table 1.

Table 1. Breaker type classification (Battjes 1974)

$0.5 > \xi_o$	Spilling breaker
$0.5 < \xi_o < 3.3$	Plunging breaker
$\xi_o > 3.3$	Collapsing or surging breaker

There exist three different mechanisms by which the longshore sand transport takes place in the breaker area. These are:

A. The boundary generated turbulence resulting from the orbital (to and fro) water motion at the sea bed stirs up the sand particles and puts them into suspension to be carried by longshore currents.

B. The turbulence and the vortex motion generated by wave breaking, if they penetrate down to the sea bed, stirs up the sand particles and puts them into suspension to be carried by longshore currents.

C. The uprush of the waves breaking at an angle with the shoreline moves the sand particles excited both by bottom and breaking generated turbulence in a path inclined with the shoreline. Then, the succeeding downrush driven by gravity carries them back in a path normal to the shoreline. As this process repeats, the sand is moved in a saw-tooth path having a net displacement in the longshore direction.

These three sand transport mechanisms and the breaker type under which they dominate, are described schematically in Fig.1.

For the spilling breakers, the turbulence generated by wave breaking is confined to a surface region near the wave crest (Sawaragi & Iwata 1974). This turbulence does not penetrate down to the sea bed. Thus, it does not provide useful energy for the transport of sand particles. The energy of turbulence is dissipated as heat by the internal friction of water particles and constitutes a total loss for the sand transport. Therefore, the excitation of the sand particles is solely due to the bed shear resulting from the orbital water particle motion similar to the nonbreaking waves (Mechanism A). The theoretical models of longshore sand transport based on the results valid for nonbreaking waves (like the model described by Madsen 1978) may thus apply for spilling breakers. The occurance of spilling breakers is associated with a wide breaker area since this type of wave breaking is favored by a mild sea bed slope. The energy of a spilling breaker is dissipated both by the bed shear and the breaking turbulence over a rather long distance. Therefore, the spilling waves arrive at shore with much reduced heights and thus cause a gentle uprush-downrush motion in a narrow swash zone.

A plunging breaker creates a violent water particle motion in a narrow zone of the breaker area just shoreward of the longshore bar (Fig.1). As the curling wave crest hits the water surface, vortices are created which migrate down to the sea bed (Sawaragi & Iwata 1974). This vortex motion and the breaking induced turbulence lift huge quantities of sand into suspension (Kana 1979) to be carried by longshore currents (Mechanism B). The width of the breaker area is much smaller than that for the spilling breakers. Also, the energy loss by the internal friction is large so that the wave almost dies away in about a wave length (Sawaragi & Iwata 1974). Therefore, the resulting uprush-downrush motion in the swash zone is not significant.

A collapsing breaker also creates turbulence which is used to lift up the sand particles into suspension. However, this is not as violent as that generated by a plunging breaker. The energy loss due to the internal friction is thus more moderate so that a stronger uprush-downrush motion takes place in the swash zone. Therefore, both the mechanisms B & C may be significant for the longshore sand transport under collapsing breakers.

In the case of surging breakers, there exists almost no breaker area. The wave reaches the shore with very little energy loss resulting from the breaking induced turbulence, and runs energetically up the beach. Therefore, a strong uprush-downrush motion takes place (Kemp 1975), and the mechanism C becomes the dominant mode of longshore sand transport. Due to the steepness of the sea bed slopes however, the loss of wave energy for the sand transport in the form of reflection may be significant.

The above qualitative descriptions of the water motion in the breaker zone and the longshore sand transport mechanisms make one feel that it would be by unbelievable luck if the breaker type did not have an effect on the quantity of longshore sand transport. This feeling is developed upon considering the differences of the transport mechanisms, the width of the breaker area, and the part of wave energy which is not used for the sand transport.

Dimensional analysis of the longshore sand transport problem provides the following relationship (Özhan 1981):

$$\frac{I_\ell}{P_{\ell,b}} = K_\ell\left(Re_{w,b},\frac{H_o}{L_o},\frac{D}{H_o},m,\frac{\rho_s}{\rho},\frac{V_c}{gT}\right) \qquad (3)$$

where: $Re_{w,b}$ is the wave Reynolds number at breaking; D, the sand diameter; ρ_s & ρ, the densities of sand and of sea water respectively; V_c, current speed parallel to the shoreline other than the one created by wave motion; g, the gravitational acceleration; and T, the wave period. As the water motion in the breaker area is highly turbulent, the Reynolds number can be neglected. Considering that H_o, L_o, D, m and ρ_s may not be independent variables (the problem of equilibrium beach profile) and thus omitting the variables D/H_o and ρ_s/ρ, and also letting $V_c = 0$, one ends up with the functional relationship:

$$\frac{I_\ell}{P_{\ell,b}} = K_\ell\left(\frac{H_o}{L_o}, m\right) \qquad (4)$$

The variables which appear to affect the proportionality parameter of the Inman - Bagnold model according to Eq.4, are the ones which determine the breaker type. Eq.4 is the relationship which is investigated in the present study.

3 LABORATORY EXPERIMENTS

Experiments were carried out in a wide laboratory channel having the dimensions: length=28.9 m, width=6.2 m, and depth=1.1 m. An auxiliary compartment (length=16.5 m and width=1.6 m) was built beside the channel at the downdrift side to allow the recirculation of the water mass carried by the littoral current. Natural quartz sand (specific gravity = 2.65) was used to build the model beach. The median diameter was 0.83 mm. Size distribution of the sand particles was nearly uniform, having the uniformity coefficient of D_{60}/D_{10}=2.35. In all test groups, the model beach had an initial slope of 1:10, and made an angle of 10° with the wave maker which was placed normal to the channel axis to avoid wave reflection from the channel walls.

A concrete ditch having the same surface slope as the beach was used as the sand trap. The surface elevation of the ditch's wall was lower than that of the beach initially by 20 cm at every point. To support the vertical sand column along the length of the trap, a perforated metal sheet was vertically placed. This set-up provided sufficient support to prevent the sand column from collapsing into the trap. The sand carried in the longshore direction during the tests was observed to pass through the perforations with no obstruction. This was ascertained by the beach profile measurements after the tests as it was seen that the beach profiles along the sand trap were not different from those measured elsewhere to indicate superfacial erosion or deposition along the perforated metal sheet support. The sand trap was placed in the auxiliary compartment at the start of the study. During the preliminary (heat-up) tests, this arrangement was not seen succesfull and the trap was moved to the main channel. In order to simulate an "infinitely long beach" and thus to obtain steady sand transport, sand was fed to the beach from the updrift side of the channel. The sand feeder used was similar to the one developed by U.S. Army Coastal Engineering Research Center (Savage 1962). The location of the sand feeder across the beach could be significant in affecting the feeding and thus the transport rates (Kampuis & Readshaw 1978). The sand feeder used in the present study was a fixed structure. In order to feed the sand at a constant relative position in all tests, in the swash zone by the action of uprush and downrush of the waves, the water depths were slightly varied to compensate for the differences in the wave run-up heights of the individual test waves which occurred as the wave height and the period were changed.

A test run was continued for a duration which was constant for a test group. At the start of a test, the tank of the sand feeder was completely filled. If the sand in the tank was depleted before the duration specified for that test group, the test was stopped. The sand collected in the trap after a test run was weighed in water to obtain the immersed weight rate of transport. The transport rates were constantly compared with the feeding rates. The feeding system was seen to function very well to feed sand at almost the same rate as it was transported. This was obviously essential to simulate a steady longshore sand transport.

As mentioned before, the model beach had a plane surface with a slope of 1:10 at the start of a test group. The equilibrium shape of the beach developed during the first test of a test group. Thus, the transport rates obtained from these first tests were not included in the analysis since they did not reflect the steady conditions.

The regular test waves were generated by a flap type wave maker. The wave profiles were recorded at the toe of the model beach by moving a resistance wire gage attached to a carriage across the channel width.

Table 2. Summary of the measured and observed characteristics of the laboratory tests

Test group	No. of tests	Duration	H_b (cm)	T (sec)	Depth (cm)	I_ℓ (N/hr)	Breaker type
1	4	41 hr 35 min	4.5-4.8	3.98-4.00	60	161-191	Surging - Collapsing
2	6	11 hr 35 min	14.2-18.8	2.17-2.23	55	2,865-3,916	Collapsing - Plunging
3	6	14 hr	12.3-13.0	1.90-2.02	55	973-1,173	Plunging - Collapsing
4	6	14 hr 5 min	15.6-16.8	1.45-1.57	57	1,063-1,314	Plunging
5	6	11 hr 35 min	9.9-10.7	2.75-2.85	53	1,599-2,893	Collapsing
6	6	11 hr 20 min	16.6-17.5	1.43-1.50	57	1,432-2,504	Plunging

Also the heights of the breaking waves were measured at both updrift and downdrift ends of the beach. The average of these two values was used as the breaking wave height in the analysis.

A total of 34 test runs in 6 test groups were conducted. With the omission of the first tests, this provided 28 transport rates for the analysis. In two of the test groups the breaking waves were the plunging type, and in one test group they were the collapsing type. The breaker type was identified as plunging-collapsing in two test groups, and as surging-collapsing in the remaining one. The characteristics of the tests are summarized in Table 2.

4 ANALYSIS AND DISCUSSION OF RESULTS

4.1 Computation of longshore wave energy flux

The longshore component of wave energy flux at breaking is written as:

$$P_{\ell,b} = (E_f)_b \sin \alpha_b \cos \alpha_b \qquad (5)$$

Here, $(E_f)_b$ is the mean wave energy flux at breaking per unit length of wave crest, and α_b is the angle between the wave crests at the breaking depth and the shoreline. A number of procedures are available in literature for the computation of $(E_f)_b$. These are:

a. To employ the expression derived from the small amplitude wave theory together with the observed breaking wave characteristics (Bruno & Gable 1976, Vitale 1981). A variation of this procedure is to use the wave celerity of solitary waves of the same height instead of the group velocity of the small amplitude wave theory (Komar & Inman 1970, Galvin & Schweppe 1980).

b. To employ the cnoidal wave theory expression and the observed breaking wave characteristics (Svendsen 1972).

c. To use the small amplitude wave theory expression for the deep water waves and to consider the conservation of wave energy as the waves move upto the water depth where they break (Das 1972).

It has been shown that the above procedures give $(E_f)_b$ values which may vary by 600 % depending on the Ursell number of the breaking waves (Özhan 1981). Thus which procedure is to be used is a crucial question for the analysis of the experimental data. In this study, the procedure "c" is favored. According to this, the longshore component of the wave energy flux at breaking is written as:

$$P_{\ell,b} = \frac{1}{32} \rho g H_o^2 C_o K_{r,b}^2 \sin 2\alpha_b \qquad (6)$$

where: C_o is the deep water wave celerity computed from the small amplitude wave theory; and $K_{r,b}$ is the refraction coefficient at the breaking depth. The corresponding deep water heights of the test waves are computed from the measured breaking heights by using two independent empirical breaking criteria (Le Mehaute & Koh 1967, Collins 1970). These are:

$$\frac{H_b}{H'_o} = 0.76 \, (m \cos \alpha_b)^{1/7} \left(\frac{H'_o}{L_o}\right)^{-1/4} \qquad (7)$$

and:

$$\frac{H_b}{d_b} = 0.72 + 5.6 \, m \qquad (8)$$

In these equations, H'_o is the unrefracted deep water wave height ($=K_{r,b} H_o$), and d_b is the depth of breaking. For the computation of the refraction coefficient, the small amplitude wave theory is used uniformly for the wave length (i.e. both in deep

Table 3. Summary of the computed characteristics of the laboratory tests

Test group	d_b (cm)	H_o (cm)	α_o (degrees)	α_b (degrees)	$H'_o/L_o \times 10^3$	$P_{\ell,b}$ (N-m/m-hr)	K_ℓ
1	3.5-3.7	1.44-1.52	27.2-27.3	2.47-2.54	0.55-0.58	109-125	1.480-1.530
2	11.1-14.7	8.60-12.4	15.9-16.2	4.76-5.44	11.2-16.1	4,469-10,602	0.321-0.876
3	9.6-10.2	7.65-8.07	14.3-15.0	4.50-4.64	12.2-13.5	2,944-3,420	0.304-0.361
4	12.6-13.1	13.10-14.10	11.7-12.3	5.37-5.56	32.5-42.0	7,855-9,532	0.121-0.164
5	7.2-8.0	3.80-5.30	20.1-20.8	3.82-4.03	2.9-4.0	883-1,746	0.916-2.789
6	13.0-13.7	13.79-14.71	11.6-11.9	5.52-6.02	39.5-45.5	9,169-10,411	0.138-0.272

water and at the depth of wave breaking). A summary of the computed characteristics of the tests are given in Table 3.

4.2 Breaker type effect on longshore sand transport rate

As seen from Table 3, the values of the proportionality constant K_ℓ of Eq.1 computed for each test run are far from being constant. The ratio of largest to smallest values is about 23.

For a constant beach slope, the wave steepness determines the breaker type according to the classification using the Irribaren number. In Fig.2, the K_ℓ values for each test run are plotted against the corresponding unrefracted deep water wave steepnesses. Also shown in the figure is the least square curve. This curve is rather strongly affected by three data points of test group 1, and is shifted away from the bulk of the remaining points. The tests in group 1 were conducted with waves having a mean wave height of 4.6 cm (the smallest among all test groups). Owing to the small size of the waves, there could exist a significant level of scale effects in the results of these tests. Indeed, according to an empirical criterion for the threshold of the sand particles under waves (Horikawa 1978:250), the critical sand diameter for the average wave characteristics of test group 1 is found as D = 0.88 mm. Since the median size of the sand used to build the model beach is D_{50} = 0.83 mm, about 40 % of the sand particles have diameters in excess of the threshold value. Thus, it is reasonable to accept a scale effect in the K_ℓ values of these tests with regard to the threshold of the larger particles. (The true values are expected to be larger).

With the omission of three data points from test group 1, the least square curve is recomputed as (Fig.3):

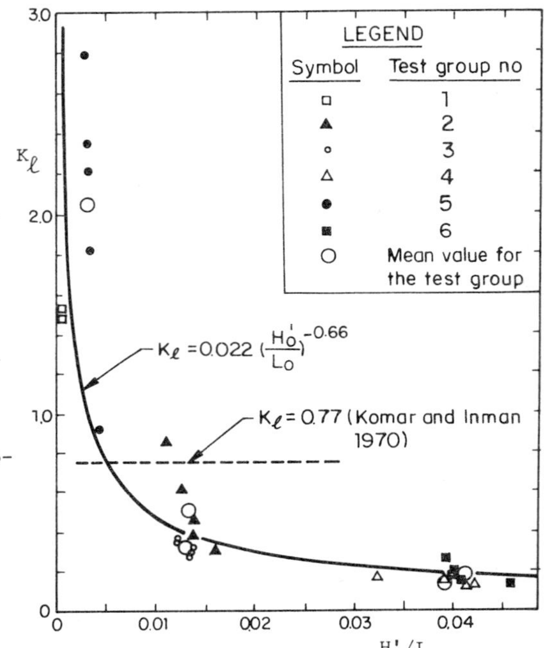

Fig. 2. The dependence of K_ℓ on the unrefracted deepwater wave steepness.

$$K_\ell = 0.0055\ (\frac{H'_o}{L_o})^{-1.03} \tag{9}$$

for $0.0029 < H'_o/L_o < 0.0455$

The power of (H'_o/L_o) in the above equation is very close to unity. The least square equation which results when this power is required as unity is:

$$K_\ell = 0.007\ (\frac{H'_o}{L_o})^{-1.0} \tag{10}$$

Eq.10 is also shown in Fig.3.

The laboratory data obtained by Saville

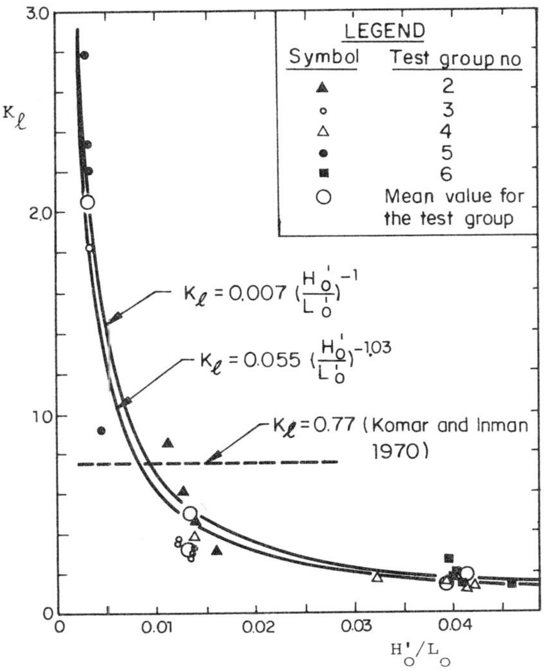

Fig. 3. The relationship between K_ℓ and H'_o/L_o. (Three data points from test group 1 are disregarded).

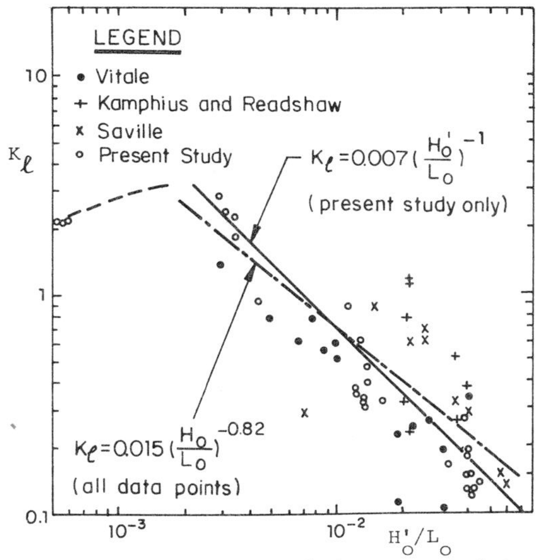

Fig.4. The comparison of the present data points with those obtained in three earlier studies.

(1950), Kamphuis & Readshaw (1978), and Vitale (1981) were reanalyzed using the method outlined in Section 4.1. The computed K_ℓ values are compared with those of the present study in Fig.4. In this figure, K_ℓ values of the test group 1 are located at $K_\ell \simeq 2.1$ by increasing the original values by 40 % to compansate approximately for the scale effect discussed earlier.

The same trend (that K_ℓ increases with decreasing H'_o/L_o) is observed in all data sets in Fig.4. The present data points fall somewhat lower then the points of Saville, and of Kamphuis & Readshaw, and somewhat higher than those of Vitale. This seems reasonable when the differences between the measurement techniques and the experimental set-ups in different studies are taken into account. The least square line in this logarithmic plot obtained for all data points has a slightly milder slope then the one representing the data points of the present study only.

The results presented in this section indicate that a collapsing breaker is much more efficient in longshore sand transport than a plunging breaker. That is to say, among two breakers having the same amount of longshore wave energy flux, the collapsing one (test group 5) carries sand in the longshore direction at a rate greater than the plunging one (test groups 4 & 6). This finding contradicts the earlier general belief that the plunging breaker is the most efficient breaker type for the longshore sand transport (see f.e. Kamphuis & Readshaw 1978, Walton 1979). According to the limited information obtained in this study, the longshore sand transport efficiency of a surging breaker may be in between those of collapsing and plunging breakers. The field measurements of Komar & Inman (1970) indicate that a spilling breaker may have a sand transport efficiency comparable with that of a plunging breaker.

4.3 Beach slope problem

A relationship between the proportionality parameter K_ℓ and the Irribaren number would be more fundamental than the $K_\ell \sim (H'_o/L_o)$ relationship discussed in the previous section. However, various segments of a beach profile may have different slopes. Therefore, which slope of the sand bed should be used to compute the Irribaren number is an important question which needs to be answered.

The typical beach profiles measured at the end of tests with plunging, collapsing and collapsing-surging breakers in this study, are sketched in Fig.5. Looking at these profiles, one may identify at least three different slopes having a connection with the breaking of waves. These are: a) the average foreshore slope m_1; b) the seaward slope of the longshore bar or the

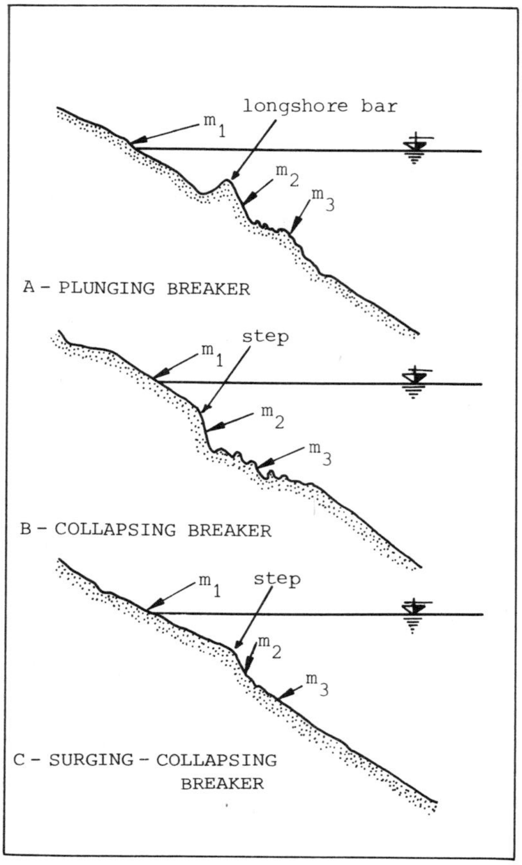

Fig.5. The typical beach profiles measured under different breaker types.

step m_2; and c) the average offshore slope m_3. In Fig.6, the values of K_ℓ are plotted against the Irribaren number, $(\xi'_o)_3$ computed by using the offshore slope, m_3. As seen, a rather high level of correlation exists between K_ℓ and $(\xi'_o)_3$, giving the least square relationship:

$$K_\ell = 1.26\ (\xi'_o)_3^3 \qquad (11)$$

valid for: $0.4 < (\xi'_o)_3 < 1.2$. For the test group 5 which were conducted with collapsing breakers, $(\xi'_o)_3$ varies from 0.86 to 1.20. These figures are much lower than the smallest Irribaren number given by Battjes(1974) for the occurance of a collapsing breaker (i.e. $\xi_o = 3.3$ in Table 1). Thus, the offshore slope alone does not appear to account for the breaker type. The waves break either on the longshore bar or just upon climbing the step. Thus, the geometries of such sea bed features should also affect the breaking of waves. This problem

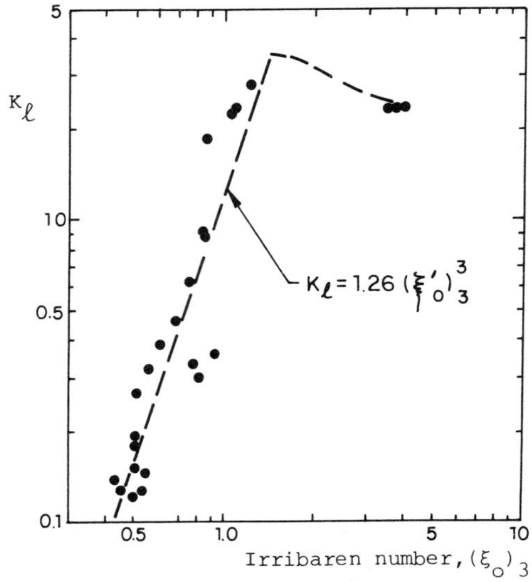

Fig.6. The relationship between K_ℓ and the Irribaren number computed by using the average offshore sea bed slope.

is certainly worthy for further investigation.

The use of the other beach slopes m_1 and m_2 in the Irribaren number cause considerable scatter of the data points (Fig.7).

The dependence of K_ℓ on the Irribaren number was also investigated by Kamphuis & Readshaw(1978), and later by Vitale(1981). However, some geometric features of the

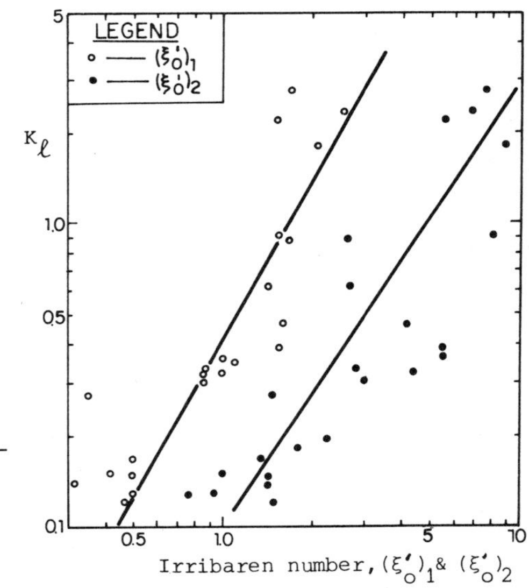

Fig.7. The scatter of data points caused by the beach slopes m_1 and m_2 when they are used in the Irribaren number.

beach profile which are used to define the beach slope by Kamphuis & Readshaw do not seem to be universally existing. (At least, they do not all exist in the profiles measured in the present study). On the other hand, Vitale uses a constant beach slope for all tests corresponding to the initial plane (stepped) beach. Therefore, his result can not be considered as more general than the $K_{\ell} \sim (H'_o/L_o)$ relationship discussed in Section 4.2.

5 CONCLUSIONS

Like other hydrodynamic phenomena in the breaker area, the breaker type has an effect also on the rate of longshore sand transport. For a unit longshore wave energy flux, collapsing breakers cause a sand transport at greater rates than the other three breaker types.

The scatter of all available data points when plotted according to the Inman - Bagnold equation for the transport rate, may largely be due to the variation of the wave steepness. The empirical relationship obtained from the results of the laboratory experiments of the present study, explains this scatter up to about 23 times. The variation of the proportionality constant K_{ℓ} of the Inman-Bagnold model with the wave steepness is also apparent in the results of three previous laboratory studies.

The proper definition of the beach slope to be used in the Irribaren number in order to obtain a more fundamental relationship: $K_{\ell} \sim (\xi_o)$, appears as vague at present.

REFERENCES

Bagnold, R.A. 1963. Mechanics of marine sedimentation. In M.N.Hill (ed.), The sea: Ideas and observations, v.3, p.507-529. New York: Interscience.

Battjes, J.A. 1974. Surf similarity. Proc. 14th Conf. coastal engineering, v.1, p.466-480. New York: A.S.C.E.

Bruno, R.O. & C.G.Gable 1976. Longshore transport at a littoral barrier. Proc. 15th Conf. coastal engineering, v.2, p.1203-1222. New York: A.S.C.E.

Collins, J.I. 1970. Probabilities of breaking wave characteristics. Proc. 12th Conf. coastal engineering, v.1, p.399-413. New York: A.S.C.E.

Das, M.M. 1972. Suspended sediment and longshore sediment transport data review. Proc. 13th Conf. coastal engineering, v.2, p.1027-1048. New York: A.S.C.E.

Dean, R.G. 1973. Heuristic models of sand transport in the surf zone. Proc. Conf. on engineering dynamics in the surf zone, Sydney, p.208-214.

Galvin, C.J. 1968. Breaker type classification on three laboratory beaches. J.Geophys.Research, 73:3651-3659.

Galvin, C.J. & C.R.Schweppe 1980. The SPM energy flux method for predicting longshore transport rate. U.S.Army Corps of Engineers, C.E.R.C. Tech. paper no.80-4 34 p.

Guza, R.T. & A.J.Bowen 1976. Resonant interactions for waves breaking on a beach. Proc. 15th Conf. coastal engineering, v.1, p.560-579. New York:A.S.C.E.

Horikawa, K. 1978. Coastal engineering: An introduction to ocean engineering. Tokyo: Univ. Tokyo Press.

Huntley, D.A. & A.J.Bowen 1975. Field observation of edge waves and discussion of their effect on beach material. J.Geol. Soc. London, 131:69-81.

Inman, D.L. & R.A.Bagnold 1963. Littoral processes. In M.N.Hill (ed.), The sea: Ideas and observations, v.3, p.529-553. New York:Interscience.

Kamphuis, J.W. & J.S.Readshaw 1978. A model study of alongshore sediment transport rate. Proc. 16th Conf. coastal engineering, v.2, p.1654-1674. New York:A.S.C.E.

Kana, T.W. 1979. Suspended sediment in breaking waves. Rept. no. 18-CRD, Univ. of South Carolina, 153 p.

Kemp, P.H. 1975. Wave asymmetry in the nearshore zone and breaker area. J. Hails & A. Carr (eds.), Nearshore sediment dynamics and sedimentation, p.47-68. London: John Wiley & Sons.

Komar, P.D. & D.L.Inman 1970. Longshore sand transport on beaches. J.Geophys. Research, 75:5914-5927.

Komar, P.D. 1971. The mechanics of sand transport on beaches. J.Geophys.Res. 76:713-721.

Komar, P.D. 1976. Beach processes and sedimentation. Englewood Cliffs:Prentice Hall.

Le Mehaute, B. & R.C.Koh 1967. On the breaking waves at an angle with the shoreline. J.Hyd.Res. 5:67-88.

Madsen, O.S. 1978. An analytical model of longshore sand transport. Proc. Workshop coastal sed. trans. Univ. Deleware, p.95-102.

Özhan, E. 1981. A study of longshore sand transport by wind waves, Unpubl. rept. Coastal Eng. Res. Inst. Middle East Tech. Univ. Ankara, Turkey. 210 p.

Savage, R.P. 1962. Laboratory determination of littoral transport rates. J.Waterways and Harbors Div. A.S.C.E. 88:69-92.

Saville, T.Jr. 1950. A model study of sand transport along an infinitely long straight

beach. Trans. Am. Geopys. Union, 31:555-565.
Sawaragi, T. & K.Iwata 1974. Wave deformation after breaking. Proc. 14th Conf. coastal engineering, v.1, p.481-499. New York:A.S.C.E.
Svendsen, I.A. 1972. Energy transport in cnoidal waves. Prog. rept. no.25, I.S.V.A. Tech. Univ. Denmark, p.25-32.
Vitale, P. 1981. Movable bed laboratory experiments to compare radiation stress and energy flux factor as predictors of longshore transport rate. U.S.Army Corps of Eng. C.E.R.C. Miscel. rept. no.81-4, 94 p.
Walton, T.L. 1979. Littoral sand transport on beaches. Phd. dissertation, Univ. of Florida, 322 p.

Euromech 156: Mechanics of Sediment Transport / Istanbul / 12-14 July 1982

Sand ripple motion under combined surface wave and current action

M.C.QUICK
University of British Columbia, Vancouver, Canada

ABSTRACT: Nearshore sediment bedwave processes in shallow water and with low bed slopes are influenced by currents, waves and combined wave and current motion. Sediment kinematics are defined for these various types of sediment behaviour and are related to laboratory measurements. A stream power relationship is used to analyse sediment transport produced by combined wave and current behaviour. The results are sensitive to the behaviour of the mass transport velocity and to the onset of wave breaking due to adverse currents.

INTRODUCTION

Near shore regions, especially river delta fronts, can have extensive sand shelf areas where water depths are moderate and bottom slopes are quite low. The sediment transport in these regions is influenced by river and tidal currents and by wave action, so that the total sediment movement is a combination of these several processes, which is well summarised and extensively referenced, Greenwood, 1975. When examining the total sediment balance of such a region it is important to identify the sediment supply and sediment loss from such a region and it is also necessary to understand the relative importance of the various sediment transport mechanics. Examination of a particular estuary delta region, namely the Fraser River delta front in British Columbia, (Luternauer 1973) indicates that there may will be sub-regions within the shallow offshore zone which can be catagorised as wave dominant, river dominant or combined wave and current dominant zones. The resulting onshore, offshore and longshore motion appear to be linked together to form cells of sediment circulation and these cells can be self-contained, or supplied or deprived of sediment according to the precise physical location. To unravel some of the complexities of the behaviour of these nearshore shallow regions, studies have been made of the individual processes, namely the action of steady flow on sediment (Quick 1978), the wave induced sediment motion (Quick 1982) and the combined wave and current sediment movement. In each of the above studies made by the writer, two major aspects of the total water-sediment behaviour have been studied, firstly, the sediment kinematics necessary for the existence of various sandwave forms have been defined and, secondly, the velocity field near the sand bed has been related to the net sediment motion using a streampower relationship similar to that used by Bagnold.

SEDIMENT KINEMATICS

Sandwaves formed from moving sediment require a quite specific pattern of sediment movement. It is possible to define this sediment movement for both unsteady and steady sediment and water motion. This sediment movement will be referred to as sediment kinematics, because the motion is defined from the demands of continuity and not from any demand of the dynamic forcing field, which is a further stage in the total theoretical development.

For steady flow, such as flow in a river or steady tidal flow, the sediment kinematics have been defined for sandwaves which are (a) developing from the flat bed, (b) fully developed and steadily progressive waves, (c) steady standing waves and (d) regressive waves, such as antidunes (Quick 1978).

The sediment continuity equation, Equation 1, relates spatial gradients of local sediment transport, q_s, to timewise variations of bedwave height, η,

$$\frac{\partial q_s}{\partial x} + \frac{\partial \eta}{\partial t} = 0 \qquad (1)$$

Equation (2) satisfies Equation 1 and relates the sediment transport to local bedwave height and the initial value of q_{so} for a wave progressing at speed nk.

$$q_s = k\eta^n + q_{so} \qquad (2)$$

Referring to Equation (2) which is based on simple kinematic wave theory (Lighthill 1955), the conditions for various types of sand bedwaves to exist in steady flow are as follows, and are derived in detail in the earlier work (Quick 1982).

1. For a steady, progressive bed wave, the exponent n is equal to one and q_{so} is zero.
2. For a stationary, or standing, steady bedwave, q_{so} is again zero and n is also zero, so that q_s is a constant.
3. For a regressive steady bedwave, which is moving upstream against the flow, such as an antidune, the exponent, n, is minus one and q_{so} can have a finite value.
4. For a sandwave to develop from the flat bed condition, the exponent n must have a value between zero and one as is illustrated in Figure 1. Conversely, for a bedwave to flatten and return to a flat bed condition, n must be greater than one. Development of such a progressive bedwave is also dependent on sediment supply q_{so}.

If the local value of sediment transport is considered to depend on the local value of streampower, as defined by Bagnold (1966), then the exponent, n, describes how local streampower must vary with bedwave height for these various modes of bedwave behavior.

Unsteady bedwave behaviour can be defined using relationships which are essentially similar to those defined for the steady flow situation, except that the q_s values now become the instantaneous $q_s(t+\tau)$ values of sediment transport, where t represents time and τ represents a phase relationship. The existence of a phase lag from one part of a bedwave to another, makes complete definition of the unsteady sediment transport difficult, but the time averaged behavior can be explicitly defined (Quick 1982), so that $q_s(t)$, the time averaged behavior, is described by Equation (2) and by the values of n and q_s discussed above. Therefore, for a progressive ripple of constant shape, Figure 2, Equation 2 becomes,

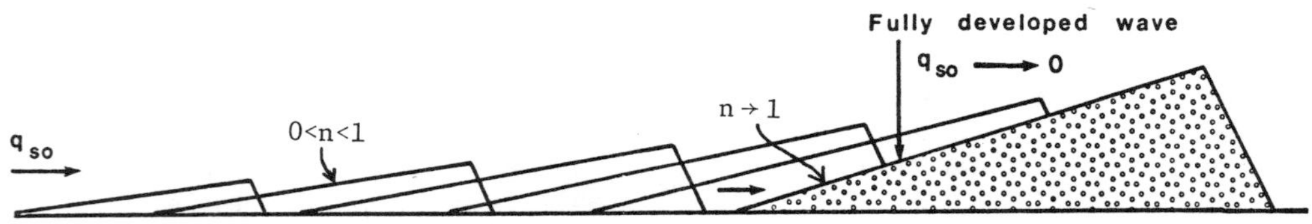

Fig. 1. Progressive Development of a Sandwave: Wave Steepens and Lengthens [Simplified Picture Using Triagular Wave].

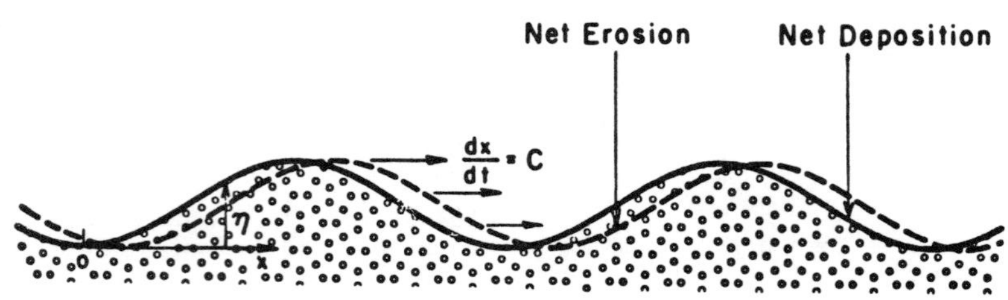

Fig. 2a. The Time-Averaged Steady Translation of Ripple Shape.

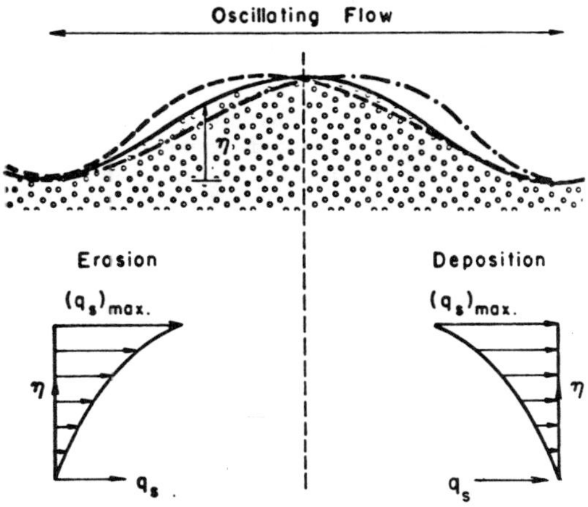

$q_s = k\eta^n$, n>1 $\qquad$ $q_s = k\eta^n$, n>1

Fig. 2b. Possible Sediment Transport Pattern for Fully Developed Ripple.

$$\overline{q_s(t)} = k\eta \qquad (3)$$

where k equals $\frac{dx}{dt}$, the speed of ripple movement. Because the time averaged $q_s(t)$ is a linear function of η, the net sediment transport rate averaged over the total bedwave, $q_s(x)$ is,

$$q_s(x) = \frac{1}{2} k\eta_{max} \qquad (4)$$

where η_{max} is the bedwave height. This relationship is used in the next section to calculate the net sediment transport rate.

WAVE MODIFICATION BY SHOALING AND CURRENTS

Offshore wave spectra are governed by the characteristics of fetch and wind duration. The development of a wave spectrum is illustrated by Figure 3 based on work by Bretschneider (1975). Energy is cascaded to the lower frequencies as time increases. The developed deep water wave spectrum is then modified as the waves move into shallow water and also as they encounter currents. Longuet-Higgins & Stewart, 1964, argue that in deep water, longer waves transfer energy to shorter waves which then break when near the crest of the longer wave. This wave breaking dissipates both short and long wave energy, so that short period wave content of the total wave spectrum decreases with time. A similar result appears to carry through to shallow water wave behaviour. Preliminary indications are that energy continues to be cascaded to the longer period waves, but energy is also dissipated by wave breaking induced by both shoaling and adverse currents. The inshore wave spectrum therefore becomes increasingly limited to the longer wave periods with heights that are limited by water depth.

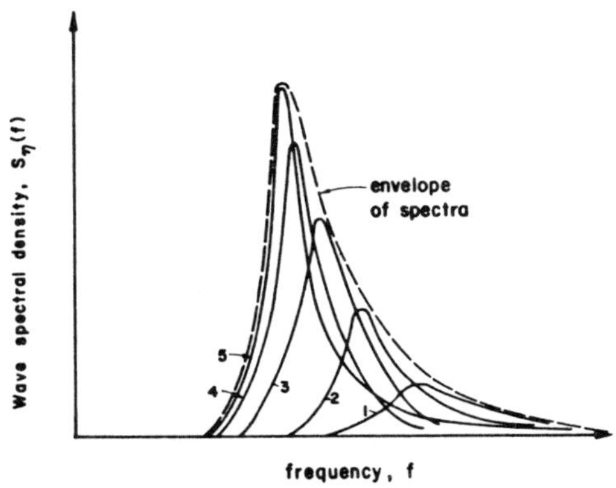

Fig. 3. Envelope of Spectra Illustrating the Development of Spectra with Time.

The shallow water wave spectrum behaviour is too complex for the present discussion, but some guide to the behaviour of the more dominant wave periods can be obtained from a simplified examination of shallow water wave behaviour. By way of illustration, one specific deep water wave will be traced into shallow water, but refraction will be ignored. To illustrate this wave limitation for one specific deep water wave, the special case of shallow water wave behaviour will be considered. For limiting wave steepness, where K depends on sea-bed slope,

$$\left(\frac{H}{L}\right)_{max} = K\left(\frac{H_o}{L_o}\right)_{max} \tanh \frac{2\pi d}{L}. \qquad (5)$$

In shallow water Equation 5 reduces to,

$$H_{max} = 2\pi K\left(\frac{H_o}{L_o}\right)_{max} \cdot d \qquad (6)$$

The equation for the orbital velocity, u, in shallow water reduces to

$$u = \frac{H}{2}\sqrt{\frac{g}{d}} \qquad (7)$$

Combining Equations (6) & (7) to find the maximum oribital velocity, u_{max},

$$u_{max} = \pi K\left(\frac{H_o}{L_o}\right)_{max} \sqrt{gd}, \qquad (8)$$

Equation (8) shows that the maximum orbital velocity in shallow water is essentially dependent on d, the water depth, and is independent of the wave period T, because wave speed in shallow water is primarily depth dependent.

A uniform current parallel to the direction of wave propogation causes a change in the wave period relative to the current. By considering the wave relative to the current, all the usual wave formulae can be used, (Dalrymple 1974) so that first and higher order orbital velocity components and mass transport velocity can be computed. The relative wave period is related to the incoming wave period, T_a, the local value of wave speed, c, and the steady current velocity V by the relationship,

$$T_r = T_a\left(1 - \frac{V}{c}\right)^{-1} \qquad (9)$$

For shallow water, c is simply $\sqrt{gd}$, so that T_r depends on the current Froude number.

The shallow water wave and current situation is therefore fairly simple, because Equations (6) & (9) can be used to calculate the orbital and mass transport velocities using Stokes second order wave theory and Longuet-Higgins (1953) modified value for mass transport velocity. The total velocity field can then be used in conjunction with a transport power relationship to define the net sediment transport at a point.

THE TRANSPORT POWER RELATIONSHIP

Using an argument similar to Bagnold's (1966) original streampower development,

$$P = \tau_o V_T = \rho V_T^{\,3} \qquad (10)$$

where τ_o is the boundary shear stress and V_T is the total velocity at the bed, associated with the sediment transport, and the shear stress, τ_o, can be written as proportional to $\rho V_T^{\,2}$. The net transport power, P_V, can be found by time averaging equation (10) and expressing V_T in terms of the various velocity components, i.e.

$$P_V \equiv \rho V_T^{\,3} \equiv \frac{3}{\sqrt{2}} u_1^{\,2} u_2 + \frac{3}{\sqrt{2}}\left(u_1^{\,2} + u_2^{\,2}\right)V_s + V_s^{\,3} \qquad (11)$$

where V_s is the steady velocity component, $V+\bar{U}$. Writing the transport power proportional to the weight of sediment movement, $\rho g q_s$,

$$q_s \;\alpha\; \frac{P_V}{\rho g} \;\alpha\; \frac{V_T^{\,3}}{g} \qquad (12)$$

where $V_T^{\,3}$ is defined in equation (11). Equation (12) has been tested against laboratory measurements of sediment transport induced by waves alone and for combined effect of waves and steady currents. Results for waves alone are presented in Figure 4, where S is essentially similar to P_v. Results on the combined wave and current behaviour are shown in Figure 5. Co-currents and adverse currents with and against the direction of wave propogation were used. The data showed good agreement if certain assumptions were made about the mass transport velocity, namely that $\bar{U}$ is zero for co-currents and is doubled for adverse currents until wave breaking

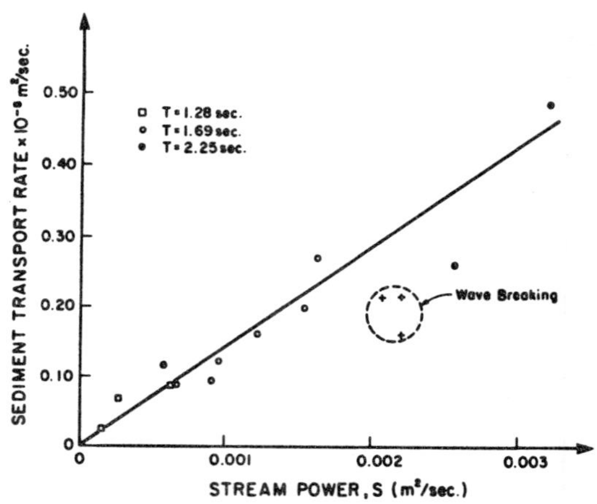

Fig. 4. Sediment Transport as a Function of Transport Power for Wave Action Alone.

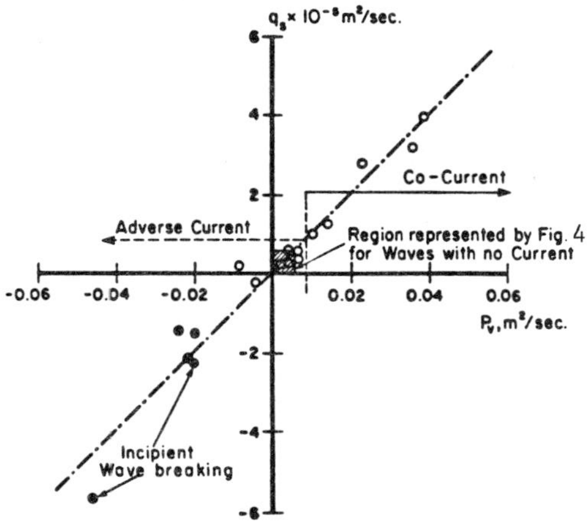

Fig. 5. Sediment Transport as a Function of Transport Power for Combined Wave and Current Action.

occurs, at which time the mass transport velocity is again assumed zero. These assumptions (Quick 1982) are discussed in terms of the vorticity generated at the bed and by the breaking waves, but is only a qualitative argument at this time. A quantitative argument would follow the lines of Longuet-Higgins 1953, which used a theoretical analysis of the mass transport velocity in the presence of vorticity generated in the bottom boundary layer. The present qualitative argument notes that shear, and hence, vorticity, is increased by adverse currents but decreased by co-currents. There should therefore be a corresponding intensification of mass transport velocity for the adverse current situation and a reduction for the co-current case. Wave breaking introduces vorticity at the surface, in addition to the shear induced vorticity at the bed, so that the mass transport velocity should be reduced.

CONCLUSIONS

Sand transport in nearshore, shallow water regions is seen to depend on several possible transport mechanisms, such as waves or currents or a combination of both waves and currents. The kinematics of sediment motion have been defined for various steady and unsteady sediment bedwave situations. This sediment kinematic behaviour can be used to calculate the net sediment transport rate for various observed bedwave behaviours.

The complex deep water wave spectrum is modified by shallow water conditions, so that the longer wave periods become dominant. A simplified shallow water analysis shows that the maximum wave induced velocity field becomes essentially dependent on water depth, so that an apparently complex situation is greatly simplified. A further stage in the simplified theory shows how steady current influences can be incorporated into the analysis. The resulting equations can be used for any type of wave, but become particularly simple for shallow water.

A transport power argument is used to relate the sediment transport rate, calculated from the kinematics, to the wave and current velocity field. This transport power relationship is for net sediment motion and is averaged over the wave period. In addition to the wave orbital velocities and the steady current velocity, the mass transport velocity is seen to play an important role. Examination of the sediment transport rate as a function of transport power indicates that the mass transport velocity is enhanced by adverse currents and reduced by co-currents or wave breaking. Qualitative arguments, based on vorticity production in the flow, are put forward to explain this mass transport velocity variation.

ACKNOWLEDGEMENTS

This study has been carried out with funding support from the Natural Sciences and Engineering Research Council of Canada. The work on the Fraser River Delta Front is being carried out in cooperation with the Geological Section, Marine Sciences, of the Canadian Ministry of Fisheries and Oceans.

REFERENCES

Bagnold, R.A. 1962 "Mechanics of marine sedimentation". The Sea, Ideas and Observations, Editors M. Munk et al. Interscience Publishers.

Bagnold, R.A. 1966 "An approach to the sediment transport problem from general physics". U.S. Geol. Survey Prof. Paper 422-I U.S. Govt. Printing Office, Washington, D.C.

Bretschneider, C.L. 1975. "The envelope wave spectrum" Proc. 3rd Int. Conf. on Port & Ocean Eng. under Arctic Conditions, Univ. of Alaska, Vol. II pp. 689-703.

Dalrymple, R.A., and Dean, R.G. 1975 "Waves of maximum height on uniform currents", Journal of theWaterways and Harbors and Coastal Engineering Division, A.S.C.E. Vol. 101, No. WW3, pp. 259-268.

Greenwood, B. and Davidson-Amoot, R.G.D. 1975, "Marine bars & nearshore sedimentary processes, Kouchibouguac Bay, New Brunswick", pp. 123-150, Nearshore Sediment Dynamics & Sedimentation, Hails & Carr, Editors, Wiley, Publishers.

Lighthill, M.J. and Whitham, G.B. 1955 "On kinetmatic waves", Proc. Roy. Soc. (London) A. Vol. 229 No. 1178 p. 281.

Longuet-Higgins, M.S. 1953 "Mass Transort in Water Waves", Proceeding of the Royal Society, A. 903, Vol. 245, pp. 535-581.

Longuet-Higgins, M.S. and Stewart, R.W. 1964, "Radiation stresses in water waves; a physical discussion with applications". Deep Sea Research Vol. 11, pp. 529-562, Pergamon Press.

Luternauer, J.L. and Murray, J.W. 1973 "Sedimentation on the western delta front of the Fraser River", Can. J. Earth Sci., Vol. 10, No. 11, pp. 1642-1663.

Quick, M.C., 1978 "Sediment transport by ripples and dunes", Dept. of Civil Engineering, In Press (June).

Quick, M.C., 1982 "Wave induced sand ripples", Canadian Journal of Civil Engineering, In Press (June).

Sediment transport due to waves and currents

D.M.McDOWELL
University of Manchester, UK

INTRODUCTION

The main thrust of research into sediment transport in the Department of Engineering of the University of Manchester is towards improvement of numerical modelling methods for coastal sediment transport. The present model (O'Connor, Fanos and Cathers 1981) has modules for estimation of far-field wave climate, current field, near-field combined current and waves, sediment transport and bed-level changes. The near-field modules extend through the breaker zone and include the effect of wave-induced longshore currents.

NUMERICAL MODELLING OF SEDIMENT TRANSPORT

The modular method was adopted to allow modification as improvements are made to modelling any of the component parts. Although decoupling is not strictly permissible, it is thought that the errors that result are smaller than the errors in other parts of the model; wave-breaking on sloping beaches on which three-dimensional flow patterns occur is one example; sediment transport is another. Only when errors in the components have been reduced will it be worth trying to reduce the consequences of decoupling. Research directed towards improvement of the models is being done in several key areas. This paper is concerned with aspects of one of these - sediment transport due to waves and currents.

Earlier work (Fanos 1979) included a study of some approaches to the study of sediment transport (Bijker 1968, Ackers and White 1973, Engelund and Hansen 1967). It was concluded that a modification of the Bijker-Frijlink equation for bed load transport combined with the Rouse equation for suspended load gave the best combination of performance over a wide range of conditions and relative simplicity. Even so, its use should be restricted to particles larger than 1000 μ, if only because estimation of bed roughness cannot yet be made for smaller sized particles. There are many shortcomings to this approach; nevertheless it can be made to give a reasonable fit to data such as was obtained during an extensive series of experiments carried out in the Delft Hydraulics Laboratory (Bijker 1967). There is urgent need for further data obtained from carefully conducted experiments that include the variables that have been neglected so far.

Bijker's pioneering work laid the foundations of much that has followed. The main features were:

1. Determination of a resultant shear stress at the bed due to combined currents and waves.
2. Modification of the Frijlink equation for bed-load transport in which a ripple factor was modified to take into account the effect of waves.

In practice, the resultant shear stress must be expressed in terms of wave and current characteristics and these depend in turn on bed resistance, which is a function of bed form and grain properties under given conditions of depth, hydraulic gradient and wave climate.

The modified Frijlink equation is:

$$q_b = AD_{50} \; V \; C^{-1} \exp\left[B\rho D_{50}(S-1)/(\mu \tau_r)\right] \; \dots \; \dots \quad (1)$$

q_b = volume transported over unit width in unit time

D_{50} = equivalent diameter of bed grains of which 50% are finer

D_{90} = " 90% "

ρ = Water density
S = Ratio of bed-grain density to water density ρ_b/ρ
V = Depth-mean current speed
C = Chezy coefficient = 18 log (12d/R)
C_{90} = " " = 18 log $(12d/D_{90})$
d = Water depth
R = Equivalent height of bed forms
A = Coefficient. If q_b is based on the equivalent in-situ volume, A may be about 5 to give a best fit to experimental values.
B = Coefficient, usually taken as 0.27 to give best fit to experimental values
μ = Ripple factor = $(C/C_{90})^{3/2}$
τ_r = Resultant shear stress at bed = $(V^2g/C + f_wU_o^2/4)$
f_w = Wave friction coefficient
U_o = Velocity amplitude due to waves at bed
SI metric throughout

Perhaps the most difficult aspect of applying this equation is determination of R, the equivalent roughness due to bed forms. It is necessary to decide whether waves or currents dominate bed form generation and then to estimate their equivalent size or the resistance they offer to flow. The author has used recent work (Yalin and Karahan 1979, Engel and Lau 1980) to obtain equivalent roughness due to steady flows over dunes but there is much less certainty about equivalent roughness when ripples are present or when there is an increasing proportion of finer, silt-sized particles. Similarly there has been a lot of work done on pure waves in laboratory channels but the effect of wave spectra on bed forms has hardly been touched. The combination of waves and currents is even more difficult, though useful experiments carried out at the Delft Hydraulics Laboratory led to estimation of the coefficients in the Bijker-Frijlink equation.

Practical application of equation 1 requires a choice to be made between situations in which bed forms are dominated by currents or waves. This may be a choice between a situation in which dunes and ripples are well developed and one in which the bed is almost plane. The rate of bed load transport in such cases may differ by an order of magnitude. Gross errors can occur if the wrong choice is made. Moreover the variability of resistance due to bed forms must have a considerable effect on the flow field in which case it becomes necessary to adopt an iterative approach to the problem.

Clearly there is much that is unsatisfactory about the present state of knowledge of sediment transport relationships; although the Bijker-Frijlink equation has been used as an illustration of some of the difficulties, the same problems arise with any other commonly used relationship.

The major weakness is that it is necessary to specify a shear stress or to provide some equivalent information before sediment transport rates can be estimated. The equations developed so far incorporate the effect of shear stress as a function of flow parameters and they try to reproduce the considerable reduction in transport rate that occurs when bed forms develop.

The research group at Manchester is working on a different approach to this problem. Sediment transport rates increase with increasing shear stress regardless of the presence of bed forms. When bed forms are present, however, the maximum forward rate of sediment transport occurs locally, at the crests and a maximum reverse rate occurs in the troughs. Much of the material moving as bed load goes towards movement of the bed forms, so the resultant net rate of transport is considerably less than the local rate. Present knowledge of development of bed forms in steady flows should enable a rational relationship to be developed. Work is beginning on a method of calculating the local velocity field and shear stress around bed forms using numerical methods used previously on study of wakes behind bluff bodies.

EXPERIMENTAL RESEARCH

An essential part of research into sediment transport is measurement of local shear stress close to the bed. Until recently, this has been an impossible task because insertion of instruments into a sediment-laden flow has interfered with the parameters being measured. The advent of the two-channel laser-doppler anemometer and recent developments in data handling have changed that. After considerable testing, the Manchester group selected a DISA instrument that could be used in forward - or backward - scattering modes. Tests with a single channel version showed that good measurements could be obtained in forward-scatter mode using a frequency tracker even when there was a moderate quantity of sediment suspension. It was also possible to use a particle counter in the back-scatter mode to measure the rate of transport of particles in suspension. These measurements might be combined in a narrow channel with low concentrations in suspension.

The information from the laser doppler anamometer and other instrumentation is stored and analysed in the laboratory using a Plessey minicomputer with 10 Mb cartridge discs, a high resolution VDU and a digital plotter.

It has taken three years to set up the whole system including traversing gear for the LDA, development of the specialised computer software and proving in practise. Much time was spent in identifying the causes of errors in measurement which were not immediately apparent when work began.

The two-channel LDA relies on polarisation of the laser light to separate the velocity components, and one troublesome error occurred because the planes of polarisation were not aligned at precisely 90°. This was easily corrected once it was identified. Another problem arose during proving under a wide range of conditions. Measurements were made over the bottom 10 mm of flow over a smooth bed, over a sand bed and over a gravel bed. It was found that the frequency tracker was not able to follow signals continuously when measuring high levels of turbulence close to the gravel bed.

For successful operation, there must be sufficient minute particles moving through the measuring volume to permit rapid fluctuations to be followed. The frequency response of the tracker must also be high enough to follow the signals.

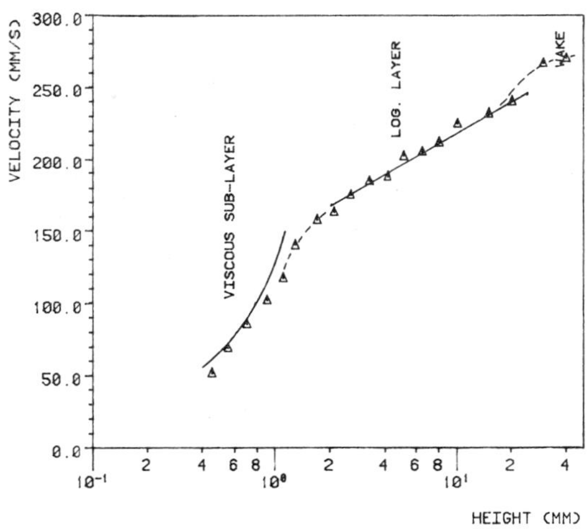

Figure 1. LDA measurements (△) compared with published velocity profiles (solid lines)

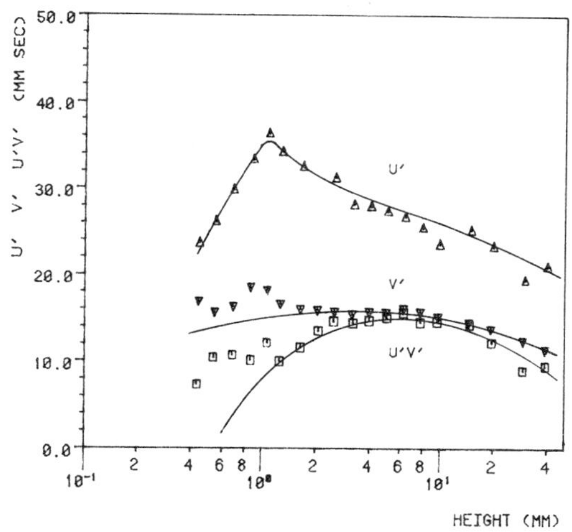

Figure 2. Turbulence and Reynolds stress (points) compared with published data (solid lines)

It was found that the frequency trackers were susceptible to spurious electronic noise and to minimise the effect of this, very careful setting up of the instrument was required with detailed checking of the output signals to ensure data integrity.

As part of the proving process, measurements of turbulence levels in steady flows were compared with measurements obtained by other workers, e.g. in wind tunnels. A comparison is shown in Figs. 1 and 2. It can be seen that both the mean and fluctuating velocities are accurately reproduced in the main body of the flow. Very close to the bed the differences are probably due to errors in the measuring techniques in regions of significant velocity gradient across the measurement volume. In the case of the LDA, the width of the measuring volume is 0.1 mm.

The main series of experiments will lead up to measurements in random waves up to 250 mm high and period 4 sec, combined with currents. Initial testing has been done in a small flume with mean depth 300 mm and with regular waves up to 110 mm high and period 1s or 1.5s. Some typical results are shown in Fig. 3.

These data were taken at 0.2 mm below the top of a 2 mm sand bed; the results are an ensemble average of 100 waves.

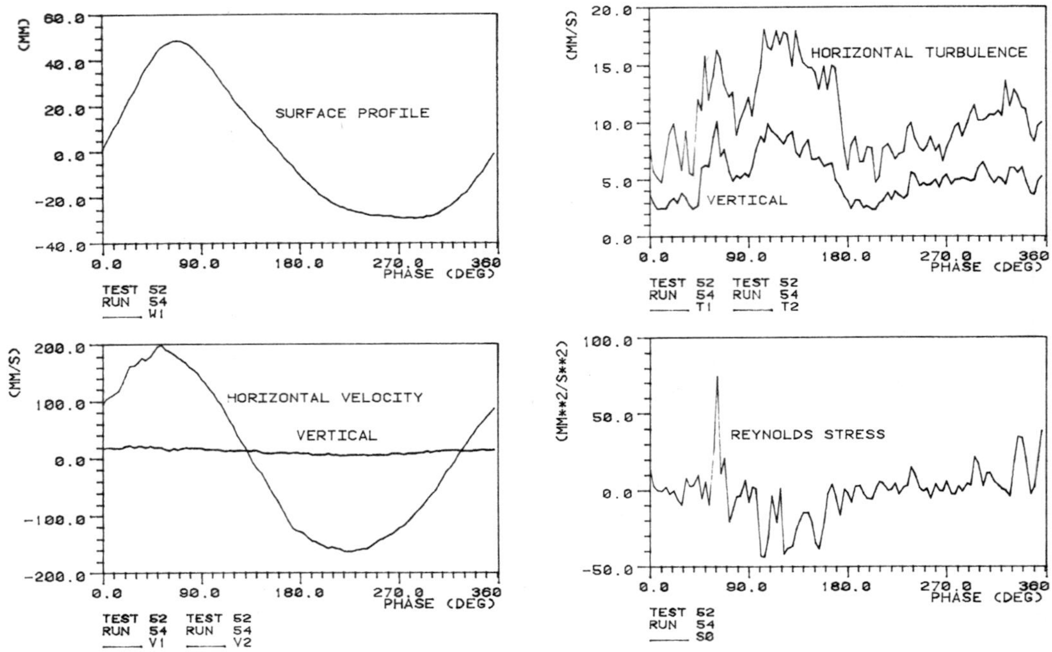

Figure 3. Typical output from each experiment showing ensemble averages of 100 waves

The main conclusion to be drawn so far is that, although fluctuations in velocity can be detected within 10 mm of the bed for a short time before the crest of each wave passes, Reynolds stresses are negligible beyond 6 mm from the bed. This indicates a limit for the turbulent boundary layer in these tests. Although the Reynolds number based on orbital velocity and displacement was in excess of 10,000 the flow was virtually laminar under the trough of the wave and the Reynolds stress was erratic elsewhere.

A series of tests was conducted to investigate this point further. The bed was 2mm sand in a layer 10mm thick. Waves of 1s period were increased in height and boundary layer thicknesses were measured. Fig. 4 shows the result which indicates a jump in thickness above a certain wave height.

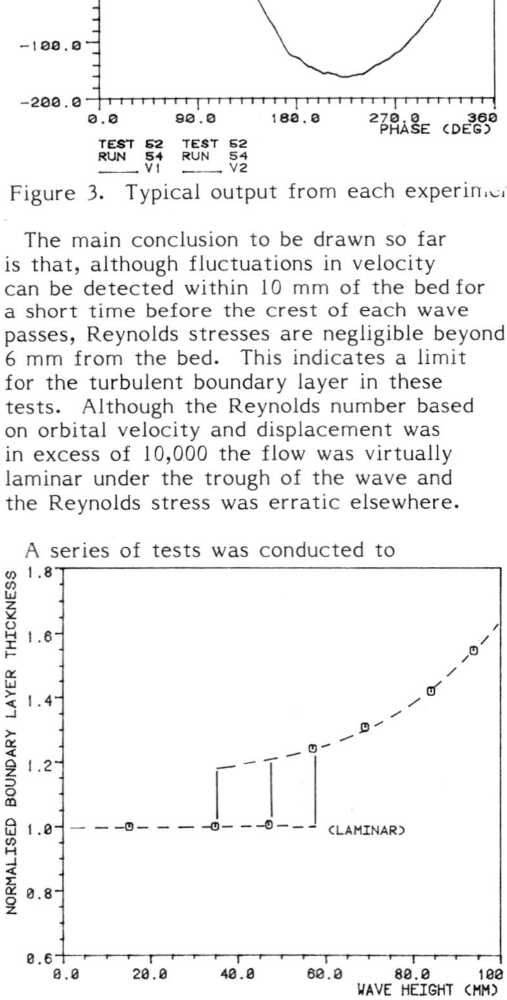

Figure 4. Boundary layer thickness relative to laminar conditions

Figure 5 shows profiles of instantaneous velocity (horizontal component) and Reynolds stress at different times (phase angles) during the passage of a wave. Each data point is an ensemble average of 100 waves. Despite this, there is considerable scatter in the measurements of Reynolds stress.

A test involving the ensemble averaging of data from 100 waves at 30 levels requires the measurement, storage and analysis of over 1,300,000 data values. Sophisticated computer-based data acquisition and processing systems are therefore essential.

Computer animation techniques were developed, using the experimental data, to study the dynamic behaviour of the velocities, turbulence and Reynolds stress during a wave cycle. Some of the results were transferred on to 16 mm cine film.

A frame from this film is shown in Fig. 6 The film showed that fluctuations of horizontal velocity just before the passage of each wave crest propogated upward into

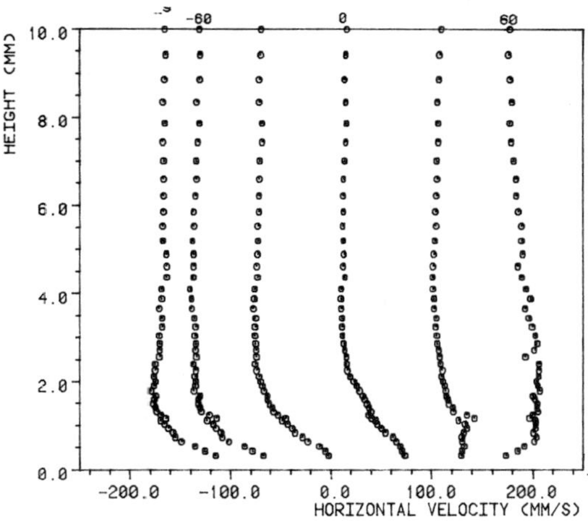

Figure 5a. Profiles of horizontal velocity

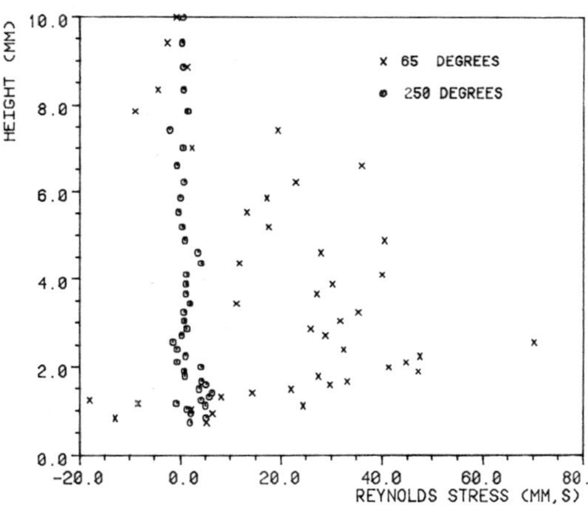

Figure 5b. Profiles of Reynolds stress

the flow. Reynolds stresses were much more restricted in their occurrence. One possible explanation is that the velocity fluctuations were caused by vortices shed from the bed (the so-called bursting phenomenon). If this is correct they were irrotational and would not contribute to the maintenance of sediment in suspension in the flow.

ACKNOWLEDGEMENTS

The team leaders are Dr. B.A. O'Connor and the author. Mr. G. Ellis built up the LDA installation and the computer system. Mr. I. Savell and Mr. G. Ellis developed the computer software and carried out the experiments on which the figures were based.

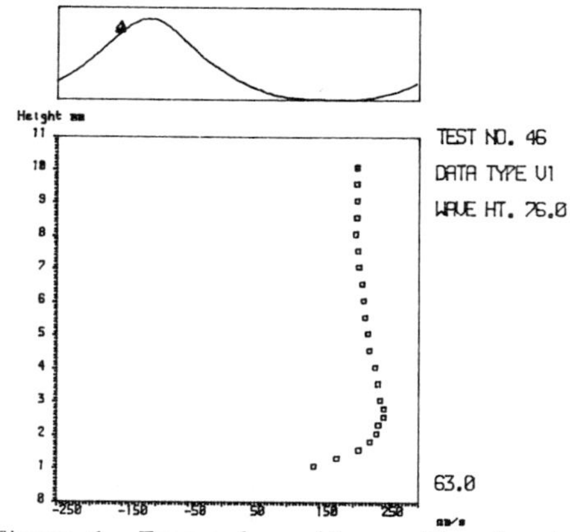

Figure 6. Frame from 16 mm film showing horizontal velocity. Other sequences showed turbulence and Reynolds stress.

The computer animation and the 16 mm film was produced by Mr. G. Ellis using the facilities of the University of Manchester Computer Graphics Unit. The work was supported by the Marine Technology Directorate of the Science and Engineering Research Council.

REFERENCES

Ackers, P. and White, W.R. Sediment transport: new approach and analysis. Journal of the Hydraulics Division, ASCE, 99, HY 11, Nov. 1973, 2041-2060.

Bijker, E.W. Some considerations about scales for coastal models with movable bed. Delft Hydraulics Laboratory Publication No. 50 1967.

Bijker, E.W. Littoral drift as a function of waves and currents. Proceedings of 11th Conference on Coastal Engineering, London, 1968, 26, 415-435.

Engel, P. and Lau, Y.L. Friction factor for two-dimensional dune roughness, Journal of Hydraulics Research, 1980, 18, No.3. 213-255.

Engelund, F. and Hansen, E. A monograph on sediment transport in alluvial streams. Teknisk Forlag, Copenhagen, 1967.

Fanos, A. Morcos. A mathematical computer model for coastal sediment transport. Ph.D. thesis, University of Manchester, 1979.

O'Connor, B.A., Fanos, A. Morcos and Cathers, B. Simulation of coastal sediment movements by computer model. Proceedings Second International Conference on Engineering Software, Section VI, March 1981.

Yalin, M.S. and Karahan, E. Steepness of sediment dunes. Journal of the Hydraulics Division, ASCE, 105, No. HY 4, April 1979.